Communication in Extension Education

THE AUTHOR

The author, the learned personality of the academic world of Social Science and related disciplines, Prof. M.M. Adhikary, is presently rendering his precious services as the Professor of Agricultural Extension, Faculty of Agriculture, Bidhan Chandra Krishi Viswavidyalaya. Prof. Adhikary has been attached with the profession of teaching, research and extension for about thirty-five years. His immense potentiality of academic activities and capability to work with the people, sense of research and extension works helped him to serve as Dean, Faculty of Agriculture, Registrar, Bidhan Chandra Krishi Viswavidyalaya, Vice-Chancellor, Bidhan Chandra Krishi Viswavidyalaya, and Director (In charge), Directorate of Extension Education in the same University as well as Head of the Department of Agricultural Extension. His hard work and the devotion in the aegis of extension affiliated him as Master Trainers of Experiential learning Cycle in ICAR-USAID programme in 1991. He is the expert of different specialized fields like Group dynamics and leadership, Training of Trainers, Project Evaluation, Participatory Project Management, Value added Agriculture, Media Management etc. He has conducted fourteen research and extension projects as investigator. To his credit there are 80 numbers of scientific research papers published in National & International Journals, a good score of chapters and 16 numbers of books in related areas. He is widely traveled all through his carrier covering China, Japan, USA, Australia, Turkey, Bangladesh and Thailand etc.

Communication in Extension Education

Author

M.M. Adhikary

2015

Daya Publishing House®

A Division of

Astral International Pvt. Ltd.

New Delhi-110002

Cataloging in Publication Data–DK
Courtesy: D.K. Agencies (P) Ltd. <docinfo@dkagencies.com>

Adhikary, M. M., author.
Communication in extension education / author, M.M. Adhikary.
pages cm
Includes bibliographical references (pages).

ISBN 978-93-5130-746-4 (International Edition)

1. Agricultural extension work. 2. Agricultural journalism. 3. Communication in agriculture. I. Title.

DDC 630.7 23

Published by : **Daya Publishing House®**
A Division of
Astral International Pvt. Ltd.
– ISO 9001:2008 Certified Company –
4760-61/23, Ansari Road, Darya Ganj
New Delhi-110 002
Ph. 011-43549197, 23278134
E-mail: info@astralint.com
Website: www.astralint.com

Laser Typesetting : **SSMG Computer Graphics,** Delhi - 110 084

Printed at : **Replika Press Pvt Ltd**

Preface

It is obviously a fact that in the archives of human civilization no single factor like communication has stormed into the human society like anything. Starting from primitive society to the post modern elite human lives, communication has become the super single mounting force. Even before the human civilization set for a journey, it was the non human communication networked into the nature, ants, bees, birds, animals different creatures and millions of floral expanses.

The extension science being pivoted on the dynamics of system behaviour and system expansion has rightly been the fulcrum with the attributes of communication and its different components. The communication process is nothing but a social chemistry creating covalence and imbalance and synergy as well.

In the recent past the farmers of India including those from South-East Asia have changed their life styles responding to quick urbanization and faster change in rural life systems. The book "Communication in Extension Education" will help in gathering knowledge about the upcoming decades choosing communication intervention out of huge choices of communication. The modern civilization has uptoned the textural configuration of rural people; so the role of prompt, precise, palpable information and its proper communication would write a new biography for million of farmers. The need of organizational communication has gone many folds because modern organizations have become more natural with family and individual lives.

So, a splendid combination of information and communication technology will be of higher symbiosis in every day life of rural people. A farmer possesses not only a pair of hands to work but also a pair of ears to listen something musical, something appealing and something mesmerizing. That is why the plurality of need and multiplicity of role of a single entity can be reshaped only by the amazing science of communication. The much desirable transformation of information into knowledge is possible by communication technology and ultimately bringing us to a cenotaph of acumen. Thus, it is felt that the creation of this book will pour a sense of satisfaction; while, a scholar of communication would find worth meaning and worth application.

M.M. Adhikary

Contents

Part: I
Communication and Audio-visual Aids

Chapter 1

Concept and Element of Communication

Introduction

The beginning and booming of human civilization, its growth and stability, its decline and demolition have all been attributed to the process of communication. Once the helpless and armless human being started scrambling across the meandering path of progress to reach the destination, the initiation of the process of communication also kept on moving on and on. Without symbol he remained culturally mute and with symbol he started creating his culture to communication.

The logical transformation of forest life into an agrarian life is the resultant of a revolutionary technological problems spearheaded by the need of survival. But both for the nomadic life and agrarian life, communication went on contributing comprehensively. The transformation of blunt stones into sharp stone itself had been the most signification development communication in the annals of human civilization. For the first time the human need had been translated through a primitive version of technology, the development communication.

Passing through ancient society, the process of communication started entering the medieval era human civilization. The script and languages, the art and artifacts, the gothic structures and the grand monuments had been glittering the grace of communication in different forms and faces. The great literally works and the paintings of all time based heralded the victory of human civilization over time and space.

The ninetieth and twentieth centuries had certainly been the era of communication. The invention of stem engine, telegraphs, telephones, radio, television and printing devices are some of the great creation of human civilization that redefined the value of life. This process lifted to a level of grandeurs when television and computer stormed into the every spell of human life. Through both pace and grace, it has enlightened the valleys of human walks and trades.

The twenty first century has experienced, what we may call, a grand metamorphosis. A commodity society has already been metamorphosed into a knowledge society. This has been possible by the great contribution of computer and ITs, empowered by satellite communication.

We have already declared the days of the distance and the death of the obscurities. We are moving with the space of photon and growing with expansion of universe. While change is the go of the day, communication has revolutionized it beyond the fence. This unbound cherishes of victory has been possible through the explosion of communization process and application.

Communication

The word 'communication' comes from the Latin word *'communis'*, meaning common. 'Communication is a process by which two or more people exchange ideas, facts, feelings, or impressions in way that each gains a common understanding of meaning, intent and use of message.

Levels of Communication

According to Thayer (1968) there are four levels of human communication. These are as following:-

1. **Intrapersonal Communication** - Interpersonal communication is defined as the communication process within an individual. Intrapersonal communication is the active internal involvement of the individual in symbolic processing of messages. The individual becomes his or her own sender and receiver, providing feedback to him or herself in an ongoing internal process. It can be useful to envision intrapersonal communication occurring in the mind of the individual in a model which contains a sender, receiver, and feedback loop. Although successful communication is generally defined as being between two or more individuals, issues concerning the useful nature of communicating with oneself and problems concerning communication with non-sentient entities such as computers have made some argue that this definition is too narrow. In communication, the Social Matrix of Psychiatry, Jurgen Ruesch and Gregory Bateson argue that intrapersonal communication is indeed a special case of interpersonal communication, as "dialogue is the foundation for all discourse." Intrapersonal communication can encompass: day-dreaming nocturnal dreaming, including lucid dreaming, Speaking aloud (talking to oneself) reading aloud and repeating what one hears. The additional activities of speaking and hearing (in the third case of hearing again) what one thinks, reads or hears might increase concentration and retention. This is

considered normal, and the extent to which it occurs varies from person to person. It should become a matter of concern only when talking to oneself occurs outside of socially acceptable situations. It aids in ordering one's thoughts. It also creates a record that can be used later again. Copying text to aid memorizing also falls in this category. Making gestures while thinking: the additional activity of body motions, might again increase concentration, assist in problem solving, and assist memory.

2. **Interpersonal Communication -** "A spoken word flies, you won't catch it". A Russian proverb.

"I will pay more for the ability to deal with people than for any other ability under the sun". John Rockefeller.

"Trust is the single most important factor in both personal and professional relationships".

If you want to become a senior manager, interpersonal communication skills are much more important than technical skills. Good communication is a very important aspect of interpersonal relationships. Listening is one part of communication which sounds fine in theory but is rare to find in practice. Empathy will help you to understand people and respond to their needs more effectively. Understanding the basic communication model should help to improve your communication skills. Experts maintain that, as much as 65 per cent of the message communicated is non-verbal or through body language. There are some practical tips you can apply to help people to like you more. Similarly, an extension worker fails to achieve results in the absence of interpersonal communication skills though he may be endowed with the best of the technology or even the best of the strategies. The ability to get along with others is immensely helpful for getting success in almost all walks of life. Whether one plays the role of extension worker, farmer, village developmental officer, scientist, research worker, the manner in which we interact with people would determine very largely if life remains buoyant or miserable.

Research reveals that there is a positive correlation between good relationship and productivity and profit. Success of an extension worker will largely depend upon an effective communication which in turns depends upon the interpersonal communication skills such as empathy, listening, positive attitudes, perception, presentation skills, written communication skills, courtesy, manners, a sense of humour, etc. In the basic communication model there is a sender, a message and a receiver. In other words, it's a speaker, words and listeners.

Sender

First the sender conceives an idea which must be translated into words. The effectiveness with which the sender conveys the message will depend on choice and appropriateness of words, vocabulary, tone and gestures. All these factors help to project the message with precision and clarity. The message may be distorted, ignored or misunderstood, if any of these is mishandled by the sender.

Message

The message consists of words and non-verbal cues, what are words? Words are symbols which represent concrete objects and concepts. It is important to remember that words can create understanding and inspire people to great deeds or they can create misunderstanding, distract, annoyance and bad feeling. So pick your words with care and discretion. Words are not necessarily the precision tools of description that they are made out to be.

For example:

- Words mean different things in different contexts. In the context of a training programme, 'course' means curriculum, but in the context of a horse or a greyhound race it means route. Therefore meaning of words lies in the people and not in the words.
- Words mean different things in different cultures.
- Words mean different things to different people. A bull to a farmer means livestock. On the other hand, to a stockbroker, it means a person who buys shares.

Receiver

The third ingredient in the communication model is the receiver. The involvement of the receiver can be considered in three ways. Firstly, the receiver recognizes what the speaker is saying and tunes in. Secondly, the receiver interprets the meaning of what is being said by mentally summarizing and reorganizing the idea. Thirdly, the receiver understands the message, which has been assimilated, associates and consolidates with exiting knowledge and only then responds to what has been said. You can now see just how complicated communication is. Mishaps can occur at any stage and interfere with the message. This is the communication cycle.

Verbal and non-verbal communication

The experts tell us that only thirty five (35) per cent of what we mean is in the verbal message. The other sixty five (65) per cent of the meaning is contained in the non-verbal communication. Facial expressions, eyes and other body movements must be considered when interpreting meaning. In any social interaction with other people you must be aware of the non-verbal cues and signals. Psychologists have been paying increasing attention to these non-verbal aspects of communication. The following are some of their findings.

Boredom

Boredom is "an unpleasant, transient affective state in which the individual feels pervasive lack of interest and difficulty concentrating on the current activity. Or in other way we can say " boredom is an emotional state experienced when an individual is left without anything in particular to do, and not interested in their surroundings". In positive psychology, boredom is described as a response to a moderate challenge for which the subject has more than enough skill. There are three types of boredom, all of which involve problems of engagement of attention.

These include times when we are prevented from engaging in wanted activity, when we are forced to engage in unwanted activity, or when we are simply unable for no apparent reason to maintain engagement in any activity or spectacle. Boredom proneness is a tendency to experience boredom of all types. This is typically assessed by the Boredom Proneness Scale. Boredom and its proneness are both theoretically and empirically linked to depression and similar symptoms. Nonetheless, boredom proneness has been found to be as strongly correlated with attentional lapses as with depression. Although boredom is often viewed as a trivial and mild irritant, proneness to boredom has been linked to a very diverse range of possible psychological, physical, educational, and social problems.

Arms folded

Folded arms usually indicate resistance. It suggests that the listener is forming a protective barrier against you and your ideas. But beware, it can also mean other things.

Frown

The frown may suggest disagreement, lack of understanding or annoyance. Several responses are possible. You can back up and restate your last point. You can pause and ask for questions, or ask you can clarify your point.

Upward glance

Watch people's eyes. The eyes are the minor to the soul. The sky ward glance may mean "I have heard that one before". Eye contact is essential for good communication.

Memory

The average person has a poorly trained memory. Both the long-term and short-term memories are important factors in communication. Research has shown that we forget 50 percent of what we read immediately unless we review. It is also believed that we forget 80 per cent of what we hear within 24 hours and 90 per cent within 48 hours. One should not rely on ones over worked memory that we tend to remember what we like and forget what we dislike.

Encouraging people to like you

There are some practical steps we can take to encourage other people to like us better. We can put the following tips into practice straight away:

Positive thinking

Many books have been written about the power of positive thinking. The problem with negative thinking is that it becomes a self-fulfilling prophecy. The good news is that positive thinking is likewise a self-fulfilling prophecy. Shakespeare said: there is nothing good or bad except that thinking makes it so. Believe in yourself. Think positive experiences. Think about your successes rather than your failures. Develop a mindset for success. Positive expectations are more likely to lead to positive outcomes. Positive thoughts proceed positive actions and positive actions produce positive results. Negative expectations are more likely to lead to negative outcomes. Positive thinking is about developing a favorable image of yourself. Remember there is often little correlation between success and a positive self-image. Many successful, Hollywood stars have a very negative self-image and often end up committing suicide. Unless you have a favorable image of yourself you cannot impress others favorably. You must be able to like yourself before you can like other people. Develop faith in your own ability to handle colleagues and bring work task to completion. The more successful you are the more confident you will become in the future. Do not underestimate yourself and never sell yourself short.

Show interest

Take a genuine interest in the people around you and in the work environment. To be interesting, be interested. Ask questions that the other person will enjoy answering. Encourage them to talk about themselves and their achievements. Disraeli said: talk to a man about himself and he will listen for hours. Develop a friendly and helpful attitude. Always try and be sincere. If you are artificial, people will eventually see through you. Be positive. If you do not know something, say: 'I don't know, but I'll certainly find out and let you know', rather than a curt: 'Don't know'. Develop a reputation for reliability. If you say you're going to do something, do it!

Remember to smile

A smile is like a piece of sunshine. You'll be surprised at the friendly way people will react. Smiling facilitates positive thinking by giving you a psychological lift. You can't smile and entertain negative thoughts at the same time. Develop a cheerful disposition.

- **Name:** People like to hear the sound of their own names. One of the best ways of crating goodwill is to remember names and use them. To remember names, relate them to the person's features and repeat the name frequently during the conversation.
- **Manners:** Good manners are a sign of good breeding. Manners help make living pleasant. Like smiling, good manners cost nothing but reap many benefits.
- **Learning is a lifelong continuous process:** Seek out opportunities for self-development, not only in your professional life but also in your personal life. Action cures fear. Don't procrastinate. Undertake that educational programme non-whether formal, such as certificate, diploma or degree programme, or informal, such as challenging recreational pursuits. It will make you a more interesting person.
- **Practice:** Practice the previous advice. It is like driving a car, if you practice these, they will become automatic confident responses. Bernard Shaw once remarked: 'If you teach a man anything, he will never learn'. Learning is an active process. We learn by doing. The knowledge that is gained through experience sticks to the mind.

Some experts maintain that the ability to handle people is three times more important than technical expertise in determining the suitability of people for senior managerial positions. So all the technical knowledge (or job knowledge) in the world is of relatively little use if you aspire to senor management, without the social skills to handle people.

Handling people

The following guidelines will help you to handle people.

Tact

Don't argue: The best way to win an argument is to avoid it. Let the other person do most of the talking. Use the feedback technique. Ask questions in a friendly voice. Use the open-ended questioning technique – Why? What? When? Where? How? And Who? Arguing encourages emotional responses such as anger. Anger leads to irrational behavior and anything that encourages anger should be avoided. If, despite everything, a person gets angry, don't anger yourself in return. 'Anger blows out the light of reason, but, after a roaring hurricane expends its fury, the full of calm sets in'. Let the angry person blow himself/herself out.

Respect

Show respect for the other person's feelings and opinions. Tact and diplomacy are the watchwords here. Never say directly, or suggest to a person, 'You're wrong'. You could say you may be right, but let's look at the facts'. Nobody likes being told directly that they're wrong especially in front of a third party. Lord Chesterfield said to his son: 'be wiser than other people, if you can; but do not tell them so'. You may praise but never criticize in front of others. If you use diplomacy and tact, they might admit to themselves that they are wrong however they would never do so, if you crudely try to ram (Force) the fact down their throats.

Empathize

See things from the other person's point of view. How would you feel if you were in their shoes? That is what empathy means. Think ahead and try and anticipate reactions. Henry Ford said: 'If there is any one secret of success it lies in the ability to get the other person's point of view and see things from his angle as well as from your own'.

Simple

Speak in simple language. Remember the KISS technique-Keep if Short and Simple. Avoid jargon or technical terms. Don't try to impress others by using big words. Always use the simple work in preference to the long work. Keep your sentences short. Big words and long-winded sentences amount to woolly thinking, confusing others as well as yourself.

Praise

Most people are very reluctant to praise others, even when it is due. We spend most of our time knocking others. Treat people as winners and they will live up to your expectations. A sincere compliment for work well done is a boost to morale and an incentive for sustained excellence in the future. 'You can always catch more bees with honey than you can with vinegar'. However, insincerity is counterproductive. If you criticize for whatever reason, always soften the criticism with praise first. Criticize the act not the persons.

Appeal to senses

Dramatize your ideas and involve the other person. Appeal to as many of the senses as possible – hearing, sight, taste, smell and touch. When making a presentation to a group use visuals- it is said, "a picture speaks a thousand words". Remember the old proverb: 'I hear, I forget; I see, I remember; I do, I understand'.

Synchronize with your listener

Watch for body language and non-verbal communication. Try and synchronize your words with the speed of receptivity of the listener (you can gauge this from the actions and expressions of the listener). Mirror your body language to that of the listener.

Sell

Make other person feel that the idea is theirs. This is the best way to implement your ideas. Use the open-ended questioning technique to obtain viewpoints. Concentrate and develop those on which there is common ground and agreement. Show what is in it for them. Most people are motivated by self-interest.

Making an impression on others

The critical points about making an impression on other people are:

- Posture
- Look, listen and learn
- Expression
- Appearance
- Speech
- Eagerness to help

Posture

Your posture can indicate clearly whether or not you are going to be friendly and helpful. Indifference, nerves, restlessness and lack of confidence can all be indicated by posture.

Look, listen and learn

For the total message listen for the words and observe the accompanying body language. Pay attention to what they say and show that you are really interested in them. People prefer to talk about themselves, rather than listen to you. Do not interrupt the speaker or change the subject or you will show lack of interest and may create resentment.

Expression

Most people look at your face, and particularly your eyes, at some point during your conversation. Shakespeare said that the eyes are a mirror of the soul. Remember, your face and eyes reveal your feelings. They show the other person not only how your yourself feel – tired, interested, uninterested – but also how you feel about the other person. It's polite and a sign of attentiveness to look at the person who is talking to you (but avoid staring). Looking at a person also gives you an opportunity to notice their expression. Use your mouth. A smile, especially when greeting someone, can be the biggest icebreaker of all. Your mouth can show friendliness as easily as it reveals boredom and hostility.

Appearance

When somebody comes up to you, both you and your work area are on view. Do you give the impression that you are friendly, neat, well-groomed, smart and organized? Remember, if you look untidy, your work may be untidy too. People act on impressions. Make sure that you are neatly turned out and look after your appearance and personal hygiene.

Speech

Your first words create an impression, which steer the other person's reply. If they are friendly and positive, they invite a smile and a thank you from the other person. Nothing is more annoying than unhelpful, negative remarks. Remember to use the person/s name. There is nothing sweeter to a person's ear than the sound of their own name. Use the name frequently during the conversation. Link some outstanding feature with the person's name to remember it better. To be heard and understood, you need to speak clearly and look at the person as you speak. Do not use improper language or slang in conversation. If you use inappropriate language you will let yourself down and people will judge you by the way you speak.

Eagerness to help

Adopt a positive attitude and helpful nature in your dealings with people. Show in a positive way that you are enthusiastic, eager and willing to help in solving problems. If you want to make friends you must be seen to be obliging, unselfish and thoughtful.

Consequences of poor personal inter communication skills are:

- Stress
- Lack of communication
- Irritation
- Close mindedness
- No team spirit
- Lack of credibility
- Poor self-esteem

- Suspicion
- Loss of productivity
- Isolation
- Poor health
- Distrust
- Anger
- Prejudice
- Un co-operative behavior
- Conflict
- Frustration
- Unhappiness

Interpersonal communication skills are much more important than technical skills if you want to become a senior manager. Good communication is a very important aspect of interpersonal relationships. Listening is one part of communication, which sounds fine in theory but is rare to find in practice. Consequences of poor interpersonal communication skills are stress; lack of communication; irritation; close mindedness; no team spirit; lack of credibility; poor self esteem; suspicion; isolation, poor health etc.

Organizational Communication

Communication is the most vital ingredient of an organization. In fact, an organization cannot be conceived of without communication. An organization is a group of persons constituted to achieve certain specific objectives. The achievement of these objectives largely depends upon a proper co-ordination and integration of human effort in an organization. The people working in an organization are interrelated; their activities are also interrelated because all activities are performed only to achieve the organizational objectives. Co-ordination and integration of various human activities are possible only if there is an effective system of communication in the organization which provides for exchange of information and sharing of various ideas. The more effective the system of communication is, the better are the relations between the workers, and between workers and the management.

Communication is the nervous system of an organization. It keeps the member of the organization informed about the internal and external happenings relevant to a task and of interest to the organization. It coordinates the efforts of the member to achieving organizational objectives. It is the process of influencing the action person or a group. It is a process of meaningful interaction among human beings to initiate, execute, accomplish, or prevent certain actions. Communication is, thus, the life source of an organization. Without communication, an organization is lifeless and its very existence is in danger.

Types of Organizational Communication

Communication in an organization carries innumerable kinds of messages which may be difficult to map out; but it may be possible to classify communications in regard to how to transmit, or who communicates to whom, or what kinds of relationships communication develops. Thus, Communication may be grouped on the following basis.

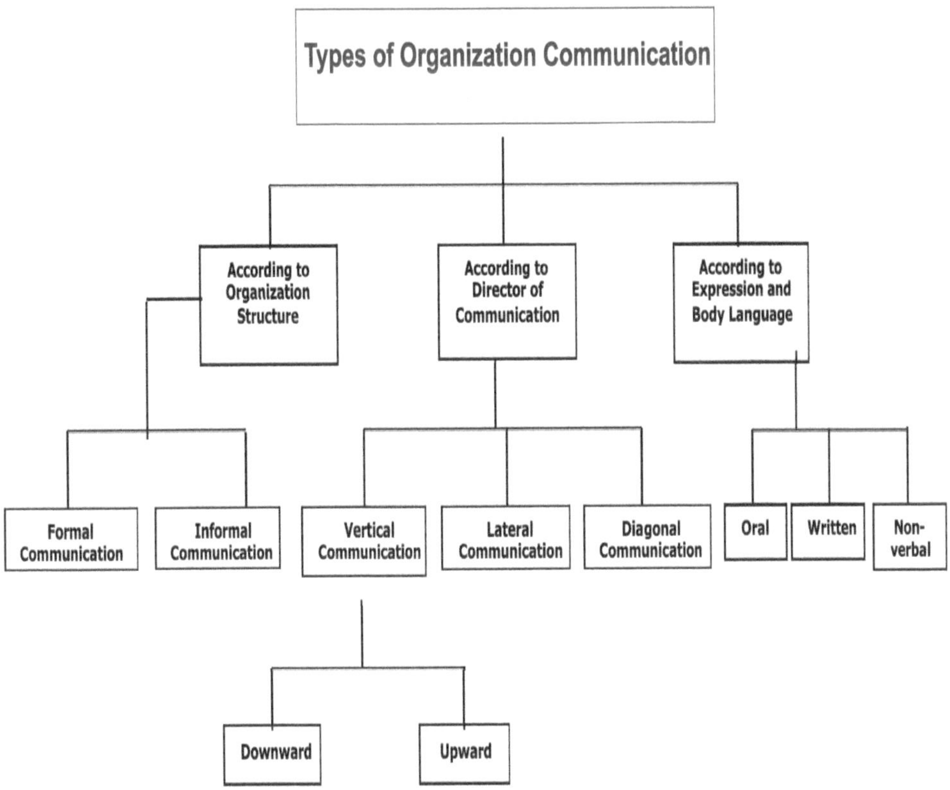

According to organizational structure

Formal communications

Such communications are those communications, which are associated with the formal organization structure. The travel through the formal channels – officially recognized positions in the organization chart. They are established mainly by the organization structure. Formal communications are mostly in black and white. We generally hear the phrase "through proper channel". It explains the essence of formal channels. Such communications include orders, instructions, decisions or intensions etc., of the superior.

Informal Communications

Informal Communications are also known as "Grapevine Communications". They are free from all sorts of formalities, because they are based on the informal relationship between the parties, such as friendship, membership of the same club or association or origin from the same place. Such communication includes comments, suggestions of any other informal reaction also. They may be conveyed by a simple glance, gesture, nod, smile or mere silence.

According to direction of communication

Downward Communication

Communications which flow from the superiors to subordinates are known as downward Communications. They include orders, rules instructions and policy directives etc. Their nature is directive. It would be impossible to manage an enterprise without downward Communications.

Upward Communications

Upward Communications are just reverse of the downward Communications. It flows from the sub-ordinates to their superiors. Such communications include reactions and suggestions from workers, their grievance etc. Contents of the upward Communication are reports, reactions, suggestions, statements and proposals prepared for the submission to the boss etc. Upward Communication is considered to be a main source of motivation in employees.

Exhibit : Vertical Communication

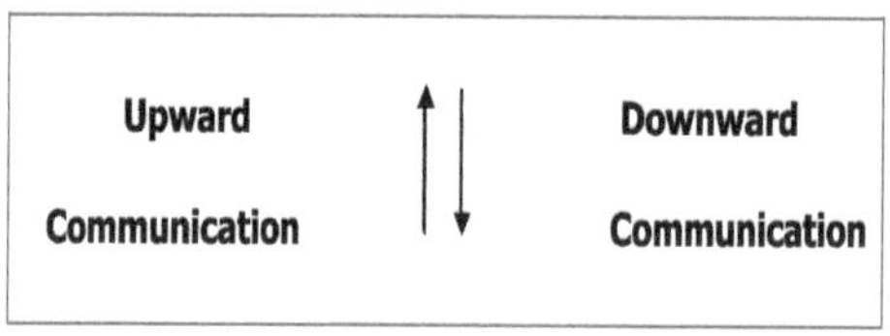

Lateral / Horizontal Communication

This type of communication refers to communication between various departments or units representing the same level, or people within the same or different departments, without having a superior – sub-ordinate relationship viz., peer groups, friends and trainees. It flows between persons at the same hierarchical level. The main object of this type of communication is to co-ordinate the efforts of different departments performing different but related activities. Such communications may be oral or written.

Exhibit: Lateral Communication

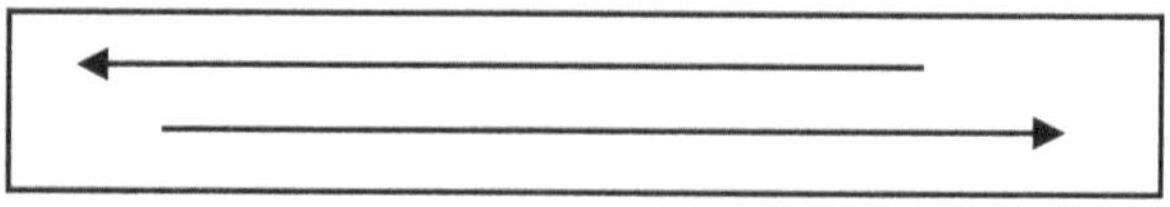

Diagonal communication

In an organization, communication does not necessarily traverse along a stipulated path. While vertical and lateral forms for transmission of messages are important, there is yet another mode to be taken cognizance of, that is, the diagonal. In this type of communications, there is no direct path chalked out for information to travel. It could, at certain stages, take on the upward path, then a lateral direction and finally, move downward, or it could even skip certain stages. This channel proves to be very affective as hierarchical bindings are done away with and communication flows irrespective of position or status. It also helps in building relationships and binding ties between the superior and the subordinate.

Inter-organizational communication

This is a system developed by each organization to communicate with another organization .Department or organizations do not intercommunicate as such rather people do .Communication always occurs between people.

The communication process

The dynamics of communication has been both explicit and implicit through the process. It retains and organizes. As a process, it is ongoing and a continuity as well. It starts from a source and reaches the ultimate receiver through a channel. While it is traveling and traversing along and across the channels, the message keeps moving in its original form or in a changed version. The source is either any human or non human entity and even the receiver. The accepted definition of communication is that it is a process by which a 'Source' is sending a 'Message' through a 'Channel' to some 'Receiver' for some 'Effect'. It is popularly known as SMCRE approach in defining communication process.

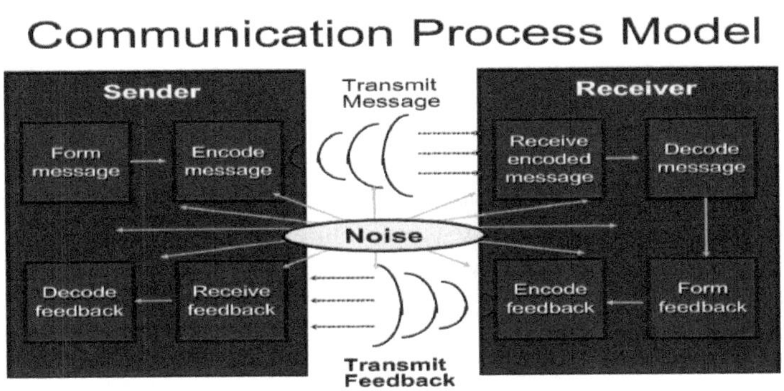

Chapter 2

Principles of Effective Communication

Interpersonal communication is the foundation of human interaction. Its importance for innovation and change can hardly be overemphasized.

Basic Principles of Effective Communication

Many definitions describe communication as a transfer of information, thoughts or ideas to create shared understanding between a sender and a receiver. The information may be written or spoken, professional or social, personal or impersonal to name a few possibilities. Basically, the communication process involves a sender, receiver, message, channel and feedback. However, this simplistic description significantly under-represents what can actually be a very complex process. Click here for a brief overview of the communication process.

Essential issues to be aware of in any communication situation are:

- Content refers to the actual words or symbols of the message that are known as language - the spoken and written words combined into phrases that make grammatical sense. Importantly, we all use and interpret the meanings of words differently, so even simple messages can be misunderstood. And many words have different meanings to confuse the issue even more.

- Process refers to the way the message is delivered - the nonverbal elements in speech such as the tone of voice, the look in the sender's eyes, body language, hand gestures and state of emotions (anger, fear, uncertainty, confidence, etc.) that can be detected. The non-verbals that we use often cause messages to be misunderstood as we tend to believe what we see more than what we hear. Indeed, we often trust the accuracy of nonverbal behaviors more than verbal behaviors. A well-known UCLA study found that only around 7% of the meaning of spoken communication came from words alone, 55% came from facial expression and 38% came from the way the words were said.
- Context refers to the situation or environment in which your message is delivered. Important contextual factors that can subtly influence the effectiveness of a message include the physical environment (eg. a patient's bedside, ward office, quiet room etc.), cultural factors (eg. international cultures, organizational cultures and so on) and developmental factors (eg. first, second or third year student, experience in similar clinical settings, stage of the practicum etc.).

The goal of communication between a sender and a receiver is primarily facilitating the understanding of the message being sent. Anything that interferes with this can be referred to as 'noise'. Communication noise can influence our interpretation of messages and significantly affect our perception of interactions with others. Read more about some examples of noise. However, there are a number of strategies that can help to help minimise communication noise and maximise communication efficiency. Read more about these strategies. Developing competence in communication, particularly in the professional context, requires ongoing practice and reflection on practice. Watch these **Videos** for practical suggestions from students and staff for enhancing communication in the workplace. While you're watching the videos, think about your own communication skills and what you can learn from the information in this section that will help you communicate even more effectively.

Principles of Communication

Communication is a two-way process of giving and receiving information through any number of channels. Whether one is speaking informally to a colleague, addressing a conference or meeting, writing a newsletter article or formal report, the following basic principles apply:

- Know your audience
- Know your purpose
- Know your topic
- Anticipate objections
- Present a rounded picture
- Achieve credibility with your audience
- Follow through on what you say

- Communicate a little at a time
- Present information in several ways
- Develop a practical, useful way to get feedback
- Use multiple communication techniques

Communication is complex. When listening to or reading someone else's message, we often filter what's being said through a screen of our own opinions. One of the major barriers to communication is our own ideas and opinions.

There's an old communications game, telegraph, that's played in a circle. A message is whispered around from person to person. What the exercise usually proves is how profoundly the message changes as it passes through the distortion of each person's inner "filter."

Environmental factors

Communication can be influenced by environmental factors that have nothing to do with the content of the message. Some of these factors are:

- The nature of the room, how warm it is, smoke, comfort of the chair, etc.
- Outside distractions, what is going on in the area
- The reputation/credibility of the speaker/writer
- The appearance, style or authority of the speaker
- Listener's education, knowledge of the topic, etc
- The language, page layout, design of the message

People remember

- 10% of what they read
- 20% of what they hear
- 30% of what they see
- 40% of what they hear and see

Communication with Decision Makers

Innovation and change often depends upon persuading potential users of the benefits of an innovation. To deal persuasively with decision makers, it is necessary to know and understand their interests and opinions. The following questions are helpful in organizing technology transfer efforts:

a) Who are the key people to persuade?

b) Who will make the decisions about innovation and change?

c) What are these decision makers' past experiences with innovation and change?

d) What are the decision makers' current attitudes toward innovation and change? Are they neutral, friendly, hostile or apathetic?

e) What is the most appropriate way to approach the decision maker?

f) What are the work styles of the decision makers? Are they highly formal people who want everything in writing and all appointments scheduled in advance? Or are they more flexible, responding favorably to personal telephone calls and informal meetings?

g) What networks or groups is the decision maker a part of?

h) What programs or services will the new innovation improve?

i) What programs or services will the new innovation cause problems with?

j) How will the innovation or change benefit the decision maker?

Principles of Effective Persuasion

Whether making a formal presentation at a meeting or writing a report or fact sheet, the following principles hold.

- Do not oversell or overstate your case. Make effective use of understatement.
- Outline the topic you are trying to cover into two parts. The first part should give broad background information, while the second part provides a detailed summary.
- Persuasion depends on clarity and simplicity. Avoid the use of jargon and buzz words.
- Be prepared to back up claims or facts immediately.
- Incorporate major anticipated objections into your program or presentation.
- Address all relevant aspects of a topic, especially those that may affect the functioning of an organization.
- Use graphics and audiovisuals appropriately.
- Consider ways to get meaningful input from people. Find out what they think about the innovation or change.

Selling New Ideas

Creating Isn't Selling

Often the creators of an innovation feel that convincing others of the idea's value is somehow superfluous to their activities. To them, conceiving the idea is enough. This combines with their inner conviction that their idea will "sell itself." Change agents provide a link between creators of new techniques and users.

Ideas Need Selling

Someone must recognize when an idea is good. It is important that when an idea is good it is sold to those who can act on it--those who have the power to evaluate and adopt it. Understanding users is an important activity for any change agent. People must be convinced that a particular idea or innovation has enough merit to warrant adoption.

Selling Ideas Takes Effort

Selling innovations requires preparation, initiative, patience, and resourcefulness. It may take more effort than originating the idea. In an age of technical complexity and information overload, new ideas seldom stand out. Information on new ideas must be targeted to the appropriate users and relate to their needs and motivations.

Once is not enough

A new idea has to be suggested many times before it will "catch on." Initial failures at promoting a new idea are to be expected, so don't get discouraged if you don't get the results you want the first time. Some ideas take years to catch on. However, first exposures are crucial to future prospects. Do it right the first time

Feedback (Listening)

Getting and giving feedback is one of the most crucial parts of good communication. Like any other activity, there are specific skills that can enhance feedback. Listening is a key part of getting feedback:

Listen to the Complete Message

Be patient. This is especially important when listening to a topic that provokes strong opinions or radically different points of view. In these situations, it's important not to prejudge the incoming message. Learn not to get too excited about a communication until you are certain of the message.

Work at Listening Skills

Listening is hard work. Good listeners demonstrate interest and alertness. They indicate through their eye contact, posture and facial expression that the occasion and the speaker's efforts are a matter of concern to them. Most good listeners provide speakers with clear and unambiguous feedback.

Judge the Content, Not the Form of the Message

Such things as the speaker's mode of dress, quality of voice, delivery mannerisms and physical characteristics are often used as excuses for not listening. Direct your attention to the message--what is being said--and away from the distracting elements.

Weigh Emotionally Charged Language

Emotionally charged language often stands in the way of effective listening. Filter out "red flag" words (like "liberal" and "conservative," for instance) and the emotions they call up. Specific suggestions for dealing with emotionally charged words include

- Take time to identify those words that affect you emotionally.
- Attempt to analyze why the words affect you the way they do.
- Work at trying to reduce the impact of these words on you.

Eliminate Distractions

Physical distractions and complications seriously impair listening. These distractions may take many forms: loud noises, stuffy rooms, overcrowded conditions, uncomfortable temperature, bad lighting, etc. Good listeners speak up if the room is too warm, too noisy, or too dark. There are also internal distractions: worries about deadlines or problems of any type may make listening difficult. If you're distracted, make an effort to clear your head. If you can't manage it, arrange to communicate at some other time.

Think Efficiently and Critically

On the average, we speak at a rate of 100 to 200 words per minute. However, we think at a much faster rate, anywhere from 400 to 600 words per minute. What do we do with this excess thinking time while listening to someone speak? One technique is to apply this spare time to analyzing what is being said. They critically review the material by asking the following kinds of questions:

- What is being said to support the speaker's point of view? (Evidence)
- What assumptions are being made by the speaker and the listener? (Assumptions)
- How does this information affect me? (Effect)
- Can this material be organized more efficiently? (Structure)
- Are there examples that would better illustrate what is being said? (Example)
- What are the main points of the message? (Summary)

Sending Messages

Messages should be clear and accurate, and sent in a way that encourages retention, not rejection.

- **Use Verbal Feedback Even if Nonverbal is Positive and Frequent**: Everyone needs reassurance that they are reading nonverbal communication correctly, whether a smile means "You're doing great," "You're doing better than most beginners," or "You'll catch on eventually."
- **Focus Feedback on Behavior Rather than on Personality**: It's better to comment on specific behavior than to characterize a pattern of behavior. For example, instead of calling a colleague inefficient, specify your complaint: "You don't return phone calls; this causes problems both in and outside your office."
- **Focus Feedback on Description Rather than Judgment**: Description tells what happened. Judgment evaluates what happened. For example, in evaluating a report doesn't say, "This is a lousy report!!" Instead, try: "The report doesn't focus on the information that I think needs emphasis," or "This report seems to have a lot of grammatical and spelling mistakes."

- **Make Feedback Specific Rather than General:** If feedback is specific, the receiver knows what activity to continue or change. When feedback is general, the receiver doesn't know what to do differently. For example, in an office situation, instead of saying "These folders are not arranged correctly," it's better feedback to say, "These should be arranged chronologically instead of alphabetically."
- **In giving Feedback, Consider the Needs and Abilities of the Receiver:** Give the amount of information the receiver can use and focus feedback on activities the receiver has control over. It's fruitless to criticize the level of activity, if the decision to grant the necessary monies for materials, personnel or technology is made at a different level.
- **Check to See if the Receiver Heard What You Meant to Say:** If the information is important enough to send, make sure the person understands it. One way of doing this is to say, "I'm wondering if I said that clearly enough. What did you understand me to say?" or "This is what I hear you saying. Is that right?"

Selecting the Best Communication Method

In communicating with decision makers, use the most appropriate communications method. One way to do this is to ask yourself the following questions.

- What is the purpose of your message? Do you plan to tell them something new? Inform? Do you plan to change their view? Persuade?
- What facts must be presented to achieve your desired effect?
- What action, if any, do you expect decision makers to take?
- What general ideas, opinions and conclusions must be stressed?
- Are you thoroughly familiar with all the important information on the innovation?
- What resources and constraints affect adoption of the innovation? How much time is available? How much money is available
- Which method, or combination of methods, will work most effectively for this situation? Personal contact--requires scheduling, time and interpersonal skills.

Chapter 3

Models of Communication

Models are symbolic representations of structures, objects or operations. They are useful theoretical constructs that are frequently used in social sciences for explanatory purposes. They may be used to show the size, shape or relationship of various parts or components of an object or process. A model may also be useful in explaining the working of a system.

Concepts of model

i. It is representation of the process of communication in the same sense that a blueprint is a representation of house.

ii. Models are symbolic representations of structures, objects or operations.

Purpose of models

Communication models serve three main purposes

i. They describe the process of communication.

ii. They visually show relationship among the variables involved in communication and

iii. They aid in finding and in correcting communication problems

Aristotle's model

According to Aristotle, communication has three ingredients

1. Speaker – the person who speaks
2. Speech – the speech that the individual produces
3. Audience – the person who listens

Aristotle's Model

The Osgood and Schramm circular model emphasizes the circular nature of communication.

The participants swap between the roles of source/encoder and receiver/decoder.

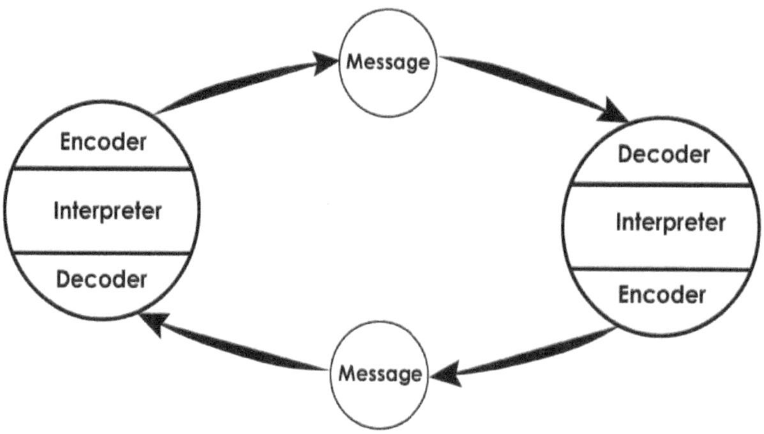

The Osgood & Schramm Circular Model
Please click a button for further information
(Massage, encodar and decoder
will take you to Shannon Weaver)

Source: CCMS-Infobase at http://www.cultsock.ndirect.co.uk/

Gerbner's General Model

Gerbner's General Model also emphasizes the dynamic nature of human communication.

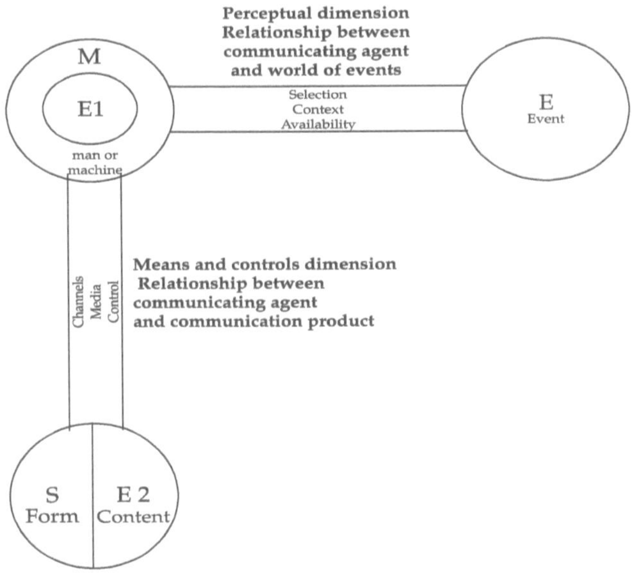

Gerbner's General Model
Please click a button for further information

Shannon-Weaver's model

The Shannon-weaver (1949) model is consistent with Aristotle's proposition. According to them, the ingredients of communication are :

1. Source
2. Transmitter
3. Signal
4. Receiver
5. Destination

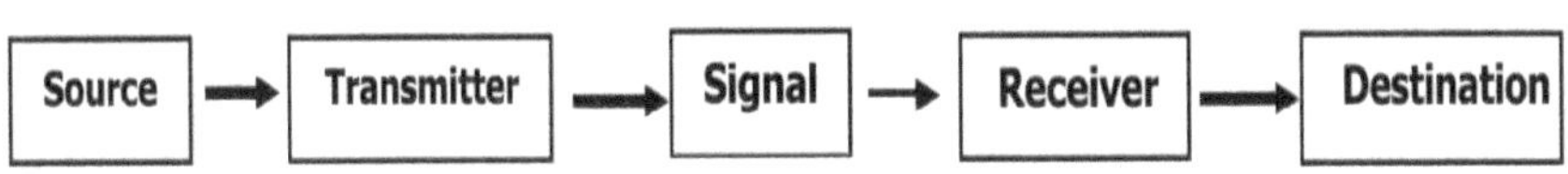

Shannon and Weaver's Model

Compared with the Aristotelian model, the source is the speaker, the signal is the speech and the destination is the audience, plus two added ingredients, it transmits which sends out the source's message and a receiver which catches the message for the destination.

Berlo's model

According to Berlo (1960) the model of communication consists of

a) Source
b) Encoder
c) Message
d) Channel
e) Decoder
f) Receiver

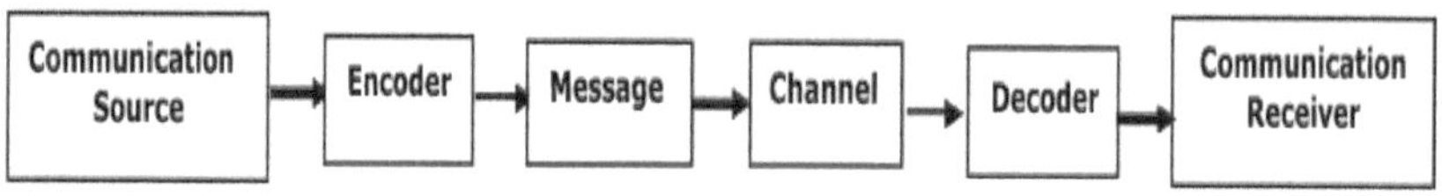

Berlo's model

Code is a system of signals for communication. Encode means to put the message into code. Channel means the medium through which the signals move, the decoder means which converts the message in the code into ordinary language which may be easily understood. He further elaborated that all human communication has some source, some person or group of persons with a purpose. The purpose has source has to be expressed in the form of message. The communication encoder is responsible for taking the ideas of the source and putting them in a code, expressing the source's purpose in the form of a message. A channel is a medium, a carrier of message. For communication to occur there must be somebody at the other end, who can be called the communication receiver, the target of communication.

Mass-media

It is a collective phrase that represents not only the press, cinema, radio, television and internet, but also to some extent, books magazines, pamphlets , direct mail literature, posters, folk media, and natural communication methods such as rumors, education and preaching. It is so termed because its reach extends to vast heterogeneous populations. Generally the mass media employ technological means to communicate to the masses. They are founded on the idea of mass production and distribution. Wiebe defined mass media as those readily available to the general public.

Selective attention

The media are full of competing messages. The process of screening vast amount of information in which one has no interest through mental filters is called selective attention, for example, an adult will be more tuned to listening to the news while a child would rather watch a cartoon show.

Selective perception

This is the tendency to interpret communication messages in terms of one's existing attitudes. People of distinct psychological character prefer same media content in different ways. This depends on factors such as age, values, family, opinions etc. Selective perception is influenced by social relationships.

Selective retention

The ability of an individual to retain certain messages in his mind while ignoring others is called selective retention. This is influenced by various psychological and physiological factors such as choice, values, culture, emotions etc.

Selective exposure

Some individuals are exposed to certain media effects/messages while some are not. This screening aspect depends on many factors such as reach of media, accessibility, age, cultural acceptability, taboos, etc.

Constructionist Models

There is an additional working definition of communication to consider that authors like Richard A. Lanham (2003) and as far back as Erving Goffman (1959) have highlighted. This is a progression from Lasswell's attempt to define human communication through to this century and revolutionized into the constructionist model. Constructionists believe that the process of communication is in itself the only messages that exist. The packaging can not be separated from the social and historical context from which it arose, therefore the substance to look at in communication theory is style for Richard Lanham and the performance of self for Erving Goffman.

Linear Model

This is a one way model to communicate with others. It consists of the sender encoding a message and channeling it to the receiver in the presence of noise. Draw backs – the linear model assumes that there is a clear cut beginning and end to communication. It also displays no feedback from the receiver.

- For example; a letter, an email, a text message or a lecture.

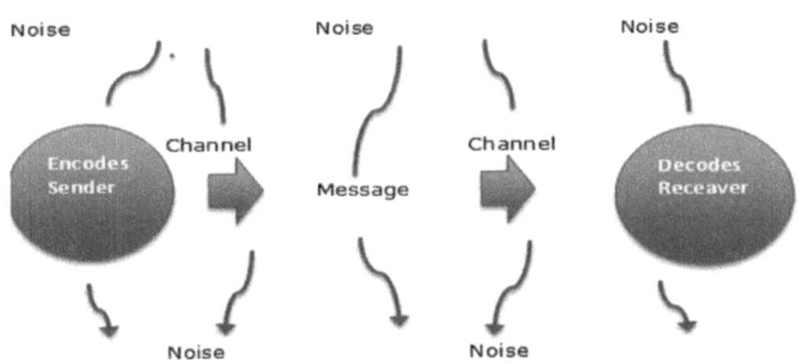

The Linear Model

Interactive Model

The two linear models are stacked on top of each other. The sender channels a message to the receiver and the receiver then becomes the sender and channels a message to the original sender. This model has added feedback, indicates that communication is not a one way but a two way process. It also has "field of experience" which includes our cultural background, ethnicity geographic location, extend of travel, and general personal experiences accumulated over the course of your lifetime. Draw backs – there is feedback but it is not simultaneous.

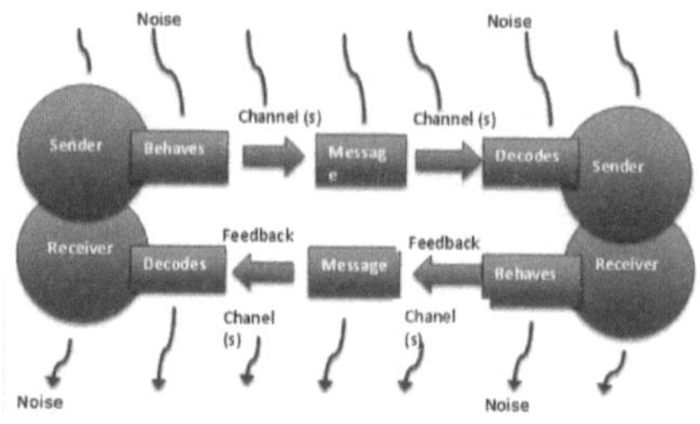

The Interactive Model

For example: Instant Messaging. The sender sends an IM to the receiver, then the original sender has to wait for the IM from the original receiver to react, or a question/answer session where you just ask a question then you get an answer.

Transactional Model It assumes that people are connected through communication; they engage in transaction. Firstly, it recognizes that each of us is a sender-receiver, not merely a sender or a receiver. Secondly, it recognizes that communication affects all parties involved. So communication is fluid / simultaneous. This is how most conversation is alike. The transactional model also contains ellipses that symbolize the communication environment (how you interpret the data that you are given). Where the ellipses meet is the most effect communication area because both communicators share the same meaning of the message.

For example – talking/listening to friends. While your friend is talking you are constantly giving them feedback on what you think through your facial expression verbal feedback without necessarily stopping your friend from talking.

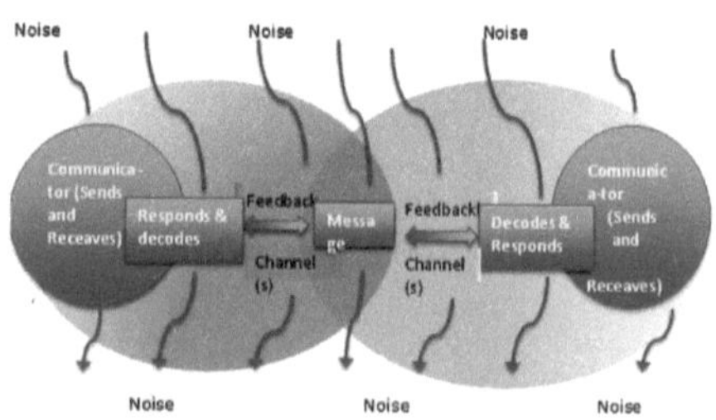

The Transactional Model

Schramm's model

According to Schramm (1961), the communication process involves –

1. Source
2. Encoder
3. Signal
4. Decoder
5. Destination

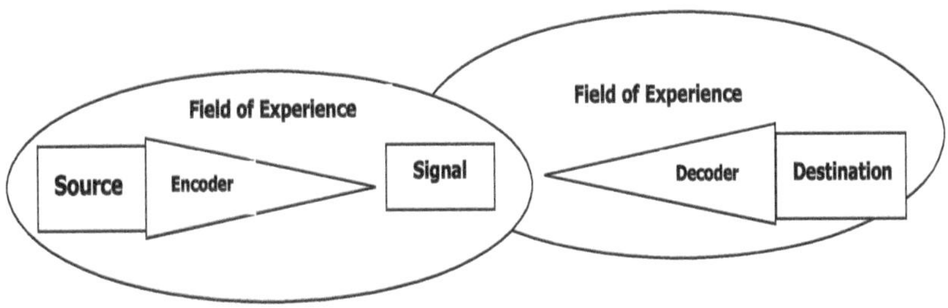

This model of communication is preferably relevant for the mass media. In human communication it is most important whether people can properly encode or decode the signal (message), and how they interpret it in their own situations.

Jean Paul Leagan's model

The communication model forwarded by Leagans (1963) has the following elements -

a. Communicator
b. Message
c. Channel
d. Treatment
e. Audience
f. Response

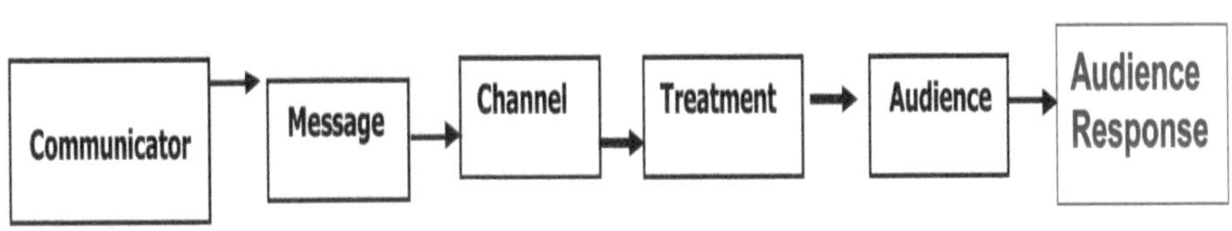

J P Leagans Model

The task of communication, according to him is to provide powerful incentives for change. Success at this task requires thorough understanding of the six elements of communication, a skillful communicator sending useful message through proper channel, effectively treated, to an appropriate audience that responds as desired.

Rogers and Shoemaker's model

Rogers and shoemaker (1971) thought of the communication process in terms of the S-M-C-R-E model, the components of which are –

a. Source

b. Message

c. Channel

d. Receiver

e. Effects

Rogers and Shoemaker's model

According to them a source **(S)** and message **(M)** via certain channels **(C)** to the receiving individual **(R),** which causes some effects **(E)** i.e., changing the existing behavior pattern of the receiver.

Westley and Maclean's model

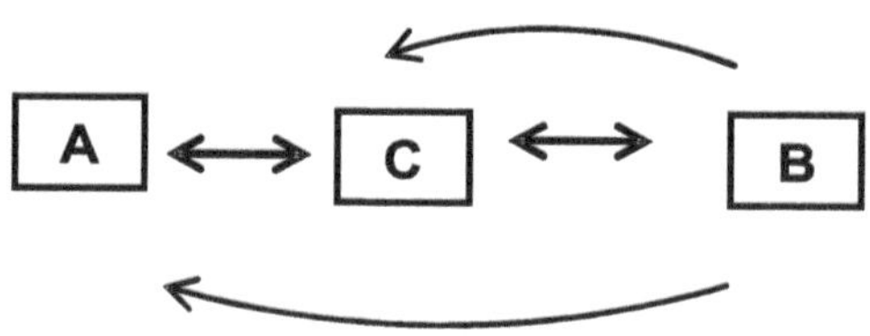

'A' is source (Eg., ANGRAU), 'B' is receiver or public, 'C' is the mass media channel (eg., E-TV Annadata), 'A' gets feedback from the public refines the message and 'C' also gets feedback and refines and the total refinements are carried out by the interpreting source that is 'C' and then refined messages are transmitted to the users. This is typical in a TV channel broadcast. eg. Annadata of E-TV does similar to what has been explained above.

Model on extension communication system

Communication in extension may also be thought of as two-way stimulus – response (S-R) situation in which the necessary stimulus is provided by the communicator, the extension agent, in the form of a message which produces certain response on the audience, the farmers and vice-versa. A favorable response by the audience reinforces learning. A diagrammatic representation of the extension communication system on the basis of the model suggested by Leagans (1963) is presented in the following figure.

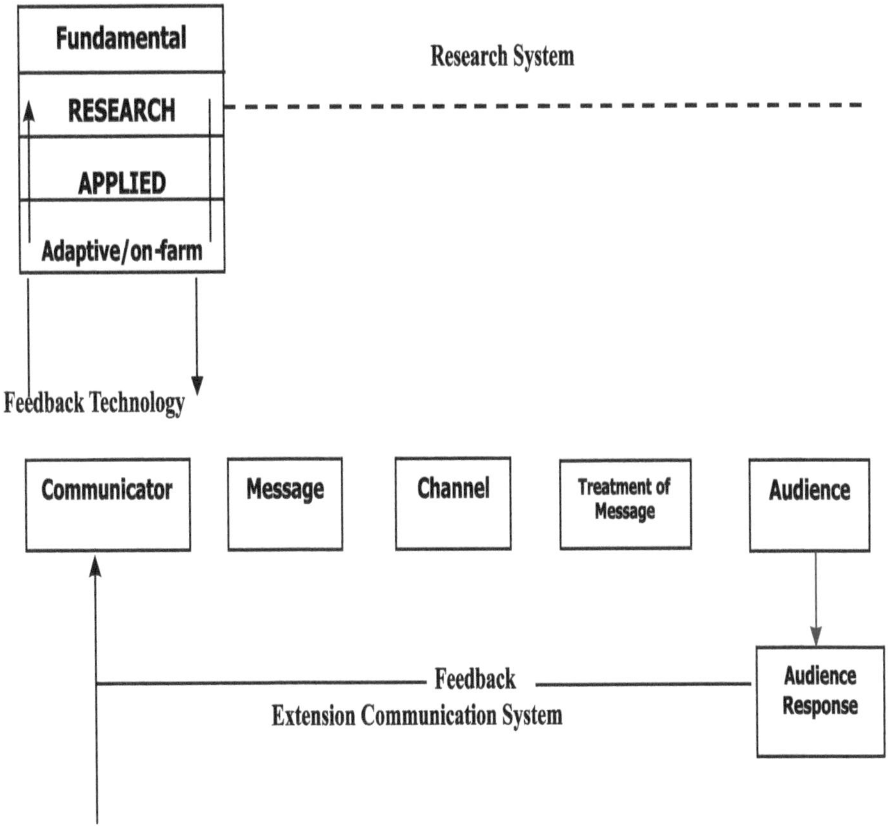
Fundamental
RESEARCH
APPLIED
Adaptive/on-farm
Research System
Feedback Technology
Communicator
Message
Channel
Treatment of Message
Audience
Audience Response
Feedback
Extension Communication System

Chapter 4
Theories of Communication

History of communication theory

Communication has existed since the beginning of human beings, but it was not until the 20th century that people began to study the process. As communication technologies developed, so did the serious study of communication? When World War I ended, the interest in studying communication intensified. The social-science study was fully recognized as a legitimate discipline after World War II.

Prior to becoming simply 'communication', or 'communication studies', the discipline was derived from three other major studies: psychology, sociology, and anthropology. Psychology is the study of human behavior, Sociology is the study of society and social process, and anthropology is the study of communication as a factor which develops, maintains, and changes culture. Communication studies focus on communication as central to the human experience, which involves understanding how people behave in creating, exchanging, and interpreting messages. Communication Theory has one universal law posited by S. F. Scudder (1980).

The Universal Communication Law states that, "All living entities, beings and creatures communicate." All of the living communicates through movements, sounds, reactions, physical changes, gestures, languages, breath, etc. Communication is a means of survival. Examples - the cry of a child (communication that it is hungry, hurt, cold, etc.); the browning of a leaf (communication that it is dehydrated, thirsty per se, dying); the cry of an animal (communicating that it is injured, hungry, angry, etc.). Everything living communicates in its quest for survival."

Communication Theory Framework

It is helpful to examine communication and communication theory through one of the following viewpoints:

Mechanistic

This view considers communication as a perfect transaction of a message from the sender to the receiver.

Psychological

This view considers communication as the act of sending a message to a receiver, and the feelings and thoughts of the receiver upon interpreting the message.

Social Constructionist (Symbolic Inter actionist)

This view considers communication to be the product of the individuals sharing and creating meaning. The Constructionist View can also be defined as, how you say something determines what the message is. The Constructionist View assumes that "truth" and "ideas" are constructed or invented through the social process of communication. Robert T. Craig saw the Constructionist View or the constitutive view as it's called in his article, as "...an ongoing process that symbolically forms and re-forms our personal identities." The other view of communication, the Transmission Model, sees communication as robotic and computer-like. The Transmission Model sees communication as a way of sending or receiving messages and the perfection of that. But, the Constructionist View sees communications as, "...in human life, info does not behave as simply as bits in an electronic stream. In human life, information flow is far more like an electric current running from one landmine to another". The Constructionist View is a more realistic view of communication because it involves the interacting of human beings and the free sharing of thoughts and ideas. Daniel Chandler looks to prove that the Transmission Model is a lesser way of communicating by saying "The transmission model is not merely a gross over-simplification but a dangerously misleading representation of the nature of human communication". Humans do not communicate simply as computers or robots so that's why it's essential to truly understand the Constructionist View of Communication well. We do not simply send facts and data to one another, but we take facts and data and they acquire meaning through the process of communication, or through interaction with others.

Systemic

This view considers communication to be the new messages created via "through-put", or what happens as the message is being interpreted and re-interpreted as it travels through people.

Critical

This view considers communication as a source of power and oppression of individuals and social groups.

Inspection of a particular theory on this level will provide a framework on the nature of communication as seen within the confines of that theory. Theories can also be studied and organized according to the ontological, epistemological, and axiological framework imposed by the theorist. Ontology essentially poses the question of what, exactly, it is the theorist is examining. One must consider the very nature of reality. The answer usually falls in one of three realms depending on whether the theorist sees the phenomena through the lens of a realist, nominalist, or social constructionist. Realist perspective views the world objectively, believing that there is a world outside of our own experience and cognitions. Nominalists see the world subjectively, claiming that everything outside of one's cognitions is simply names and labels. Social constructionists straddle the fence between objective and subjective reality, claiming that reality is what we create together.

Epistemology is an examination of how the theorist studies the chosen phenomena. In studying epistemology, particularly from a positivist perspective, objective knowledge is said to be the result of a systematic look at the causal relationships of phenomena. This knowledge is usually attained through use of the scientific method. Scholars often think that empirical evidence collected in an objective manner is most likely to reflect truth in the findings. Theories of this ilk are usually created to predict a phenomenon. Subjective theory holds that understanding is based on situated knowledge, typically found using interpretative methodology such as ethnography and also interviews. Subjective theories are typically developed to explain or understand phenomena in the social world.

Axiology is concerned with what values drive a theorist to develop a theory. Theorists must be mindful of potential biases so that they will not influence or skew their findings. A discipline gets defined in large part by its theoretical structure. Communication studies often borrow theories from other social sciences. This theoretical variation makes it difficult to come to terms with the field as a whole. That said, some common taxonomies exist that serve to divide up the range of communication research. Two common mappings involve contexts and assumptions. Many authors and researchers divide communication by what they sometimes called "contexts" or "levels", but which more often represent institutional histories. The studies of communication in the US, while occurring within departments of psychology, sociology, linguistics, and anthropology (among others), generally develop from schools of rhetoric and from schools of journalism. While many of these have become "departments of communication", they often retain their historical roots, adhering largely to theories from speech communication in the former case, and from mass media in the latter. The great divide between speech communication and mass communication becomes complicated by a number of smaller sub-areas of communication research, including intercultural and international communication, small group communication, communication technology, policy and legal studies of communication, telecommunication, and work done under a variety of other labels. Some of these departments take a largely social-scientific perspective, others tend more heavily toward the humanities, and still others gear themselves more toward production and professional preparation. These "levels" of communication provide some way of grouping communication theories, but inevitably, some theories and concepts leak from one area to another, or fail to find a home at all.

Communication theory as a field

Another way of dividing up the communication field emphasizes the assumptions that under gird particular theories, models, and approaches. While this approach also tends to have as its basis institutional divisions, theories within each of the seven "traditions" of communication theory that Robert T. Craig suggests tend to reinforce one another, and retain the same ground epistemological and axiological assumptions. His traditions include:

- Rhetorical - practical art of discourse
- Semiotic – inter subjective mediation through signs in order to mediate between different perspectives
- Phenomenological - experience of otherness, dialogue
- Cybernetic - information processing and explains how all kinds of complex systems, whether living or nonliving, macro or micro, are able to function, and why they often malfunction
- Socio-psychological - expression, interaction and influence
- Critical - discursive reflection
- Socio-cultural - reproduction of social order
- Craig finds each of these clearly defined against the others, and remaining cohesive approaches to describing communicative behavior. As a taxonomic aid, these labels help to organize theory by its assumptions, and help researchers to understand why some theories may seem incommensurable.

Human communication is understood in various ways by those who identify with the field. This diversity is the result of communication being a relatively young field of study, composed of a very broad constituency of disciplines. It includes work taken from scholars of Rhetoric, Journalism, Sociology, Psychology, Anthropology, and Semiotics, among others. Cognate areas include bio-communication, which investigates communicative processes within and among non-humans such as bacteria, animals, fungi and plants, and information theory, which provides a mathematical model for measuring communication within and among systems.

Generally, human communication is concerned with the making of meaning and the exchange of understanding. One model of communication considers it from the perspective of transmitting information from one person to another. In fact, many scholars of communication take this as a working definition, and use Lasswell's maxim, «who says what to whom in which channel with what effect,» as a means of circumscribing the field of communication theory. Among those who subscribe to the transmission model are those who identify themselves with the communication sciences, and finds its roots in the studies of propaganda and mass media of the early 20th century.

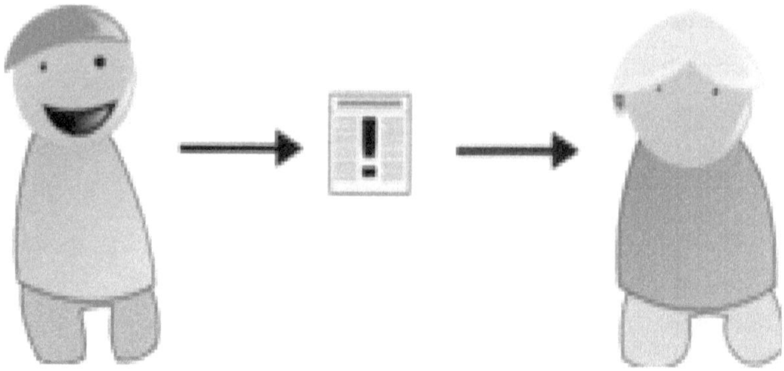

A simple communication model with a sender transferring a message containing information to a receiver

Other commentators claim that a ritual process of communication exists, one not artificially divorceable from a particular historical and social context. This tradition is largely associated with early scholars of symbolic interactionism as well as phenomenologist.

The theories of communication can be classified into three groups namely:

Theories about any phenomenon in general will highlight the key concepts involved in it. These theories of communication will be useful for a person who is trying to understand a given communication event by way of providing a framework which suggests what he should look for. These theories should be viewed based on the basic assumptions that communications is a transitive, multidimensional, multipurpose process. Obviously, no theory or model is going to do all we ask of it. All the models/theories to be surveyed, at least fail one or more of the criteria. It is the purpose of a theory or model to reduce reality to a manageable number of elements and relationships among those elements. The students of communication should be familiar with a variety of communication theories and models so that he or she can use that one which is appropriate for the analysis of a given situation.

In general, the theories of interpersonal communication explain it as the face to face interaction between people who are consistently aware of each other. Here the mutual pressure of the message generator and receiver is required. Each person assures the rules of sender and receiver of message who are independent and interchangeable. The behavior of the participants in interpersonal communication is so intertwined in the process that it is difficult to separate the sender from receiver.

Attribution theory gives stress to the psychological state, motives and intensions of the individuals involved in communication.

Game theory views interpersonal communication as the process of transaction between the people to balance possible gains and losses. Social comparison theory states that persons communicate in order to verify their perceptions of the world assumptions about themselves. New comb's theory explains the thinking processes of people who are reacting to one another simultaneously – linguistic theory deals with a description of language systems. Psycholinguistic theory focuses attention on the behavior of an individual in the use of language. Socio-linguistic theory explains patterns of variation in language across different people, stimuli relationship, context and competence.

Theories of mass communication (Message flow models)

Several models have been conceptualized to explain the flow of message from the communicator to the mass audience. Based on the number of ways involved in the flow of message between sender and receiver theories like hypodermic needle, one step flow, two step flow, and multistep flow are conceived. The response to mass media message is said to occur in line with the psychological makeup of the individuals was taken as the assumption as per the theory of individual differences. Psychodynamic model explains about the relationship between internal processes and manifest overt behavior of the persuader. Theory of social category emphasizes that people who have a number of similar characteristics are likely to have similar orientation and behavior resulting in a fairly uniform response to mass media. Free press theory stresses that the act of publication and distribution should be open to the audience and no restriction should be placed on the publication of new items, social responsibility theory states that media should accept and fulfill certain obligations to society. Development media theory states that media should accept and carryout positive development tasks in line with nationally established policy. The central point of the democratic participant media theory lies with the needs, interests and aspiration of receiver in a political theory.

Theories of communication distortion

The first one states that larger the volume of communication, larger would be the communication distortion. Increasing the volume of communication increases more chances for misunderstanding, misinterpretation, less concentration and low priorities.

Theory of communication distortion in transit

The second one emphasizes that messages pass through many layers in an organization. Therefore, repetition and reproduction of the same message is multiplied into many which reduces the efficiency of communication. The overview of communication theories suggests that communication can be described in terms of networks, which give the observer some indication of who is communicating with whom it can also be described in terms of the types of interactions which are occurring among communicators. Further, it can be described in terms of message transactions, which indicate how the elements of communication combine to produce a unique unrepeatable event. It can be also viewed from the point view of individual gathering information as he passes through his environment.

Communication process and the key communicator

"To be a leader you have got to lead human beings with affection" J.R.D. TATA. "The key is to select good people, be accessible to them when need you but leave them free to operate" Russi Modi. Key communicators or opinion leaders are persons in any social systems that are sought out for information and advice on general or specific topics. Rogers and Kincoid (1981) define opinion leadership as the degree to which an individual is able to informally influence other's knowledge, attitudes, or overt behavior in a desired way with relative frequency. It is assumed that such persons are respected persons in each social system to whom some people look for advice and information and who, through such consultations, influence their behavior and actions. The concept of key communicator was first developed by Katz & Lazarfield et al. (1944). Since then they have been variously called as fashion leaders, influencers, information leaders, opinion leaders, spark plugs, style-setters, taste-makes etc. However, all these names have been used more or less for individuals operating in a social system who are more important in the communication of information than others. Research evidence suggests that each social system has its own opinion leaders.

Identification methods of key informants

Three methods are in common use for identifying key-informants. These methods are (i) Socio-metric method, (ii) self-designation technique and (iii) information ratings.

Socio-metric method

In this method, members of a social system are asked to name the person(s) to whom they go for advice and information on a particular topic. It has several advantages over other methods in so far as socio-metric questions are easy to administer and are adaptable to different situations. This method has also relatively higher validity. However, the use of this method requires a large number of respondents and complex statistical analysis. This is concerned primarily with obtaining choices in interpersonal relations, such as with whom one would like to work, play, etc. or to whom one would go for advice on farming or other problems. It attempts to describe that social phenomena is a quantitative term. It may be used in selecting both professional and lay leaders, but a greater use of it is made in the latter case. It is necessary that the persons involved in a socio-metric test known to one another. It is also to be emphasized that the grouping of individuals upon the basis of socio-metric tests is in terms of choices relative to specific popularity. In a sense, the persons chosen by several others are popular, but it is popularity or acceptance in terms of specific activities. Socio-grams for the same individuals will manifest differences when the choices are in relation to different activities.

This method is very useful to the Extension Workers in finding out the natural or local or informal leaders in the villages, who are the influential persons that help in the introduction and popularization of new, improved practices in their communities or neighborhoods. As extension workers goes into a given area and asks the farmers to indicate whom they ordinarily consult for advice on farming (or any particular aspect of farming in which the extension worker wants to introduce some improvement). Usually after a few interviews, it becomes apparent which farmer is the influential person or natural leader. Figure (socio-gram) illustrates this type of test.

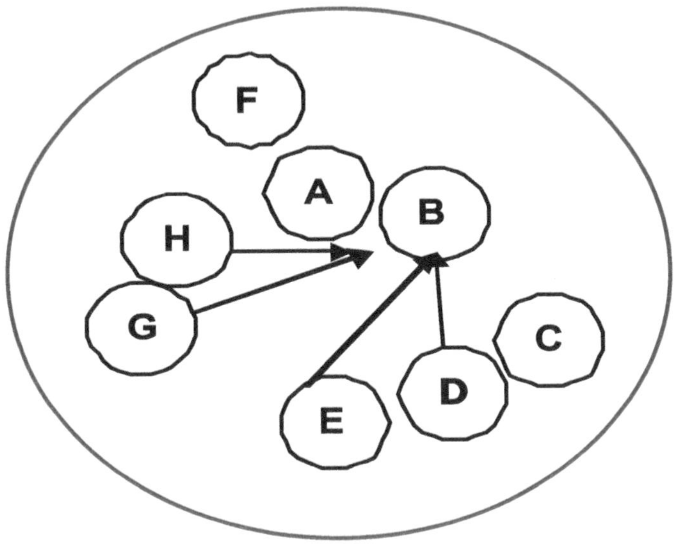

Socio-gram

Socio-gram

When 'H' is interviewed he may indicate that he generally goes to 'B' for advice on farming. 'G', 'F' and 'D' may also say that 'B' is the one whose advice they take on farming. Then 'B' is the operational or potential, 'natural' leader for these people and therefore if the extension worker succeeds in inducing 'B' to take up certain new practice, it is quite likely that others will be influenced by his behavior. It should be remembered that 'B' is the natural leader or the one who initiates action for the others. 'B' may or may not hold an office or leadership position in organizations in this area. He may not even think of himself as a leader and may insist that he is not a leader. However, so far as these farmers are concerned, 'B' is the operational leader in relation to farming practices.

Information ratings

In this method, some members of a social system are purposively or randomly selected and asked to designate persons giving advice and information on a general or particular topic. This method is economical and quick as it saves cost and time. However, it is limited to the extent each informant is thoroughly familiar with the social system.

Self-designating method

In this method, each selected person is asked a series of questions designed to determine the degree to which he perceives himself to be a key-communicator. This method has one important advantage in so far as it also measures the individual's perception of his being a key-communicator, which in turn influences his behavior. However, its accuracy is limited to the extent the respondents can identity and reports their self-image correctly.

Characteristics

Key-communicators have been found to have distinctive characteristics. On the basis of a review of a large number of research studies, Rogers and Shoemarker (1971) drew the following conclusions regarding characteristics of key-informants.

(i) External communication:

a. Key-communicators have greater exposure to mass media than their followers

b. Key-communicators are more cosmopolitan than their followers.

c. Key-communicators have greater change agent contact than their followers.

(ii) Accessibility:

Key-communicators exercise relatively greater social participation than their followers in a social system.

(iii) Social status:

Key communicators enjoy a relatively higher social status than their followers as far as conditions in that social system exists.

(iv) Innovativeness:

a. Key-communicators are more innovative than their followers.

b. When the social system's norms favors change, key-communicators are more innovative but otherwise not especially so.

c. When the norms of a system are more modern, key-communicators are more monographic. Monography is the tendency of a key-communicator to act as such for only one topic.

(v) Relationship to the social systems:

According to Shankaraiah (1969), the key-communicators under Indian conditions are:

a. Perceived as the best farmers when the social system is actively modern.

b. Social stars when the social system is relatively traditional.

On the basis of a review made by Sandhu (1970) of a large number of research conducted under Indian conditions, no specific conclusions could be drawn regarding age, education, farm size, etc., although, key-communicators tend to belong to the middle-age group i.e. 30 to 50 years, be somewhat better educated and own relatively larger farms than their followers.

Role of key communicators

1. Communication of the new technology to other villages.
2. Keeping in touch with the scientists, other institutions, media and sources of information.
3. Coordinating the functions of the village organizations and the institutions and channelizing them towards the adoption of technologies for greater production.
4. Assisting the villagers in securing the supplies and services required by them.
5. Guiding and helping people in getting assistance from government.
6. Helping the fellow villagers in the preparation of the action plans.
7. Giving constant guidance and acting as the local consultants.
8. Focusing on the problems of villagers and helping the extension workers and
9. Serving as the demonstrator in the case of agricultural innovations.

Chapter 5

Functions of Communication

The term communication is freely used by everyone in modern society, including members of the general public, scholars and management practitioners. Communication is defined as the interaction, giving and taking of information, sending and receiving of messages through verbal and non verbal means. Functions of communication vary as one must determine the function of the communication. Known as the primary function and in certain circumstances the situation or position may have one, two or three other secondary functions.

Inform functions

An organization needs a vast amount of information to function and operate a business. The top management would require timely and accurate information for the various departments to make effective decisions.

Control functions

The management of any organization will always have plans with long, medium or long term objectives for the months and years ahead. To achieve these objectives, the daily & monthly activities must proceed as planned in order to achieve the objectives for the period.

Persuading functions

Persuasion usually involves the 'selling' of an idea, product or services to a person or group. With persuasive words or actions, there is a better chance of the person or group accepting it.

Coordinating functions

Every employee plays a part in controlling inputs, outputs, job scope and head count. In any form of communication, there is a sender and receiver of the message. The question of whether the message is sent and how the message is received is of vital importance in communication. Communication is successful only when the receiver receives the intended message of the sender.

SENDER ---> ENDCODING ---> MESSAGE ---> DECODE ---> RECEIVER ---> FEEDBACK

The above is the communication process which has to be repeated until both parties finish expressing themselves. First, the sender has an idea then he/she transforms the idea into a message. After which he/she transmits the message and the receiver gets the message. The receiver then interprets the message (decoding) which in turn reacts and sends feedback to the sender. In conclusion, there are many functions and characteristics of communication and the above listed are just some of them.

According to the Management Study Guide website, communication involves a mutual exchange of understanding and is the essence of the basic functions of business management, which include staffing, directing, planning, organizing and controlling. Communication is necessary to effectively function, both inside an organization and when dealing with the outside world. Business communication can be both oral and written in nature.

i) Information function

The basic requirement of adapting and adjusting oneself to the environment is information. There must be some information about what is going on in the environment which concerns the people. The receiving or giving of information underlines all communication functions, either directly or indirectly.

ii) Command or instructive function

Those who are hierarchically superior in the family, society or organization, often initiate communication either for the purpose of informing their subordinates or for the purpose of telling them, what to do, how to do when to do etc. The command and instructive functions of communication are more observable in formal organizations than in informal organizations.

iii) Influence or persuasive function

According to Berlo (1960), the sole purpose of communication is to influence people. Persuasive function of communication i.e. to induce people is extremely important for extension in changing their behaviour in the desirable direction.

iv) Integrative function

A major function of communication is integration or of continuously offsetting any disintegration at the interpersonal or at the organizational level. This helps in maintaining individual, societal or organizational stability and identity.

Answer

Communication caters to 'change.' It is through the involvement with and the development of communication that, along the periods of ages, human beings evolved from the caves of the Stone Age to the present world of sophistication and glamour.

Information

The accessibility to the information brought about by communication helps in changing the society. When people are informed about the on-going problems of the world, the society can change to make them stronger to fight those menaces.

Socialization

Communication can help people socialize and come together in a common platform understanding and helping one another in times of need realizing the common ground of similarity we share being humans. Socialization can be possible only through communication.

Motivation

Managers use communication to motivate workers to achieve peak performance. By clarifying the expectations of employees and providing incentives for meeting or exceeding expectations, communication can help companies reach specific objectives. For example, by communicating to salespeople that they'll receive a 10 percent bonus if they reach their annual sales goal, it helps the company reach its overall sales goals.

Control

A company uses communication as a way to maintain control over employees and their work environment. Written human resources policies and procedures dictate how employees are permitted to act in the workplace. Job descriptions outline the parameters of an employee's job functions. Performance reviews control whether an employee receives a raise or attains a promotion.

Interaction

Communication allows employees to interact with customers and each other. A customer service department communicates with customers to help them resolve issues. A business letter can be used to introduce a company to a potential customer. Communication is essential for employees who work together on a project or during the training process. Employees may interact socially both at work and outside the workplace.

Providing Information

Information is dispersed throughout an organization through written or verbal communication. A human resources representative or business owner may send out a memo explaining a change in the company's health plan. A business meeting may be used as a way to communicate a new office procedure.

Providing Feedback

Communication allows for employees, managers and business owners to give and receive feedback on changes that are being considered or have already been implemented. For example, if a small business owner is considering the purchase of a new computer system, he may first consult with his employees to determine what features the system should include and what help they may need in learning the system.

Discussion and Debate

The fact that communication allows us to discuss and debate an issue holds extremely important, for, by doing so we can resolve any politician or social issue for the betterment of the society.

Education

Education is the backbone of the society. The development and the social and economic status is very much determined by the level of education that the citizens of the country have been able to acquire. This can pave the way to the social advancement and economic self-sufficiency.

Cultural Promotion

Communication also helps people to be informed about the different cultural varieties that countries and the world enjoys. Knowing and understanding them can help us understand one another in the common frame of reference as humans and live with peace and prosperity. Communication, thus, is an exclusive ladder to reach the ultimate destination, the destination being societal harmony and peace. According to the Management Study Guide website, communication involves a mutual exchange of understanding and is the essence of the basic functions of business management, which include staffing, directing, planning, organizing and controlling. Communication is necessary to effectively function, both inside an organization and when dealing with the outside world. Business communication can be both oral and written in nature.

Communicative Functions or Purposes of Communication

There are many different classification systems available for looking at the purposeful uses of communication. Models are primarily of interest to researchers and professionals within the field of speech language pathology. This article, however, is a hybrid listing of aspects of communication that are especially important to assess and track with children/students with an autism spectrum disorder (ASD). The article is designed to give a basic orientation to people outside of the speech pathology profession because these are the people who will spend more time with the child than the speech pathologist. Awareness of what to observe can contribute volumes to the development and implementation of a program for a child with ASD. No one person has to feel that he or she must master all the information and become an expert observer or data taker. Everyone is part of a team and the objective is to have a variety of observations that will help develop a more comprehensive picture of a child's skills.

Intentionality

Before looking at types or purposes of messages from the child, it will be important to observe and understand about how intentional the child's daily communication is. By intentional one means:

- Does the child EVER deliberately signal a message to others? It is easier to answer this question if the child engages in very direct, observable behavior. For example, does he look toward people, even briefly, before signing, pointing, gesturing or saying something? He could also take someone to the location of the item that he wants. If he seems vaguely aware of the other person who needs the message, however, he may still be intentional. Even if he is not looking directly at the person, he may have used peripheral vision to check if someone is around. More observations might be needed in case of the latter.
- Do the adults and other children in the child's life predominately need to observe what he is doing and make a guess at what he may or may not want? This means the communication partner does all the work and has to learn to "read" the situation. This is not intentional communication since it is not deliberately signaled to someone.
- What percentage of the child's messages each day is intentional?
- Is there any pattern to the types of messages that are intentional vs. non-intentional?
- Is the child more or less intentional with specific people, in specific locations, within specific activities, or at specific times of the day?

Examples of Intentional (I) Vs. Non-intentional (N) Communication

- **(I)** Iman sees that the teacher has a new toy on the table. He goes to the table, looks briefly at her, she says "try it" and he grabs it.
- **(N)** Namita sees the new toy on the table and goes and grabs it.
- **(I)** Iman takes his teacher by the hand, walks to the cupboard, and points upward. (The teacher knows his favorite toy is kept in that cupboard).
- **(N)** Namita wanders around the room and goes over to tug at the handles of the cabinet; she does nothing to indicate he needs help; she is determined to meet her own need of getting a specific toy.
- **(I)** Iman gives a "break" card to his teacher to request some down time.
- **(N)** Namita screams and throws things after an intense morning; her teacher thinks she needs a break and directs her to a quiet corner of the room.

Means of Communication

The child will have to use some way of sending a message to others. It does not mean the messages are always clear or there is no guessing. The intentional messages are deliberately targeted for someone whereas non-intentional are not. Things to observe and share include the following:

- How does the child communicate intentionally?
- What form or means of communication is used when the message is non-intentional (e.g., a teacher, aide or parent notices a behavior pattern, attaches meaning to it, and goes to intervene or take action)?
- Does he or she use one intentional/non-intentional means more than others? What else is used? (It is good to have a variety of ways of communicating).
- Are any intentional/non-intentional means more effective than others?
- What is the least effective intentional/non-intentional means used?
- Does he vary the intentional/non-intentional means by person, situation, or location?

Various Means of Communication and Examples

- Vocalizations sounds, grunts, unintelligible speech, shouts
- Understandable appropriate speech or echolalia (repetition of the words of others)
- Behavioral pacing, self injurious behavior, picking at sores, stripping off clothes, aggression
- Body language facial expression, going limp or rigid
- Gestures such as a yes/no headshake, point, push away, or made-up gestures
- Sign language whether correctly signed or not
- Communication display or single picture/words a point to, or exchange of picture, or word card
- Communication device electronic display that produces voice output or not
- Handwriting or computer typed messages

Examples of Intentional (I) and Non-Intentional (N) + Means of Communication

- (I) Signs "help" as he looks toward aide.
- (N) Signs "help" when no one is in the room; does not look around for a person.
- (I) Gives picture card to teacher in order to get popcorn.
- (N) Flips picture card in repetitive manner, discards it, and reaches for popcorn.
- (I) Presses button of electronic communication device with voice output to request puzzle. Looks toward teacher and walks to help himself since there was no indication that this was not OK.

- (N) Presses the button repeatedly on an electronic communication device and fixate on the action. After playing with it, he tires, and gets up to get something else.

Purposes of Communication

The topic of purpose can be looked at in two ways. First, what does the child accomplish by communicating a message TO others through some form of communication and secondly, how does he respond to the same purposeful message when provided to him FROM others? So, for example, can a child request an object? What happens if a person requests an object from him? While the second question is important, this article will only focus on the first question (i.e., the impact of the child's messages to others).

It may be easier for the non-speech pathology person to have noted the characteristics of intentionality and means of communication before addressing the element of purpose. Sometimes the purposes are very obvious and other times, it may take trial and error and/or consultation with others, to figure out the exact message. The following two categories are not inclusive of every purpose for communication. They basically reflect the common types of communication that are part of a repertoire of a young child under the age of five. A few other references at the end can be used to add more detail. It is important to know what communicative purposes a child with ASD is using so new purposes can be taught. While requesting objects is the frequent goal of early intervention programs, requesting is not the primary function expressed during conversations. Expansion of the child's repertoire of purposes can occur while teaching use of some of the means of communication outlined in the previous section.

Group 1: Overall Purpose - Regulation

When communication is used for regulation, there are two elements: 1) the child is interested in meeting his needs, and 2) he will attempt to regulate the actions of others to help him meet his needs or to help keep himself in his personal comfort zone. In the latter instance, he may protest violations of his routines, lack of information about pending changes of activities, refusing tasks that he considers difficult or boring, and so forth. The teacher/parent/or whosoever will want to observe things such as: 1) What outcome is the child trying to achieve? 2) Did he initiate the action or did someone cue or prompt him; if so, how much prompting was needed, and 3) did he achieve his desired outcome? Examples of often desired outcome include:

Group 2 Social Interaction shared attention with others

In this second category, the child is not acting solely to meet his own needs. There is a social aspect and an interactive component rather than a controlling component. The interaction can be initiated by the child with ASD or by another adult or peer

Purpose or Desired Outcome Regulation of Others	*Example of Regulation*
Wants tangible object.	Desires cheese. Stands by refrigerator and unconsciously assumes Mom will know his desire.
Wants action, event, or activity.	Wants push on swing. Signs "more."
Wants attention.	Wants help. Turns person's face to him.
Wants permission.	Wants approval to go watch TV; holds up the remote for 1 second.
Wants information.	Wants to know if time to go home. Gets jacket ready.
Wants reason/clarification.	"Why no grandma?" Bites self; screams.
Wants stop/avoid/refuse.	"No blue shirt." Hits, screams.
Commands, directs, demands any of the above.	"Jon-Jon go." (away from me). Pushes child away.

Purpose or Desired Outcome-Social	*Example of Social Interaction/ Shared Attention*
Calling	Says, "Mom" to get her attention.
Greeting	Says "Hi" or waves to neighbor.
Requests interaction/interactive activity	Hits peer so he might chase the child with ASD (*i.e.*, bid for chase game).
Comment/statement	Says "hot" to Mom as he touches his car-seat.
Naming (if not for personal enjoyment)	Says "Mario" to parent when he sees familiar With game in store.
Provides information/report	Tells teacher "Sara hurt" so teacher can provide assistance to the child who is crying.
Agree	Gives positive head shake when asked if he likes cheese.
Express feelings	Says "Sad. No more Grandma" as Dad helps him get into his car-seat after her funeral.
Social routines	Says "Thank you" when given an apple to eat.
Answers question	Points in a direction when Mom asks "Where's your brother?"
Assertiveness	Says "Mine" when peer attempts to grab his toy.
Exclamation	Says "Uh-oh." or "Oh, no!" to his teacher when he drops his pencil for the 10th time.

Summary

This article is an introduction to understanding the notion of purposes of communication. As described in detail, there are three main aspects to consider: Is the message directed to someone or not? What means of communication is used? What might be the function of the message? The functions listed above do not follow a particular model but represent borrowing from several sources. The functions reflect types of communication found in children under five years of age. More sophisticated categories and descriptors can be used to describe the functions of the communication of children older than five. It will be important to consider if the child initiates the functions listed. The listing is not meant to be a comprehensive checklist for readers of this article. Instead it can provide the framework from which to work with others who are looking at expressive communication skills and provide guidance for observations by a variety of team and family members.

Pulling it all together Intention + Means + Function

The following examples show how situations will involve all three categories of intent, means used for communication, and the function of the message.

Observation Message	*Intentional*	*Means*	*Function of*
Holds crotch and says "Teacher" until told OK to proceed to the bathroom.	yes	speech + gesture	Wants permission.
Begins to throw things when told Mom will be late for pickup today.	no	behavioral	Protests change; needs information.
Announces "trick or treat" as he wanders over to the candy dish.	yes	echolalia	Wants Candy.
Jumps up and down with excitement	no	behavioral	Expresses emotion.
Gives his aide a picture of a book using Picture Exchange (PECS).	yes	communication book	Wants activity-reading.
Says, "What's your name?" and other questions of stranger; this routine occurs frequently.	yes	speech routine	Wants interaction, not information.

Four Functions of Mass Communications

Mass media is increasingly prevalent in 21st century society. From Facebook to Twitter and everything in-between, small business owners are connected to their stakeholders more than ever before. It is no surprise then that small businesses are utilizing these contemporary forms of mass communication to share news, information and opinion. These functions of mass communication enable your small business to reach new customers and keep existing customers engaged.

News and Information

While websites such as Face book and Twitter are known as "social networking" platforms, they are also important outlets in disseminating news and factual information to wide audiences. Small businesses can take advantage of this by using the services of television news networks, newspapers and blogs to reach a large audience of interested and engaged readers. Iman and Namita started their portable generator company from a dorm room. They were featured on CNN Money's entrepreneur's blog and have been able to increase website traffic and marketing inquiries significantly. Small businesses can also release "breaking" news stories directly to their fans through status updates, photo posts and messages. This function of mass communication is important in developing a sense of community for your business. Customers feel connected and informed by using online mass communication tools.

Discussion of Opinions

Besides providing factual information about a business' products and services, mass communication tools also allow your small business to shape the opinions of consumers. Social media users are able to easily and quickly post their responses regarding their experience with your business. There is a risk attached to this ease of use, as opinions can become marred by a bad experience with your company. Business owners can capitalize on this, however, by getting directly involved in the debate. Mass communication is about combining marketing with know-how; when you share your opinion-oriented business messages or respond to negative feedback, customers are exposed to different viewpoints and keen to consider your products after all.

Cultural Transmission

Mass communications are also important for a sense of culture and community. People who keep engaged in social networking sites, blogs and online forums are developing relationships in ways that weren't possible even 10 years ago. Small businesses are a driving factor in this cultural transmission. It is now easy to speak to people hundreds of miles away, thus creating a "global" small business. No longer does your client base need to be restricted to a radius near a physical location. Small businesses can connect with, sell to and deliver products and services to people around the world. Businesses that offer intangible services like life coaching, branding and consulting are well-suited to take advantage of this virtual space. Mass communications function to keep diverse communities and businesses virtually connected even when they are physically far apart.

Entertainment

While social networking is a proving to be an excellent source of business, the entertainment function of these forms of mass communication is arguably just as important. YouTube, Vimeo and Twitter provide a form of relaxation and escapism from an otherwise demanding world. Music videos and TV shows are instantly available to users all across the world and people are able to participate directly within the media by posting their own videos, art and other creative works. All this adds up to the entertainment value of mass communications refusing to be overlooked. Small businesses can take this opportunity to further their essentially free access to marketing. There is no need for your business to have a large advertising budget now; as you can simply and rather inexpensively create and post electronic commercials and virtual billboards that have the potential to reach thousands of potential customers.

Four Major functions of Communication

Communication serves four major functions within a group or organization: control, motivation, emotional expression and information. Communication acts to control member behaviour in several ways. Organizations have authority hierarchies and formal guidelines that employees are required to follow. When employees, for instance are required to first communicate any job related grievance to their immediate boss, to follow their job description, or to comply with company policies, communication is performing a control function. But informal communication also controls behaviour. When work groups tease or harass a member who produces too much (and makes the rest of the group look bad) they are informally communicating with, and controlling the member's behaviour.

Communication fosters motivation by clarifying to employees what is to be done, how well they are doing and what can be done to improve performance if it's subpar. We saw this functioning in our review of goal-setting and reinforcement theories. The formation of specific goals, feedback on progress toward the goals, and reinforcement of desired behaviour all stimulate motivation and require communication.

For many employees, their work group is a primary source for social interaction. The communication that takes place within the group is a fundamental mechanism by which members show their frustration and feelings of satisfaction. Communication therefore provides release for the emotional expression of feelings and for fulfilment of social needs.

The final function that communication performs relates to its role in facilitating decision making. It provides the information that individuals and groups need to make decisions by transmitting the data to identify and evaluate alternative choices. No one of these four functions should be seen as being more important than the others. For groups to perform effectively, they need to maintain some form of control over members, stimulate members to perform, provide a means for emotional expression, and make decision choices. You can assume that almost every communication interaction that takes place in a group or organization performs one or more of these four functions.

The communication Process

Before communication can take place, a purpose, expressed as a message to be conveyed is needed. It passes between a sender and a receiver. The message is encoded (converted to a symbolic form) and passed by way of some medium (channel) to the receiver, who retranslates (decodes) the message initiated by the sender. The result is transference of meaning from one person to another. The key parts of this model are: (1) the sender, (2) encoding, (3) the message, (4) the channel, (5) decoding, (6) the receiver, (7) noise, and (8) feedback.

The sender initiates a message by encoding a thought. The message is the actual physical product from the sender's encoding. When we speak the speech is the message. When we write, the written narrative is the message. When we gesture, the movements of our arms and the expressions on our faces are the messages. The channel is the medium through which the message travels. It is selected by the sender, who must determine whether to use a formal or informal channel. Formal channels are established by the organization and transmit messages that are related to the professional activities of members. They traditionally follow the authority chain within the organization. Other forms of messages, such as personal or social, follow the informal channels in the organization. These informal channels are spontaneous and merge as a response to individual choices. The receiver is the object to whom the message is directed. However before the message can be received, the symbols in it must be translated into a form that can be understood by the receiver. This step is the decoding of the message. Noise represents communication barriers that distort the clarity of the message. Examples of possible noise sources include perceptual problems, information overload, semantic difficulties or culture differences. The final link in the communication process is a feedback loop. Feedback is the check on how successful we have been in transferring our messages as originally intended. It determines whether understanding has been achieved.

The basic function of Communication

The most basic functions of communication in an organization is to inform, persuade, and motivate. Yummy Cakes is a manufacturer of desserts that utilizes all the functions of communication in order to be successful.

Inform

Let's take a look at how Yummy Cakes uses communication to inform within their organization. The idea of informing within an organization is to provide data and information so that employees can effectively complete their jobs. Information ensures that an employee is aware of the rules and procedures of an organization. It also eliminates job uncertainty for workers when they are fully informed. There are three kind of ways in which information can travel in a company. Downward moving communication is information that moves from superiors to subordinates and usually consists of directives or updates. Yummy Cakes' CEO recently sent out an email to all of his managers that there would be mandatory layoffs that needed to be implemented by the end of the month. This would be considered an example of downward-moving communication.

Upward-moving communication

Upward-moving communication is information that moves from subordinates to superiors and usually consists of feedback or reports.

Horizontal-moving communication

Horizontal-moving communication is information that flows from worker to worker or manager to manager (peer to peer) and consists of reports or data. An example of this type of communication would be one co-worker telling another co-worker about the company's stock price.

Persuasion

Another essential function of information would be the element of persuasion. Yummy Cakes uses persuasion as an element of their communication plan. It is the ability to change the attitude or behaviour of employees, vendors, clients, etc. The three parts of persuasion are source of credibility, emotional appeal, and social and ego needs. In general, persuasion is used in an organizational environment effectively when the source of the information is a trusted figure. For example, if a co-worker at Yummy Cakes starts telling people that it would help the company to buy another small dessert company, most people will not trust the source. If, however, the CEO sends out an email outlining the new acquisition plan of a small dessert company, employees will trust that this is a good business decision.

Emotional appeal

Emotional appeal is when emotions or arguments are used to persuade others instead of facts or logic. Organizations can use emotional appeals when delivering bad news. Last year, the CEO spoke to the entire company at an emergency meeting. He explained how devastated he was over the need to have a corporate downsizing. He used emotion to explain that it was better for the overall security of the company to eliminate some positions. There is a need to meet the Social and egoistic needs in order for persuasion to work as a communication tool in an organization. Social needs are employees' need to belong or feel accepted, while egoistic needs are employees' self-esteem and pride.

Motivate

Motivation is different from persuasion in that it stimulates an action, whereas persuasion is to change an attitude or behaviour. Motivation is the ability to stimulate desire in employees and it consists of appreciation, recognition, and support. Motivation is an important function of communication because managers need to motivate their employees daily in order to get results. All Yummy Cakes managers have been trained to show appreciation to their subordinates for a job well done. For example, the managers can award points to each subordinate as a reward. The subordinate can save the points and cash them in for prizes, free lunches, or even a day off from work.

The functions of communication in an organization are to inform, persuade, and motivate. Informing provides data and information to employees so that they can make educated decisions. Upward, downward, and horizontal informing are three ways that workers can acquire information. The second function of communication is the persuasion of employees through source credibility, emotional appeal, and social and egoistic needs. The last function of communication is to motivate employees by appreciation, recognition and support. Communication performs variety of functions within a group or organization. It helps in transmitting information and exchanging opinions and ideas among the employees of an organization.

Basic communication functions: Informing messages

An important function of communication is to inform any subject matter to others. Communicator informs any subject matter, information or data to the receiver through communication. If the receiver understands the real meaning of the message, communication becomes complete.

Persuasion

Another important function of communication is to persuade employees so that they can perform duties and responsibilities according to the desire of the management. Communication helps employees to know the techniques of performing their jobs well.

Directing

Communication helps in giving direction to the employees to accomplish their assigned duties and responsibilities. It transfers orders and instructions from the superiors to the subordinates. It helps in explaining the ways how jobs are to be done.

Solving problems

An important function of communication is to help in organizational problems. It helps to find out the problem and creating an atmosphere to discuss the problem and solve them. It helps in exchanging opinions between the contrasting parties and to arrive at a solution.

Helping planning and decision making

Another function of communication is to facilitate in decision making. Communication helps in decision making by supplying relevant information. Moreover it supplies information for the purpose of making plans.

Forming opinions

An important function of communication is to help in forming opinion. It assists in exchanging ideas, thoughts, beliefs and information among human beings and helps in forming favorable opinions.

Coordination and cooperation

Communication performs another important function in coordinating the activities of various work groups and departments within the organization. It helps in removing the inconsistencies and brings harmony in the work field.

Maintaining relations

Communication helps in developing and maintaining good relations between different parties. It helps in minimizing misunderstanding and difference of opinion and creates a healthy atmosphere in the organization which facilitates in developing human relations.

Publicity of products

Modern age is the age of publicity. It is very difficult to attract customers and create market without proper publicity in modern age. Communication plays a very vital role in this connection.

Exchange of information

The main function of communication is the exchange of facts and information between human beings. It transfers massages form the sender to the receiver.

Controlling

It helps in controlling the activities of the subordinates by providing communicational measures at different levels of the organization. It helps in monitoring and evaluating the performance of the subordinates and taking corrective measures.

International relation

Another function of communication is to build up international relations. No country of the world is self-sufficient at present. Every country is to take other country's support and help for its own interest. So there is a necessity to develop good relations. Improved communication helps in developing such relations.

Building image

Strong image of the organization helps to gain competitive advantage in the market. Communication helps to build up image of the organization through timely contact with the stakeholders, advertising Advertising agencies etc.

Chapter 6

Application of Communication Technology in Rural Development

Introduction

Governments in developing country are increasingly aware that they have a major responsibility towards rural development and food security, but lack the capacity and solutions to meet the challenge. In 1996, the world's heads of state meeting in Rome committed their countries towards eradicating hunger and reducing the number of undernourished people by 50% by the year 2015 (FAO 1998). However, while on the one hand the agricultural yields in developing countries continue to decline, despite technological innovations, their populations continue to expand beyond food production capacities. This poses a great challenge for developing countries, especially sub-Saharan Africa, where it has been estimated that 50% of the total population would go hungry by the year 2000 (FAO, 1998). Policy and decision makers thus have to identify possible and appropriate solutions that ensure rural development and food security.

Food security can only be achieved, "when all people at all times have access to sufficient food for a healthy and productive life, and has three main components: food availability, food access, and food utilisation" (Haddad, 1997). This entails creating effective and efficient agricultural systems that supply food and foster utilisation of natural resources in a sustainable manner. Although "agriculture is the principal engine of growth in many low income developing countries, where it accounts for 60 to 80 percent of all employment" (CGIAR, 1995), food security should also be linked to environmental, social, cultural, political and institutional aspects of society (Forno, 1999). Rural people constitute the greater part of the population of developing countries and often lack access to basic needs such as water, food, education, health care, sanitation and security, leading to low life expectancy and high infant mortality. These conditions, considered harsh by the majority of the rural population, result in their migration into urban areas, often in search of formal employment, as the only option for survival. The urban slums in which these people find themselves are often not food secure either. Finding solutions calls for revolutions, including the biological revolution, this is one intervention that has been identified as having the potential to meet the increased food demands and to contribute to alleviating food insecurity problems (McCalla and Brown, 1999). Although this whole area is still controversial, some of its techniques are not controversial and developing countries should reap the benefits of biotechnology.

The information revolution is another intervention with the potential to ensure that knowledge and information on important technologies, methods and practices are put in the right hands. The relevance of this revolution is supported by Balit (1996) who pointed out that the least expensive input for rural development is knowledge. Knowledge and information are basic ingredients of food security and are essential for facilitating rural development and bringing about social and economic change. The purpose of rural development is "to improve the standard of living of the rural population is multi-sectoral including agriculture, industry, and social facilities". Rural communities require information inter alia on supply of inputs, new technologies, early warning systems (drought, pests, and diseases), credit, market prices and their competitors. The success of the Green Revolution in Asia and the Near East indicates that giving rural communities access to knowledge, technology and services will contribute to expanding and energising agriculture.

Traditional media and new ICTs have played a major role in diffusing information to rural communities, and have much more potential. There is need to connect rural communities with research and extension networks to provide access to the much needed knowledge, technology and services. Studies on information systems serving rural communities have focused on specific sectors such as agriculture or health, instead of covering the rural community needs in a holistic manner. Rural information systems must involve rural communities and local content must be of prime importance. Traditional media have been used very successfully in developing countries, and rural radio in particular has played a major role in delivering agricultural messages. Print, video, television, films, slides, pictures, drama, dance, folklore, group discussions, meetings, exhibitions and demonstrations have also been used to speed up the flow of information. New ICTs, however, have the potential of getting vast amounts of information to rural populations in a more timely, comprehensive and cost-effective manner, and could be used together with traditional media.

Potential of new ICTs for rural development

While the term 'ICTs' can be interpreted as including a wide range of media, 'new ICTs' is used to denote "the use of computers and communication systems between computers". These new ICTs are becoming more accessible, and users can obtain information from various sources, and one computer could meet the needs of a large rural community. A CTA seminar on "The role of information for rural development in ACP countries" concluded that "these modern technologies offer new and multiple perspectives, such as faster and better-focused access to information". Electronic mail is the most commonly used new ICT and has caused a cultural revolution in the way individuals and organisations interact, in terms of time, cost and distance. The second most significant use of new ICTs is the World Wide Web, which enables people to access information on millions of other computers.

Although the Internet is not a panacea for food security and rural development problems, it can open new communication channels that bring new knowledge and information resources to rural communities (Bie, 1996). Traditional communication channels have been used successfully but these have been monologic and have not allowed for much interaction with users. Radio for example has been very effective for disseminating information to all types of audiences, but broadcasting times are sometimes not appropriate for most people. But radio could be linked to the Internet, and a few initiatives have been started on this concept, such as the project Internet Radio in Sri Lanka. This enables users to access programmes on the web at a convenient time, and send feedback through e-mail or chat. Broadcasters could then disseminate the latest information promptly. Some examples of areas where ICTs could play a catalytic role in developing rural areas include:

Decision making process

Sound decision making is dependent upon availability of comprehensive, timely and up-to-date information. Developing countries facing food security problems demonstrate the need for informed researchers, planners, policy makers, development workers and farmers. Information is also needed to facilitate the development and implementation of food security policies. The e-mail and the Internet could also be used to transmit information to and from rural inaccessible areas.

Market outlook

Farmers could promote their products and handle simple transactions such as orders over the web while payment transactions for the goods can then be handled off-line (O'Farrell *et al*, 1999). It has been shown to be cheaper and faster to trade online than on paper-based medium, telephone or fax. Electronic-commerce could, therefore, enable entrepreneurs to access global market information and open up new regional and global markets that fetch better prices and increase farmers' earnings.

Empowering rural communities

ICTs can empower rural communities and give them "a voice" that permits them to contribute to the development process. With new ICTs, rural communities can acquire the capacity to improve their living conditions and become motivated through training and dialogue with others to a level where they make decisions for their own development (Balit 1998). Giving rural people a voice means giving them a seat at the table to express their views and opinions and become part of the decision making process. The approach should be participatory and could lead to improved policy formation and execution. Improved policy formulation and strategies, however, require "an educated and informed populace ... to reduce poverty, excessive population growth, environmental degradation and other factors that are most often the direct causes of hunger" (FAO, 1998). New ICTs have the potential to penetrate under-serviced areas and enhance education through distance learning, facilitate development of relevant local content and faster delivery of information on technical assistance and basic human needs such as food, agriculture, health and water. Farmers can also interact with other farmers, their families, neighbours, suppliers, customers and intermediaries and this is a way of educating rural communities. The Internet can also enable the remotest village to access regular and reliable information from a global library (the web). Different media combinations may, however, be best in different cases - through radio, television, video cassettes, audio cassettes, video conferencing, computer programmes, print, CD-ROM or the Internet. Rural areas also get greater visibility by having the opportunity to disseminate information about their community to the whole world.

Targeting marginalized groups

Most rural poor people lack the power to access information. ICTs could benefit all stakeholders including the civil society, in particular youth and women (UNDP 2000). Other disadvantaged groups that could be targeted include the disabled and subsistence peasants.

Creating employment

Through the establishment of rural information centres, ICTs can create employment opportunities in rural areas by engaging telecentre managers, subject matter specialists, information managers, translators and information technology technicians. Such centres help bridge the gap between urban and rural communities and reduce the rural-urban migration problem. The centres can also provide training and those trained might become small-scale entrepreneurs.

Innovative application of ICTs in rural areas

There are examples of initiatives in several developing countries (e.g. Bangladesh, Chile, India, Mauritius, Mexico, Peru, South Africa and Uganda) supported by FAO, ITU, IDRC, IFAD, UNESCO, DFID, British Council, GTZ and others.

Examples from Latin America and the Caribbean

- FAO has developed a number of methodologies for applying ICTs to promote rural development. The "Pedagogia Audiovisual" project was set up to promote rural development and effective peasant participation through an improved communication system and has brought together "peasants, government representatives, researchers, technicians, banking services, construction companies and marketing and processing institutions". The project is based on the saying "What I hear, I forget. What I see, I remember. What I do, I know" and was started in Chile and later spread to Peru. Since human resources are a key factor to ensuring food security, video-based packages and other traditional communication media were used to train farmers and intermediaries and to exchange ideas and information. A computer based information system was later established, which distributed "the bulletin" with information farmers needed - market, technical and weather, by fax. The information system also provided linkages to international sources and external databases. An evaluation of the bulletin indicated that it was an important tool but most of the information was not pertinent to specific user needs and it was not thought to be reaching enough farmers.
- FAO has embraced new ICTs with funding from the Government of Italy. The Internet-based project (1994-97) - "Communication para el Desarrollo en América Latina", of sharing knowledge and skills with small subsistence farmers developed national communication systems in Latin America. The projects also trained a critical mass of national staff in the production and use various communication channels preferred by peasant farmers and semi-literate rural populations. The project proved very successful and soon spread throughout the region and beyond to China, Mali, Indonesia and the Republic of Korea (Balit 1998).
- An evaluation of this project indicated that the methodologies and ICTs used were appropriate and cost effective. Farm families were the main beneficiaries of training and information activities. Intermediaries and extension workers had acquired more knowledge, skills and experience to train farmers and provide them with technical information. All stakeholders had access to better communication tools to facilitate transfer of knowledge and skills to the wider community, to access market information and to achieve an integrated approach to economic and social development, leading to inter organizational efficiencies (Richardson 1997). The project had developed national capacity in Argentina, Costa Rica, Honduras and Mexico and was operating in a sustainable manner and generating money for national and local institutions (Balit *et al*, 1996).
- The International Fund for Agricultural Development (IFAD) is supporting an Internet based system in Latin America and the Caribbean - FIDAMERICA - that has the objective of strengthening local capacities of poor rural communities and improving their quality of life. The system has used ICTs to assist the rural communities to access agricultural, market

and technical information and to improve access to financial systems. It offers facilities for knowledge and information exchange through electronic conferencing, e-mail, databases and web sites. FIDAMERICA, now in its second phase, has 41 projects and programmes in the region and involves about 3600 community organizations and 500,000 families.

Examples from Africa

- Successful experience with "audiovisual pedagogy" from Latin America has led FAO to apply the participatory video approach to agricultural development projects in Africa to empower rural communities. The methodology mediates between the needs of rural people, sources of information and expertise to respond to the expressed needs through the production and use of documentaries and training presentations (FAO 1998).
- The Technical Centre for Agricultural and Rural Co-operation (CTA) works towards improving the dissemination of information for the benefit of farmers through improved adoption of new technologies. CTA has a programme - Rural Radio Support Programme that supports rural radio and development of audiovisual aids in ACP (Africa, Caribbean, and Pacific) countries. The programme was launched in 1990 and distributes information packages consisting of taped interviews accompanied by a complete transcript and radio talks.
- In Uganda, the National Agricultural Research Organisation (NARO) and CAB International (CABI) are implementing a project titled "Electronic Delivery of Agricultural Information to Rural Communities in Uganda". The project, funded by the International Development Research Centre (IDRC) under the Acacia Initiative, aims to improve access to agricultural information by rural communities through use of selected traditional media and modern ICTs such as radio, video, television, print media, e-mail, CD-ROMs and The Internet to increase agricultural production. The project will accelerate the delivery of appropriately packaged agricultural information using existing telecentres at Nabweru, Buwama and Nakaseke as information resource centres. Information needs and ICT preferences will be identified with special focus on women, community leaders, youth, frontline workers, NGOs and CBOs. Local content will be developed and building the capacity of the local communities will be a priority. The evaluation and learning systems approach (ELSA) developed by the Acacia initiative will be used to ensure continuous learning and evaluation.
- FAO has recently prepared a proposal for a National Agricultural and Rural Knowledge and Information System (NARKIS) and network for Uganda, to increase "farmer access to information, knowledge and technology through an effective, efficient, sustainable and decentralized extension with increasing private sector involvement in line with government policy". The project hopes to assess user needs and facilitate

agricultural and rural information flows and knowledge building, as well as capacity building. The programmes to be implemented include the building and management of an agricultural Web portal, packaging of multi-media publications, rural networking and ICT deployment and support services.

- The Faculty of Agriculture of the University of Mauritius has developed a computer based information system – "The Potato Extension & Training Information System (PETIS)". PETIS uses the Internet and tests whether rural communities can use the web to access information. The system, destined principally for the small-scale potato growers, is equipped with audio files that provide information in English. Illiterate users have an option that provides the summary of the content in Creole and Bhojpuri, and icons and pictures that enable most rural users to easily navigate the basic levels on the site.

Examples from Asia

- In Bangladesh, ICTs have transformed the lives of rural people and village women. The women have started small-scale enterprises through small loans from the Grameen Bank. They used the loan to buy mobile cell phones to provide telephone services and earned good income. "Much of the voice traffic over the cell phones is commerce directed - access to agricultural market prices, access to agricultural trade information, facilitation of remittances from foreign workers, information on work opportunities, using the phone to reduce substantial travel costs" (Don Richardson, personal communication 1999). The Grameen Bank has been hugely successful in providing jobs to poor rural entrepreneurs and connecting the community to the world and, "is being replicated in dozens of other countries". Most of the borrowers are women, who organize weekly meetings to discuss loan payment and other health and development issues.
- In India, the British Council and partners, including bilateral and multilateral donor agencies, NGOs and government departments, are planning to establish India's Development Information Network (INDEV). A proposal drafted by 'One World' has been discussed with partners, and the project hopes to disseminate different forms of information to different target audiences using the Internet and web technology. Project outputs will include a web site, e-mail digests, printed reports, exhibitions, CD-ROMs and discussion lists (Global Knowledge).
- UNICEF developed the Meena Communication Initiative in South Asia and the governments of Bangladesh, India, Pakistan and Nepal have come out in support of a mass communication project aimed at changing perceptions and behaviours that hamper the survival, protection and development of female children in the region. The Initiative involves the production of multi-media packages, including animated films, videos, radio series, comic books, posters, discussion guides, folk media,

calendars, stickers and other materials. The package aims to put across gender, child rights and educational messages using the medium of popular entertainment. Topics for the animated film episodes and other multi-media materials were identified through field research.

ICTs for improving linkages

Weak linkages between researchers, frontline workers and farmers have been a major constraint that has resulted in research findings not being applied by poor rural farmers. ICTs can improve and strengthen these linkages and ensure knowledge and information, which are essential for improving food security are communicated to all stakeholders.

- FAO's "Research, Extension and Training Division" and the World Agricultural Information Centre (WAICENT) recently developed a prototype network - the Virtual Extension Research Communication Network (VERCON) concept that aims at improving linkages between agricultural research and extension institutions. The Internet tool allows network members to capture and develop local content, share, store, retrieve and disseminate information and connect geographically dispersed people from research and extension institutions, faculties of agricultural education, NGO workers and agricultural producers. The tool can also facilitate communication, sharing information and supporting improved agricultural production and can further broaden and strengthen collaboration through facilitating co-ordination of rural, local, national and regional development programmes.
- In Latin America, FAO has applied ICTs in a project to establish farmer information networks FARMNets - involving agricultural producers and farmer associations, extension services and NGOs in Chile and Mexico. Essential information on inputs, prices, markets, weather and credit are exchanged through the electronic network (via the Internet) to farmer organisations, co-operatives and local government. The project also provided training on how to analyze, retrieve and disseminate information of local relevance using ICTs (Van Crowder and Fortier, 2000).
- An evaluation of the project indicated that the Internet was found appropriate for transmitting information across the network. Farmers and their associations were able to sell their produce at much higher prices than they could fetch in the local market and strategize on what quantities and when to produce. Users of the centres found the Internet to be a cheaper, timely and appropriate communication option (Balit 1998, Masias, 1996). Knowing the future prices for cereals and oilseeds also enabled the association to better plan the quantities to plant. In addition vegetable producers reported that the information on weather conditions informed them of climatic conditions faced by competitors in other states and countries.
- The Farmers' Information & Technology Service (FITS - Techno-Pinoy) is a web-based information service initiative in the Philippines. FITS aims

to contribute to the empowerment of farmers, processors, entrepreneurs and traders and provides information and technology services that facilitate decision-making by rural communities. This leads to improved production, processing, trading and marketing. The initiative intends to link with organisations, networks and technology services and existing resources into a centre near the farmers and uses the Internet, traditional media, and face-to-face information delivery and access at the local level.

- The Mango Information Network (MIN) is a web-based information service that provides information on market outlook, pest management, directory of players in the mango industry, a virtual meeting place in extension/research activities. MIN is relevant to individual farmers, farmer organizations, co-operatives, frontline agents and entrepreneurs. It also provides question and answer service.

Constraints in the establishment and management of community-based ICT projects

Some general problems that have been experienced in a number of ICT-based rural projects and initiatives include:

1. Policy considerations

In most developing countries, especially in Africa, the formulation and implementation of policies in the ICT sector is still very rudimentary and calls for an integrated set of laws, regulations and guidelines that shape the generation, acquisition and utilization of ICTs (Marcelle, 2000). Most countries lack policies and strategies that facilitate the harnessing of new ICTs for rural development and where policies have been formulated, proper implementation plans are needed (Fillip, 2000). In addition, review strategies are often lacking. At present, the regulations are rigid and telecommunication tariffs and import duties on ICT equipment are high. The situation is compounded by lack of political good will.

2. High telecommunication costs in some developing countries

The cost of basic Internet remains a strong deterrent in many developing countries such as Angola, Kenya, Cuba and India. Although market liberalization has led to the entry of several private sector ISPs, service provision is through government phone companies, whose service is inadequate in terms of robustness, low bandwidth, congestion and noisy lines. Others make international calls to ISPs in the United States or elsewhere in Europe, which renders the service rather expensive.

3. Infrastructure

The telecommunication and electricity infrastructure in developing countries is lacking or is poorly developed in rural areas. Satellite and wireless technologies are now in use in some developing countries, but these are largely developed around urban cities and even here, the infrastructure is often inadequate. There are problems of low bandwidth and there is a need for strengthening the Internet backbone.

4. Lack of local content and language barrier

Information available through ICTs is mostly in English, which the majority rural communities of developing country cannot read. There is a marked shortage of relevant material in local languages that responds to their needs and this calls for "significant investment and support for local content" (O'Farrell et al 1999). Bartechi has described one example in India – the Village Earth Project and its Consortium for Sustainable Village Based Development. The project works with NGOs and individuals to help connect rural villages and develops appropriate content for individuals or organizations using ICTs and living in rural areas. The project has a library of books, microfiche and CD-ROMs that provide appropriate information resources but most of these materials are in English, and may not be appropriate for every region or culture. Thus, there is a need to develop more local content in appropriate languages. Various local, national, regional and international organizations have useful information resources (statistical, bibliographic, factual and full text) of relevance to rural communities, packaged as print, audio tapes, radio and television programmes, videos, CD-ROMs and on the World Wide Web. For many of these, there is lack of bibliographic control and there are no commonly agreed standards. At the international level, FAO, through its WAICENT web site, makes available many of FAO's own published sources, including FAO's Global Information and Early Warning System (GIEWS), the AGRIS and CARIS databases, and the FAO Statistical database. A CD-ROM – "FAO on the Internet" - makes this important information accessible to users that have no Internet facilities. Other bibliographic database producers such as CABI, Royal Tropical Institute (KIT), The Centre de Cooperation Internationale en Recherche Agronomique pour le Développement (CIRAD), the US National Agricultural Library (NAL) (which also hosts the gateway AgNic site), and the Inter American Institute for Cooperation on Agriculture (IICA) in Latin America provide data in a variety of formats, including via the Internet. CABI and others have also developed multimedia knowledge bases such as the Crop Protection Compendium, the Forestry Compendium and the forthcoming Animal Health and Production Compendium which can reach beyond the research community. CIRAD has produced similar products: Entodoc, Cotondoc and le Dromedaire. The Consultative Group for International Agricultural Research (CGIAR) has produced several multimedia products and institutional databases which could be very useful to intermediaries and literate farmers in rural areas. One important web resource designed to help the exchange of information for Internet users in the agriculture community is the Agriculture Internet Users Association. However the challenge in the present times is, to find ways of making these and other sources accessible to rural communities.

5. High rate of illiteracy in rural areas

"Illiteracy is a fundamental barrier to participation in knowledge societies" (Mansell, 1998). A large proportion of the rural population of developing countries - a majority of which are women - is illiterate. Most of the pictographic and audio -visual information usually has some text that goes with it. This means that these individuals are disadvantaged and lack the basic skills required to harness the benefits of ICTs. The assistance of intermediaries may thus be required.

6. Gender insensitivity

Men and women of developing countries play a different productive and community role in rural development and have different needs and preferences. Women produce more than half the world's food (World Bank 2000, FAO 1999) and face many problems in addressing food security and rural development. These include weak extension services, non-adoption of technologies, low status and therefore non-involvement in decision- and policy-making, varied and heavy workloads, poor access to credit; and lack of access to education and training (Munyua 2000). When new technologies are introduced, they are seen as a domain for men, and women have often been left out of initiatives associated with new ICTs. Rural women, however, have wisdom and indigenous knowledge that is rooted on culture, traditions, values and experience. Their methods of communication and information exchange should thus be harnessed and be complemented with new ICTs (FAO 1999). Women also spend most of their income on family welfare and have a greater impact on increasing agricultural productivity and improving the quality of family life. The youth are another special group that deserves special mention. In Kenya for example, out of the present 28 million people, about 50% are youth -below 18 years of age. The youth have been given little opportunity to contribute to rural development issues, despite their numbers, fresh and innovative ideas. The School Net model has been used in some developing countries (South Africa, Uganda,) to connect the youth to the Internet to enable them to move into information and knowledge based societies and develops young talent. Women, youth and the disabled will, therefore, require special treatment (affirmative action), and should be deliberately integrated into all ICT projects and initiatives through gender sensitive project development and implementation (FAO 1999). It is only by doing so that a critical mass of people well aware of all the information and leaders of tomorrow can be developed to ensure that their countries realize food security and rural development goals.

7. Inadequate human resources

To ensure more meaningful participation in rural development, and to pave way for the creation of a critical mass of people that effectively harness ICTs in developing countries, training and capacity building must be an integral part of all ICT projects. It has been observed that "a critical factor in meeting the challenge of ensuring food security in Africa is human resource development through knowledge building and information sharing" (Forno, 1999). Users of ICTs have to be trained in the use, application and maintenance of ICTs before they become capable and confident enough to use them (Richardson and Rajasunderam, 1999). Most of the staff managing ICT-based projects lack adequate training that would enable them to take full advantage of the new technologies. There is need to invest in training and advisory services for information intermediaries, telecentre staff, frontline workers and women's groups. Norrish, (1999) has pointed out the need to identify the best training approaches for rural communities targeting different user categories and different technologies. Such training could be done through conferences, workshops or training of Introductory and sensitizing workshops could be organised for different categories of users and local experts could provide ongoing on-line support. People in rural communities of developed countries are already reaping the benefits of electronic distance education and developing countries could offer similar services through selected centres (Richardson 1999). Focus should be on such skills like how to use ICTs through practical and participatory approaches. The 'Self Employed Women's Association' in India, for example, has trained rural women in the production and use of videography thought with which they can generate income, disseminate new skills and also advocate changes in policy (Balit, 1999). Some key players in the training activity have been FAO, CTA, IDRC, IFAD, UNESCO and ministries of agriculture. Traditional media will remain important and should continue being used alongside new ICTs. "It is through the use of a variety of media and their integration with local communications networks that more people throughout Africa can be heard and can be reached" (FAO, 1998).

8. Sustainability of projects

Most projects established with external funding face major challenges after the project period has ended. Sustainability of these projects should be considered right from the outset and, wherever possible, should have the support of government, private sector and community. Users should also pay for services but the cost will depend on how much they can afford. There are a few examples of success in attaining such sustainability, and there is an urgent need for viable models to be developed and tested.

Role of FAO in ensuring appropriate use of ICTs

FAO is an international and membership organisation with intergovernmental status. It is recognised as one of the leading agencies in alleviating poverty and hunger by promoting agricultural development, improved nutrition and the pursuit of food security. This means that FAO has the potential to assist governments of developing member countries to attain food security. Rural people will, in turn, themselves then have to adopt new technologies. FAO could assist developing countries to harness the benefits of ICTs to facilitate sharing and exchanging of relevant knowledge and information. FAO can, however, not achieve this on its own and international, regional, national as well as local partners will need to work together to ensure that rural communities participate in the ICT revolution. Some areas where FAO could play a role include:

1. Influencing policy in the food security/rural development context

Weak and inadequate policies have led to lack of development (social, economic, political) and to problems of food security in many developing countries. This underlines the need to address policy constraints (Jensen 1997). FAO could play a much greater role in facilitating agricultural and ICT policy formulation and implementation in developing countries. Its status makes it much easier for it to seek audience with key decision- and policy-makers within the government sector and the international community. FAO could work with both the public and non-public sector (involving all possible stakeholders concerned with food security) to advocate for the formulation and adoption of policies that have gender considerations to support and empower rural people, and in particular women and youth groups. FAO could also assist developing countries in strengthening policies and institutions. There are a number of success stories in the ICT sector, and these could be shared and discussed with policy makers and be considered as best-practice guidelines for formulating ICT policy. There is also a need to advocate the institutionalization of communication as a vital component of rural development policies of developing countries.

2. Telecentres

There is evidence that telecentres have played a major role in mobilizing communities to address their development problems. Tele-centers can be used as information hubs that capture, repackage and disseminate information to rural communities (FAO 1998). Bie (1996) recommends "an Internet and development strategy focused on rural and agricultural communities and the intermediary agencies that serve those communities with advice, project support, research, extension, and training". FAO could take the lead and work with like-minded partners to further develop the concept of multi-purpose community information tele-centers into information and knowledge systems, aimed at meeting food security and rural development goals. FAO and partners could establish pilot projects in selected countries to demonstrate the power of ICTs in addressing rural development and food security problems. A mix of media - traditional and new ICTs could be used to meet different user categories, needs and preferences. Lessons learned by FAO and other partners from the use of different communication media could be applied to deal with rural development and food security problems in developing countries (FAO, 1998).

3. Capacity building and training

One major constraint for delivery of food security initiatives in rural areas is weak institutional capacity and insufficient co-ordination. Non-governmental organizations and the private sector in particular possess a vast but often untapped potential. FAO could seek partners to fund capacity building activities. The partnership could also assist with building the required human and institutional capacities at national and regional level to provide training and education to rural communities on how to manage local knowledge and information, using ICTs. Training materials produced could be availed as electronic archive of training resources, and could be repackaged in preferred media such as the successful video-based model used in Latin America. The resources could then be translated into major languages to ensure that most developing countries benefit.

4. Harmonization of standards

Common standards are a prerequisite for sharing information. FAO could work with relevant producers and disseminators of knowledge and information to develop standards for managing information and knowledge targeted at rural communities. Some tools and methodologies such as Participatory Rural Communication Appraisal (PRCA) have already been developed and have been used to "uncover local skills and knowledge" (Anderson et al. 1999) and to fully understand the information and knowledge needs of rural communities. These could be promoted for adoption by other actors in rural areas. FAO could for example, provide simple guidelines on how to develop local content using the VERCON model to capture indigenous knowledge. Other useful tools that have the potential to enhance networking, such as FARMnets could also be promoted. FAO's multilingual thesaurus (AGROVOC), categorization/ classification scheme and guidelines for bibliographic analysis have been widely adopted and adapted in developing countries for managing agricultural information. These could be harmonized with other existing standards (such as CABI's and UNEP's thesauri), for processing and indexing agricultural information at national level and on the World Wide Web. Standardization would facilitate the pooling of relevant resources from different rural communities on the Internet, as successfully demonstrated in the case of AGRIS. Intermediaries could then tap this resource and repackage the information in different media to suit their needs.

5. Repackaging and local content development

The African Development Forum (ADF'99) underscored the need to harness indigenous knowledge for development. This is further supported by 'One World', who have stated that "a country's knowledge base needs to be developed and fostered to both improve its competitive position and to contribute to human and sustainable development goals" (One World, 2000). FAO has played a major role in helping developing countries to manage local scientific and technical agricultural information and could in the same vein; provide technical assistance to national and regional institutions managing rural knowledge and information systems. Special emphasis could be placed on developing and disseminating local content, "improving the relevance of the information to local development" (FAO, 1998), as well as capturing and auditing all relevant local resources using ICTs. Resources produced should involve the participation of local communities and be packaged in local languages, to make the services offered more valuable and accessible. An electronic co-operative agricultural information resource could be established and be hosted on the WAICENT site. This resource could also be repackaged in media such as CD-ROM, to ensure access by communities whose "last mile" has yet to be connected.

Use of FAO's own resources

FAO has a vast body of knowledge and experience and can now build on its many successes by championing, facilitating and enhancing the use of new ICTs to support this new information revolution. These include the David Lubin Memorial Library resources, FAOSTAT, GIEWS, CARIS and AGRIS databases on the WAICENT site and other invaluable resources such as the Global Plant and Pest Information system (GPPIS). The challenge to FAO is to tap all the organisation's resources in a holistic manner, and add a component of information dissemination using appropriate ICTs to all FAO programmes. These products could be packaged by subject, commodity or by country, and be adapted to a less technical level for wider dissemination through both traditional communication media and new ICTs. Regional and national institutions could then access these resources and provide continued flow of information and knowledge by remote and impoverished rural communities and provide answers to many questions.

Promoting use of traditional and new ICTs

FAO and partners could play a key role in raising awareness of the power of appropriate traditional and new ICTs in facilitating rural development and food security. This could be done through workshops, visits to telecentre models, radio, video, television and print. FAO and member countries could also produce and disseminate documentation on development and use of ICTs and electronic information management systems.

Conclusion

Richardson (1997) stated that "achieving an integrated rural Internet development approach in a given nation or region requires the collaborative participation of agencies, organisations and government services". The Strategy for Co-operation would thus be to work through international co-operation to harness synergies of the respective partners. FAO could, therefore, forge alliances and coalitions with other international, regional, national, donor, multilateral and development agencies, public and non-public institutions and rural groups. The partnership could then work jointly in planning and implementing initiatives that seek to harness ICTs for food security and rural development. The partners could include IDRC, CABI, UNDP, USAID, IFAD, CTA, World Bank, DfID, IICD, British Council, GTZ, national governments, regional organisations, private sector, NGOs, farmer associations and specialised ICT bodies such as ITU and telephone companies. Following a recommendation from one of its seminars that a 'technology watch' be established to track the evolution of ICTs, CTA has established an electronic "Observatory" with experts from ACP countries and Europe, that has the task of informing the agricultural community on appropriate ICTs (CTA 1998). Similarly, ITU has created a Global Case Library of reports on on-going projects using ICTs, planned projects using satellite and wireless technologies and examples of equipment adapted for use in remote disadvantaged rural environments. FAO could work with these groups and others, to build on the existing body of knowledge on ICTs, provide wider resource bases, increase economies of scale and put agricultural growth on centre stage. Multipurpose community telecentres could play a key role in the "information renaissance" in developing countries and ensure universal access. Locations for telecentres must be carefully selected, and should take into consideration the "level of potential demand for communication and information services from a large number and wide range of users", its proximity to other organisations and institutions, infrastructural considerations and socio-cultural issues (Anderson et al 1999). The information systems established should be multi-sectoral (agricultural research, extension, training and education, and health) and use a mix of appropriate traditional media and new ICTs depending on preferences of the users. All relevant stakeholders should facilitate the evolution of appropriate ICT policies in developing countries and work towards a common goal of ensuring rural development and food security. Where the infrastructure is not yet developed, the Internet could be used from a central point (telecentre) for online broadcasting and for exchanging relevant information from developing countries. The telecentres will also provide a stage for rural communities to address their training and development needs and vision (Anderson et al 1999). Local content and expert systems developed in different countries could also be shared through an electronic agricultural network over the Internet. New initiatives should avoid fragmentation and duplication of costly infrastructure against the challenge of an ever-diminishing resource base and projects developed should respond to needs of small scale farmers and entrepreneurs. There should also be more consultation and active involvement of private sector and civil society organisations that produce and use knowledge and information.

Food security attainment is, however, unlikely to be realised by provision of information alone and developing country governments must invest much more in impoverished rural areas where the greater population lies. "Governments are much better placed to formulate overall objectives and priorities, and to articulate a coherent strategy at the national level" (CTA, 1999), as they have the machinery to bring about food security and rural development through a wide range of mechanisms. Organisations advocating for the use and application of ICTs for information sharing and exchange must therefore work closely with developing country governments.

Chapter 7

Barriers to Communication

There are many reasons why interpersonal communications may fail. In many communications, the message (what is said) may not be received exactly the way the sender intended. It is, therefore, important that the communicator seeks feedback to check that their message is clearly understood. The skills of 'Active Listening', 'Clarification' and 'Reflection' may help but the skilled communicator also needs to be aware of the barriers to effective communication and how to avoid or overcome them.

A. Active Listening

Active listening is a skill that can be acquired and developed with practice. However, active listening can be difficult to master and will, therefore, take time and patience. 'Active listening' means, as its name suggests, actively listening. That is fully concentrating on what is being said rather than just passively 'hearing' the message of the speaker. Active listening involves listening with all senses. As well as giving full attention to the speaker, it is important that the 'active listener' is also 'seen' to be listening - otherwise the speaker may conclude that what they are talking about is uninteresting to the listener. Interest can be conveyed to the speaker by using both verbal and non-verbal messages such as maintaining eye contact, nodding your head and smiling, agreeing by saying 'Yes' or simple verbal non-verbal gestures like, 'Mmm hmm' to encourage them to continue. By Receiving this 'feedback' the speaker will usually feel more at ease and therefore communicate more easily, openly and honestly. Listening is the most fundamental component of interpersonal communication skills. Listening is not something that just happens (that is hearing), rather, listening is an active process in which a conscious decision is made to listen and to understand the messages of the speaker. Listeners should remain neutral and non-judgmental, this means trying not to take sides or form opinions, especially early in the conversation. Active listening is also about patience - pauses and short periods of silence should be accepted. Listeners should not be tempted to jump in with questions or comments every time there are a few seconds of silence. Active listening involves giving the other person time to explore their thoughts and feelings, they should, therefore, be given adequate time for that. Active listening not only means focusing fully on the speaker but also actively showing verbal and non-verbal signs of listening. Generally speakers want listeners to demonstrate 'active listening' by responding appropriately to what they are saying. Appropriate responses to listening can be both verbal and non-verbal:

Science of active listening

This is a generic list of non-verbal signs of listening, in other words people who are listening are more likely to display at least some of these signs. However these signs might not be appropriate in all situations and across all cultures.

1. Smile

Small smiles can be used to show that the listener is paying attention to what is being said or as a way of agreeing or being happy about the messages being received. Combined with nods of the head, smiles can be powerful in affirming that messages are being listened to and understood.

2. Eye Contact

It is normal and usually encouraging for the listener to look at the speaker. Eye contact can however be intimidating, especially for more shy speakers – gauge how much eye contact is appropriate for any given situation. Combine eye contact with smiles and other non-verbal gestures to encourage the speaker.

3. Posture

Posture can tell a lot about the sender and receiver in interpersonal interactions. The attentive listener tends to lean slightly forward or sideways whilst sitting. Other signs of active listening may include a slight slant of the head or resting the head on one hand.

4. Mirroring

Automatic reflection/mirroring of any facial expressions used by the speaker can be a sign of attentive listening. These reflective expressions can help to show sympathy and empathy in more emotional situations. Attempting to consciously mimic facial expressions (i.e. not automatic reflection of expressions) can be a sign of inattention.

5. Distraction

The active listener will not be distracted and therefore will refrain from fidgeting, looking at a clock or watch, doodling, playing with their hair or picking their fingernails. Although a strong signal of attentiveness, caution should be used when using positive verbal reinforcement. Although some positive words of encouragement may be beneficial to the speaker the listener should use them sparingly so as not to distract from what is being said or place unnecessary emphasis on parts of the message. Casual and frequent use of words and phrases, such as: 'very good', 'yes' or 'indeed' can become irritating to the speaker. It is usually better to elaborate and explain why you are agreeing with a certain point.

6. Remembering

The human mind is notoriously bad at remembering details, especially for any length of time. However, remembering a few key points, or even the name of the speaker, can help to reinforce that the messages sent have been received and understood i.e. listening has been successful. Remembering details, ideas and concepts from previous conversations proves that attention was kept and is likely to encourage the speaker to continue. During longer exchanges it may be appropriate to make very brief notes to act as a memory jog when questioning or clarifying later.

7. Questioning

The listener can demonstrate that they have been paying attention by asking relevant questions and/or making statements that build or help to clarify what the speaker has said. By asking relevant questions the listener also helps to reinforce that they have an interest in what the speaker has been saying.

8. Reflection

Reflecting is closely repeating or paraphrasing what the speaker has said in order to show comprehension. Reflection is a powerful skill that can reinforce the message of the speaker and demonstrate understanding.

B. Clarification

Clarifying involves asking questions of the speaker to ensure that the correct message has been received. Clarification usually involves the use of open questions which enables the speaker to expand on certain points as necessary.

Summarisation

Repeating a summary of what has been said back to the speaker is a technique used by the listener to repeat what has been said in their own words. Summarising involves taking the main points of the received message and reiterating them in a logical and clear way, giving the speaker chance to correct if necessary. In communication, clarification involves offering back to the speaker the essential meaning, as understood by the listener, of what they have just said. Thereby checking that the listener's understanding is correct and resolving any areas of confusion or misunderstanding. Clarification is important in many situations especially when what is being communicated is difficult in some way. Communication can be 'difficult' for many reasons, perhaps sensitive emotions are being discussed - or you are listening to some complex information or following instructions. This page provides dialogue and examples of clarification and how you can use this simple technique to improve your communication skills.

The purpose of clarification is to:

- Ensure that the listener's understanding of what the speaker has said is correct.
- Reassure the speaker that the listener is genuinely interested in them and is attempting to understand what they are saying.
- As an extension of reflecting, clarifying reassures the speaker that the listener is attempting to understand the messages they are expressing. Clarifying can involve asking questions or occasionally summarising what the speaker has said.

A listener can ask for clarification when they cannot make sense of the speaker's responses. Sometimes, the messages that a speaker is attempting to send can be highly complex, involving many different people, issues, places and/or times. Clarifying helps you to sort these out and also to check the speaker's priorities. Through clarification it is possible for the speaker and the listener to make sense of these often confused and complex issues. Clarifying involves genuineness on the listener's part and it shows speakers that the listener is interested in them and in what they have to say. See our page: Active Listening for more about attentive listening skills.

Some examples of non-directive clarification-seeking questions are:

- "I'm not quite sure I understand what you are saying."
- "I don't feel clear about the main issue here."
- "When you said .... what did you mean?"
- "Could you repeat ...?"

Clarifying involves:

- Non-judgemental questioning.
- Summarising and seeking feedback as to its accuracy.

Clarification Questions

When you are the listener in a sensitive environment, the right sort of non-directive questioning can enable the speaker to describe their viewpoint more fully. Asking the right question at the right time can be crucial and comes with practice. The best questions are open-ended as they give the speaker choice in how to respond, whereas closed questions allow only very limited responses.

Open Questions

If your role is to assist a speaker to talk about an issue, often the most effective questioning starts with 'when', 'where', 'how' or 'why'. These questions encourage speakers to be open and expand on their thoughts. For example:

"When did you first start feeling like this?"

"Why do you feel this way?"

Closed Questions

Closed questions usually elicit a 'yes' or 'no' response and do not encourage speakers to be open and expand on their thoughts. Such questions often begin with 'did you?' or 'were you?' For example:

"Did you always feel like this?"

"Were you aware of feeling this way?"

Guidelines for Clarifying

Clarification is the skill we use to ensure that we have understood the message of the speaker in an interpersonal exchange. When using clarification follow these guidelines to help aid communication and understanding.

- Admit if you are unsure about what the speaker means.
- Ask for repetition.
- State what the speaker has said as you understand it, and check whether this is what they really said.
- Ask for specific examples.
- Use open, non-directive questions - if appropriate.
- Ask if you have got it right and be prepared to be corrected.

Summarising

As a further extension to clarification a summary involves reviewing what has taken place during the whole conversation. It is important to keep only to the essential components of the conversation, and it must be given from the speaker's frame of reference, not an interpretation from the listener's viewpoint. The aim of a summary is to review understanding, not to give explanation, to judge, to interpret or provide solutions.

Summarising should be done at the end of a conversation, although sometimes it may be appropriate midway through as a way of drawing together different threads. At the start of a conversation, it is useful to summarise any previous discussions or meetings as it can help to provide focus. Whilst the summary is likely to be the longest time a listener will be speaking during a conversation, it is important to be as concise and straightforward as possible.

Summary of Clarification

In reflecting, clarifying and summarising, speakers must be allowed to disagree with, and correct, what the listener says. They should be encouraged to express themselves again, if necessary, giving the listener another chance at understanding, and to check understanding until agreement is reached.

Reflecting, clarifying and summarising are all tools used by active listeners to enable them to demonstrate understanding and encourage a speaker to talk openly. It is essential that the listener and speaker both have the same understanding of the discussion and the speaker must have the opportunity to correct the listener's understanding. These are the tools of any good interpersonal relationship and are therefore important interpersonal skills.

However good you think your listening skills are, the only person who can tell you if you have understood correctly or not is the speaker. Therefore, as an extension of good listening skills, you need to develop the ability to reflect words and feelings and to clarify that you have understood them correctly. It is often important that you and the speaker agree that what you understand is a true representation of what was meant to be said. As well as understanding and reflecting the verbal messages of the speaker it is important to try to understand the emotions - this page explains how to use reflection effectively to help you build greater understanding of not only what is being said but the content, feeling and meaning of messages.

C. Reflection

Reflecting is the process of paraphrasing and restating both the feelings and words of the speaker. The purposes of reflecting are:

- To allow the speaker to 'hear' their own thoughts and to focus on what they say and feel.
- To show the speaker that you are trying to perceive the world as they see it and that you are doing your best to understand their messages.
- To encourage them to continue talking.

Reflecting does not involve asking questions, introducing a new topic or leading the conversation in another direction. Speakers are helped through reflecting as it not only allows them to feel understood, but it also gives them the opportunity to focus on their ideas. This in turn helps them to direct their thoughts and further encourages them to continue speaking.

Two Main Techniques of Reflecting:

Mirroring

Mirroring is a simple form of reflecting and involves repeating almost exactly what the speaker says. Mirroring should be short and simple. It is usually enough to just repeat key words or the last few words spoken. This shows you are trying to understand the speakers' terms of reference and acts as a prompt for him or her to continue. Be aware not to over mirror as this can become irritating and therefore a distraction from the message.

Paraphrasing

Paraphrasing involves using other words to reflect what the speaker has said. Paraphrasing shows not only that you are listening, but that you are attempting to understand what the speaker is saying. It is often the case that people 'hear what they expect to hear' due to assumptions, stereotyping or prejudices. When paraphrasing, it is of utmost importance that you do not introduce your own ideas or question the speaker's thoughts, feelings or actions. Your responses should be non-directive and non-judgemental. It is very difficult to resist the temptation to ask questions and when this technique is first used, reflecting can seem very stilted and unnatural. You need to practice this skill in order to feel comfortable.

Reflecting Content, Feeling and Meaning

The most immediate part of a speaker's message is the content, in other words those aspects dealing with information, actions, events and experience, as verbalised by him or her. Reflecting content helps to give focus to the situation but, at the same time, it is also essential to reflect the feelings and emotions expressed in order to bring them into sharper focus. This helps the speaker to own and accept their own feelings, for quite often a speaker may talk about them as though they belong to someone else, for example using "you feel guilty" rather than "I feel guilty." A skilled listener will be able to reflect a speaker's feelings from body cues (non-verbal) as well as verbal messages. It is sometimes not appropriate to ask such direct questions as "How does that make you feel?" Strong emotions such as love and hate are easy to identify, whereas feelings such as affection, guilt and confusion are much more subtle. The listener must have the ability to identify such feelings both from the words and the non-verbal cues, for example body language, tone of voice, etc. As well as considering which emotions the speaker is feeling, the listener needs to reflect the degree of intensity of these emotions. For example:

	Intensity	*Emotion*
"You feel	a little bit	sad/angry?"
"You feel	Quite	helpless/depressed?"
"You feel	very	stressed?"
"You feel	extremely	embarrassed?"

Reflecting needs to combine content and feeling to truly reflect the meaning of what the speaker has said. For example:

Speaker

"I just don't understand my boss. One minute he says one thing and the next minute he says the opposite."

Listener

"You feel very confused by him?"

Reflecting meaning allows the listener to reflect the speaker's experiences and emotional response to those experiences. It links the content and feeling components of what the speaker has said.

Guidelines for Reflecting

- Be natural.
- Listen for the basic message - consider the content, feeling and meaning expressed by the speaker.
- Restate what you have been told in simple terms.

- When restating, look for non-verbal as well as verbal cues that confirm or deny the accuracy of your paraphrasing. (Note that some speakers may pretend you have got it right because they feel unable to assert themselves and disagree with you.)
- Do not question the speaker unnecessarily.
- Do not add to the speaker's meaning.
- Do not take the speaker's topic in a new direction.
- Always be non-directive and non-judgemental.

It is common, to formulate a reply while listening to someone else, even when the other person is still talking. However, this means that we are not really listening to all that is being said. Even good listeners are often guilty of critically evaluating what is being said before fully understanding the message that the speaker is trying to communicate. The result is that assumptions are made and conclusions reached about the speaker's meaning that might be inaccurate. This and other types of ineffective listening lead to misunderstandings and a breakdown in communication. Even if we are not formulating a response whilst listening, we may still be thinking of other things, albeit subconsciously. During a conversation, how often have thoughts such as "What am I going to have for my dinner", "Will I have time to finish that report?" or "I hope I am not late picking the kids up" crossed your mind? At such times, we are distracted and not giving our full attention to what is being said. In other words we are not actively listening to the speaker. We can easily pick up bad habits when it comes to listening - this page examines some of the barriers and bad habits of listening - enabling you to address and correct them. Listening is a key interpersonal skill and a prerequisite to many other communication skills – by learning to listen more effectively one can improve the quality of your professional and personal life.

Common Barriers to Listening

There are many things that get in the way of listening and one should be aware of these barriers - many of which are bad habits - in order to become a more effective listener. Barriers and bad habits obstructing effective listening can include:

- **Trying to listen to more than one conversation at a time**: This includes having the television or radio on while attempting to listen to a speaker; being on the phone with one person and talking to another person in the same room and also being distracted by some dominant noise in the immediate environment.
- **Finding the speaker/communicator attractive/unattractive:** Sometimes people find the speaker/communicator less attractive and thus pay more attention to how you feel about the communicator and their physical appearance than to what they are saying. Perhaps you simply don't like the speaker - you may mentally argue with the speaker and be fast to criticise, either verbally or in your head.

- **Lack of interest:** Listening is affected if one is not interested in the topic/issue being discussed and become bored.
- **Lack of Focus:** Not focusing and being easily distracted, fiddling with ones hair, fingers, a pen etc. or gazing out of the window or focusing on objects other than the speaker also obstructs effective listening.
- **Unfavourable physical conditions:** Feeling unwell or tired hungry, thirsty or needing to use the toilet.
- **Identifying rather than empathising**: Understanding what you are hearing but not putting yourself in the shoes of the speaker. As most of us have a lot of internal self-dialogue we spend a lot of time listening to our own thoughts and feelings - it can be difficult to switch the focus from 'I' or 'me' to 'them' or 'you'. Effective listening involves opening your mind to the views of others and attempting to feel empathetic.
- **Sympathising rather than empathising**: Sympathy is not the same as empathy, you sympathise when you feel sorry for the experiences of another, to empathise is to put yourself in the position of the other person.
- **Prejudice or Biasness:** If one is prejudiced or biased by race, gender, age, religion, accent, and/or past experiences.
- **Nurturing preconceived ideas or bias**: Effective listening includes being open-minded to the ideas and opinions of others, this does not mean you have to agree with everything but one should listen and attempt to understand.
- **Being Judgemental:** One makes judgements thinking, for example, that a person is not very bright or is under-qualified so there is no point listening to what they have to say.
- **Previous experiences**: We are all influenced by previous experiences in life. We respond to people based on personal appearances, how initial introductions or welcomes were received and/or previous interpersonal encounters. If we stereotype a person we become less objective and therefore less likely to listen effectively.
- **Preoccupation:** When we have a lot on our minds we can fail to listen to what is being said as we're too busy concentrating on what we're thinking about. This is particularly true when we feel stressed or worried about issues.
- **Having a Closed Mind**: We all have ideals and values that we believe to be correct and it can be difficult to listen to the views of others that contradict our own opinions. The key to effective listening and interpersonal skills more generally is the ability to have a truly open mind - to understand why others think about things differently to you and use this information to gain a better understanding of the speaker.

Non-Verbal Signs of Ineffective Listening

Although with all non-verbal signals a certain amount of error has to be expected, generally signs of inattention while listening include:

- **Lack of eye contact with the speaker**: Listeners who are engaged with the speaker tend to give eye contact. Thus lack of eye contact can, be a sign of shyness.
- **An inappropriate posture**: Louched, leaning back or 'swinging' on a chair, leaning forward onto a desk or table and/or a constantly shifting posture. People who are paying attention tend to lean slightly towards the speaker.
- **Being distracted**: Fidgeting, doodling, looking at a watch and yawning.
- **Inappropriate expressions and lack of head nods**: Often when a listener is engaged with a speaker they nod their head, this is usually an almost subconscious way of encouraging the speaker and showing attention. Lack of head nods can mean the opposite – listening is not happening. The same can be true of facial expressions, attentive listeners use smiles as feedback mechanisms and to show attention.

Further Signs of Ineffective Listening

Other common traits of ineffective listening include:

- **Sudden Changes in Topic:** When the listener is distracted they may suddenly think about something else that is not related to the topic of the speaker and attempt to change the conversation to their new topic.
- **Selective Listening:** This occurs when the listener thinks they have heard the main points or have got the gist of what the speaker wants to say. They filter out what they perceive as being of key importance and then stop listening or become distracted.
- **Daydreaming:** Daydreaming can occur when the listener hears something that sets off a chain of unrelated thoughts in their head – they become distracted by their 'own world' and adopt a 'far-away' look.
- **Advising:** Some people want to jump in early in a conversation and start to offer advice before they fully understand the problem or concerns of the speaker.

Ineffective Listening is Very Common

You can probably think of examples when you have listened ineffectively or not been listened to over the last 24 hours. You can probably recognise the frustration and irritation when you know the person you are talking to is not listening to you. As listening is so fundamental to the communication processes it is important to try to avoid ineffective listening.

Common Barriers to Effective Communication

There are many barriers to communication and these may occur at any stage in the communication process. Barriers may lead to your message becoming distorted and you therefore risk wasting both time and/or money by causing confusion and misunderstanding. Effective communication involves overcoming these barriers and conveying a clear and concise message.

- The use of jargon. Over-complicated, unfamiliar and/or technical terms.
- Emotional barriers and taboos. Some people may find it difficult to express their emotions and some topics may be completely 'off-limits' or taboo.
- Lack of attention, interest, distractions, or irrelevance to the receiver.
- Differences in perception and viewpoint.
- Physical disabilities such as hearing problems or speech difficulties.
- Physical barriers to non-verbal communication. Not being able to see the non-verbal cues, gestures, posture and general body language can make communication less effective.
- Language differences and the difficulty in understanding unfamiliar accents.
- Expectations and prejudices which may lead to false assumptions or stereotyping. People often hear what they expect to hear rather than what is actually said and jump to incorrect conclusions.
- Cultural differences. The norms of social interaction vary greatly in different cultures, as do the way in which emotions are expressed. For example, the concept of personal space varies between cultures and between different social settings.

Barriers to organizational communication

When a communication is sent to the receiver, it must be effective and it is effective only when it has been understood by the receiver in the same sense in which the sender intended it. If it is not properly understood by the receiver, the very purpose of communication is lost and the communication proves ineffective. There may be so many obstructions in the way of its being effective. The message may not reach the receiver; there may be some flaws in encoding and decoding the message; the channel used may be defective or there may be some noise in the channel, or the channel used may be wrong, there may also be personal problems. Thus, for various reasons, the message may not be received by the receiver as was intended by the sender.

Types of barriers

Barriers, obstructions, and interruptions in communication may broadly be categorized into the following groups:

- External barriers
- Organizational barriers
- Personal barriers

However, this classification does not suggest that these categories are mutually exclusive.

(i) External barriers

External barriers are those which are caused by factors other than organizational and personal factors. Such external barriers may be (a) semantic barriers, (b) emotional or psychological barriers.

(ii) Organizational barriers

An organization is a deliberate creation of management for the attainment of certain specific objectives. The day-to-day functioning of the organization is regulated in such a way as to contribute to the attainment of these objectives in the most effective manner. For this purpose, a variety of official measures are adopted such as designing of the structure, arrangement of activities, formulation of various policies, rules and regulations, and procedures, laying down of norms of behaviour, instituting a reward and punishment system etc.

(iii) Personal barriers

As communication is basically an interpersonal process, many personal factors inherent in the two parties to communication, the sender and the receiver, influence the flow of communication and present many hurdles in the way of effective communication.

A Categorisation of Barriers to Communication

Language Barriers

Clearly, language and linguistic ability might act as a barrier to communication. However, even when communicating in the same language, the terminology used in a message may act as a barrier if it is not fully understood by the receiver(s). For example, a message that includes a lot of Technical jargon and abbreviations will not be understood by a receiver who is not familiar with the terminology used. Regional colloquialisms and expressions may be misinterpreted or even considered offensive. See our page: Effective Speaking for more information.

Psychological Barriers

The psychological state of the receiver will influence how the message is received. For example, if someone has personal worries and is stressed, they may be preoccupied by personal concerns and not as receptive to the message as if they were not stressed. Stress management is an important personal skill that affects our interpersonal relationships. See our pages Stress: Symptoms and Triggers and Avoiding Stress for more information. Anger is another example of a psychological barrier to communication, when we are angry it is easy to say things that we may later regret and also to misinterpret what others are saying. See our pages: What is Anger?, Anger Management and Anger Management Therapy for more information. More generally people with low self-esteem may be less assertive and therefore may not feel comfortable communicating - they may feel shy about saying how they really feel or read negative sub-texts into messages they hear. Visit our pages on Improving Self-Esteem and Assertiveness for more information.

Physiological Barriers

Physiological barriers may result from the receiver's physical state: for example, a receiver with reduced hearing ability may not grasp to entirety of a spoken conversation especially if there is significant background noise.

Physical Barriers

These barriers are those that separate people from each other and mark territories. This type of barrier can often be seen in the workplace where offices and closed doors stop communication An example of a physical barrier to communication is geographic distance between the sender and receiver(s). Communication is generally easier over shorter distances as more communication channels are available and less technology is required. Although modern technology often serves to reduce the impact of physical barriers, the advantages and disadvantages of each communication channel should be understood so that an appropriate channel can be used to overcome the physical barriers.

Systematic Barriers

Systematic barriers to communication may exist in structures and organisations where there are inefficient or inappropriate information systems and communication channels, or where there is a lack of understanding of the roles and responsibilities for communication. In such organisations, individuals may be unclear of their role in the communication process and therefore not know what is expected of them.

Attitudinal Barriers

Attitudinal barriers are behaviours or perceptions that prevent people from communicating effectively. Attitudinal barriers to communication may result from personality conflicts, poor management, resistance to change or a lack of motivation. Effective receivers of messages should attempt to overcome their own attitudinal barriers to facilitate effective communication.

Effective communication requires messages to be conveyed clearly between communicators, but along the way there are many communication barriers that can create misunderstandings and misinterpretations of your message. Successful communication requires knowing what barriers to communication exist and how to navigate around these roadblocks.

Language Barriers

Not using words another can understand will certainly stop your message from being conveyed. This not only applies to actual languages, but that of expressions, buzz words, and other jargon. If one is not familiar with your language, misinterpretation will occur.

Gender Barriers

Variation exists among masculine and feminine styles of communication. While women often emphasize politeness, empathy, and rapport building, male communication is often more direct. Meshing these two styles without awareness could be becomes a barrier.

Interpersonal Barriers

These are barriers are created to distance themselves from others. These can be done through withdrawal, meaningless rituals which keep one devoid of real contact, superficial activities through pastimes, and more.

Perceptual Barriers

Different world views can create misunderstanding. Without thinking, one might only view a message from their mindset rather than looking to see it from another viewpoint.

Cultural Barriers

Ethnic, religious, and social differences can often create misunderstandings when trying to communicate. These differences can also affect perceptual factors, as mentioned above.

Emotional Barriers

Troubled listening can occur if one is consumed with emotion. Hostility, anger, fear, and other emotions make it hard to hear outside of one's self. Overcoming these barriers to communication is no easy task. It takes great awareness and a willingness to adapt and look at communication from new perspectives. But, if you begin to focus on how these communication barriers are affecting your everyday conversation, you will be well on your way to becoming an effective communicator.

Rules for overcoming barriers

Eliminating differences in perception

The organization should ensure that it is recruiting right individuals on the job. It's the responsibility of the interviewer to ensure that the interviewee has command over the written and spoken language. There should be proper Induction program so that the policies of the company are clear to all the employees. There should be proper trainings conducted for required employees (for eg: Voice and Accent training).

Use of Simple Language

Use of simple and clear words should be emphasized. Use of ambiguous words and jargons should be avoided.

Reduction and elimination of noise levels

Noise is the main communication barrier which must be overcome on priority basis. It is essential to identify the source of noise and then eliminate that source.

Active Listening

Listen attentively and carefully. There is a difference between "listening" and "hearing". Active listening means hearing with proper understanding of the message that is heard. By asking questions the speaker can ensure whether his/her message is understood or not by the receiver in the same terms as intended by the speaker.

Emotional State

During communication one should make effective use of body language. He/she should not show their emotions while communication as the receiver might misinterpret the message being delivered. For example, if the conveyer of the message is in a bad mood then the receiver might think that the information being delivered is not good.

Simple Organizational Structure

The organizational structure should not be complex. The number of hierarchical levels should be optimum. There should be a ideal span of control within the organization. Simpler the organizational structure, more effective will be the communication.

Avoid Information Overload

The managers should know how to prioritize their work. They should not overload themselves with the work. They should spend quality time with their subordinates and should listen to their problems and feedbacks actively.

Give Constructive Feedback

Avoid giving negative feedback. The contents of the feedback might be negative, but it should be delivered constructively. Constructive feedback will lead to effective communication between the superior and subordinate.

Proper Media Selection

The managers should properly select the medium of communication. Simple messages should be conveyed orally for instance through, face to face interaction or meetings. Use of written means of communication should be encouraged for delivering complex messages. For significant messages reminders can be given by using written means of communication such as: Memos, Notices etc.

Flexibility in meeting the targets

For effective communication in an organization the managers should ensure that the individuals are meeting their targets on time without skipping the formal channels of communication. There should not be much pressure on employees to meet their targets.

Chapter 8
Some Relevant Terminologies in Communication

Opinion leaders/change agents

The opinions of people in a group are influenced by what they hear from "opinion leaders". An individual who is a member of a group manifests certain characteristics in his thinking and behavior that contributes to the formation of "public opinion". The opinion of the leader is based on rational thinking due to education and experience. They weigh the pros and cons of the information they receive and then give their judgment on it.

Encoder

In the process of communication, the sender or source of the message is referred to as the encoder.

Decoder

The person receiving the message and decodes it is referred to as the decoder.

Sender

The initiator and encoder of a message

Receiver

The one that receives the message (the listener) and the decoder of a message.

Channel

The medium through which the message travels such as either through oral communication (radio, television, phone, in person) or written communication (letters, email, text messages)

Message

This is the verbal and nonverbal components of a language which is sent to the receiver by the sender, which conveys an idea.

Feedback

Feedback - term form cybernetics - is the study of messages. It refers to an inquiry, response or experiment. Feedback can be positive (when the required result is achieved) or negative; instantaneous (when the response is immediate) or delayed. Feedback is used to gauge the effectiveness of a particular message put forth or situation that has taken place the receivers verbal and nonverbal responses to a message such as a nod for understanding (nonverbal), a raised eyebrow for being confused (nonverbal), or asking a question to clarify the message (verbal).

Positive feedback

This feedback shows the communicator that his message was understood.

Negative feedback: This feedback shows the communicator that his message was not understood. It does not imply disagreement just a lack of understanding.

Evaluation: Measuring the effectiveness of the message.

Noise

In all communication, there is a sender, a message/communication and a receiver. The meaning of a message is greatly dependent on the culture in which it is transmitted. The sender encodes a message, the receiver decodes it. Between the sender, the message and receiver, noise gets in the way and complicates the process. A noiseless communication does not exist. There always is some kind of noise entering the communication. Noise can be physical noise for example static or psychological i.e. when culture, taboos or values come into play to disrupt the normal transmission process of communication. Misunderstanding of a particular message i.e. distortion of meaning is a form of noise, example, the game of Chinese Whisper "a person starts off with a particular message and the original message may be distorted by the time it comes to the final player.

Communication fidelity: Faithful performance of communication process by all its elements *i.e.* communicator, message, channel and receiver.

Problem: Here is something that stands between you and person with whom you are communicating which interferes the complete transfer of information.

Interaction and communication

The concept of interdependence is itself complex and can be illustrated by defining the possible relationship between any two concepts, such A and B. A and B are independent if and only if neither affects the others. For example, the color of a person's here (A) and his left-handedness are independent. They don't affect each other. Interdependence can be defined as reciprocal or mutual dependence. If A affects B and B affects A, then A and B are interdependent. For example, in this country the farmer and the grocer are interdependent.

Communication between two or more people requires an interdependent relationship. However, the level of communicative interdependence varies from situation to situation. How do these levels differ?

Levels of communicative interdependence

a) Definitional-physical interdependence

The functions of the source and receiver are physically interdependent, although the functions may be performed at different points in time and space. When two people are communicating, they rely of the physical existence of the other for the production or reception of messages. Occasionally, this is the only kind of mutual interdependence involved to any appreciable extent.

b) Action-reaction interdependence

Communication often involves action-reaction interdependence. The action of the source affects the reaction of the receiver, the reaction of a receiver affect the subsequent reaction of the source, etc. The source or the receiver can make use of the reaction of the other. Reactions serve as feedback. They allow the source or receiver to check up on himself, to determine how well he is doing in accomplishing his purpose. Feedback also affects subsequent behaviour, if the source and receiver are sensitive to it. Communication sources and receiver are mutually interdependent, for existence and feedback. Each of them continually exerts influence over him and others by the kind of responses that he makes to the messages he produces and receives. A news paper affects its readers by selecting the news, they allowed reading. On the other hand, the readers also affect the news paper. If readers don't buy the news paper (negative feedback), it may change its selection and presentation of news.

c) Interdependence of expectations: empathy

Every communicator carries around with him an image of his receiver. He takes his receiver into account when he produces his message. He anticipates the possible responses of his receiver and tries to predict them ahead of time. These images affect his message behavior. The development of expectations of the receiver by the source has its counterpart in the development of expectation of the source by the receiver. The receivers have the expectations about the sources as well.

Empathy: Empathy is "I feel how you feel".

Theories of empathy

a) Inference theory of empathy

An inference theory of empathy is psychologically oriented. It argues that man can observe his own physical behavior directly, and can relate his behavior symbolically to his own internal psychological states-his feelings, thoughts, emotions, etc. Through this process, man comes to have meanings (interpretations) for his own physical behavior. He develops a concept of self, by himself, based on his observations and interpretations of his own behavior.

b) Role taking theory of empathy

Role taking theories argue that the new born infant cannot distinguish himself from other people, can't tell one person from another. In order to develop the concept of self, the infant must first look at himself as an object and must act toward himself as he acts toward other objects and other people. In other words, the concept of self does not precede communication. It is developed through communication.

Interaction: The Goal of Human Communication

One necessary condition for human communication is an interdependent relationship between the source and the receiver. Each affects the others. At one level of analysis, communication involves only a physical interdependence; i.e source and receiver are dyadic concepts, each requires the other for its very definition, each requires the other for its existence.

At a second level of complexity, interdependence can be analyzed as an action-reaction sequence. At a third level of complexity, communication analysis is concerned with empathic skills, the interdependence produced by expectations about how others will respond to a message. Empathy names the process in which we project ourselves into the internal states or personalities of others in order to predict how they will behave. The final level of interdependence complexity is interactions. The term interaction means the process of reciprocal role taking, the mutual performance of empathic behavior. If two individuals make inferences about their own roles and take the role of the others at the same time and if their communication behavior depends on the reciprocal taking of roles, then they are communicating by interacting with each other.

Types of communication

Communication is something that cannot be escaped at any point in our lives. It is a very wide aspect of our lives which can be divided into various types. The different types of communication are what we'll be dealing with, in this article. Communication is a process that involves an exchange of information, thoughts, ideas and emotions. The components of a communication process are: a sender who encodes and sends the message, the channel through which the message is sent, and the receiver who decodes the message, processes the information and sends an appropriate reply via the same channel back to the sender. Communication can occur via various processes and methods, and depending on the channel used and the style, there can be various types.

Communication Channels

Based on the channels used for communicating, the process of communication can be broadly classified as verbal and non-verbal communication. Verbal communication includes written and oral communication whereas non-verbal communication includes body language, facial expressions and visuals diagrams or pictures.

Oral Communication

Oral communication refers to the spoken words in the communication process. It can either be face-to-face or a conversation over the phone or on the voice chats over the Internet. Spoken conversations or dialogs are influenced by voice modulation, pitch, volume and even the speed and clarity of speaking.

Oral Communication

The other type of verbal communication is written communication. It can be either via snail mail, or email. The effectiveness of written communication depends on the style of writing, vocabulary used, grammar, clarity and precision of language.

Written Communication

Nonverbal

Non-verbal communication includes the overall body language of the person who is speaking, which will include the body posture, the hand gestures, and the overall body movements. Facial expressions also play a major role in communication. Expressions on a person's face say a lot about his/her mood. Gestures like a handshake, a smile or a hug can independently convey emotions and do not need to be accompanied by words. Non verbal communication can also be in the form of pictorial representations, signboards, or even photographs, sketches and paintings.

Nonverbal Communication

Categories of non-verbal communication

Elements such as physique, height, weight, hair, skin color, gender, odors, and clothing too send non-verbal messages during interaction. The various sources and categories of Non-verbal communication have been described below:

Arbitrariness

While much Non-verbal communication is based on arbitrary symbols, which differ from culture to culture, a large proportion is also to some extent iconic and may be universally understood. Paul Ekman's influential 1960s studies of facial expression determined that expressions of anger, disgust, fear, joy, sadness and surprise are universal.

Clothing and bodily characteristics

Clothes have both a functional and a communicative purpose. A villager's clothes alone can identify him as a well-to-do farmer having a respectable position in the village community. Or the attire of a village headman will differ from an ordinary farmer. The Sarpanch will in all probability have a turban on in most part of the country. In fact, in Rajasthan, the volume of the cloth in one's turban determines his position in many rural communities.

Physical environment

Environmental factors such as furniture, architectural style, interior decoration, lighting conditions, colors, temperature, noise, and music affect the behavior of communicators during interaction.

Proxemics: Physical space in communication

Proxemics is the study of how people use and perceive the physical space around them. The space between the sender and the receiver of a message influences the way the message is interpreted. The perception and use of space varies significantly across cultures and different settings within cultures. Space in non-verbal communication may be divided into four main categories: intimate, social, personal, and public space.

The distance between the speaker and the audience has a profound effect on the acceptance of the message being delivered. The speaker has to be one with the audience and should try to single with them in order to get his/her message over to the recipients. Speaking from the dais erected too far from the audience or too near can hamper delivery of the message. Sitting amongst the listeners can encourage a positive environment for learning. It should be kept in mind that the physical distance of the speaker and the audience should be such that the intended message should be audible to everyone present and should be decipherable.

Chronemics: Time in communication

Chronemics is the study of the use of time in Non-verbal communication. The way we perceive time, structure our time and react to time is a powerful communication tool, and helps set the stage for communication. Time perceptions include punctuality and willingness to wait, the speed of speech and how long people are willing to listen. The timing and frequency of an action as well as the tempo and rhythm of communications within an interaction contributes to the interpretation of Non-verbal messages. Gudykunst and Ting-Toomey (1988) identified two dominant time patterns.

1. Monochronic Time

A monochronic time system means that things are done one at a time and time is segmented into precise, small units. Under this system time is scheduled, arranged and managed. Monochronic cultures place a paramount value on schedules, tasks and "getting the job done". These cultures are committed to regimented schedules and may view those who do not subscribe to the same perception of time as disrespectful.

2. Polychronic Time

A polychronic time system is a system where several things can be done at once, and a more fluid approach is taken to scheduling time. These cultures are much less focused on the preciseness of accounting for each and every moment. As Raymond Cohen notes, polychronic cultures are deeply steeped in tradition rather than in tasks – a clear difference from their monochromic counterparts. "Traditional societies have all the time in the world. The arbitrary divisions of the clock face have little saliency in cultures grounded in the cycle of the seasons, the invariant pattern of rural life, and the calendar of religious festivities". Instead, their culture is more focused on relationships, rather than watching the clock. They have no problem being "late" for an event if they are with family or friends, because the relationship is what really matters. As a result, polychronic cultures have a much less formal perception of time. They are not ruled by precise calendars and schedules. Rather, "cultures that use the polychronic time system often schedule multiple appointments simultaneously so keeping on schedule is impossibility". Polychronic cultures include India, Saudi Arabia, Egypt, Mexico, Philippines and many in Africa.

Movement and body position

In 1952, Ray Birdwhistell, an anthropologist studied how people communicate through posture, gesture, stance, and movement. Part of Birdwhistell's work involved making film of people in social situations and analyzing them to show different levels of communication not clearly seen otherwise. The study was joined by several other anthropologists, including Margaret Mead and Gregory Bateson.

Posture

Posture can be used to determine a participant's degree of attention or involvement, the difference in status between communicators, and the level of fondness a person has for the other communicator. Studies investigating the impact of posture on interpersonal relationships suggest that mirror-image congruent postures, where one person's left side is parallel to the other's right side, leads to favorable perception of communicators and positive speech; a person who displays a forward lean or a decrease in a backwards lean also signify positive sentiment during communication. Posture is understood through such indicators as direction of lean, body orientation, arm position, and body openness.

Gesture

A gesture is a non-vocal bodily movement intended to express meaning. They may be articulated with the hands, arms or body, and also include movements of the head, face and eyes, such as winking, nodding, or rolling ones' eyes. The boundary between language and gesture, or verbal and non-verbal communication, can be hard to identify.

Although the study of gesture is still in its infancy, some broad categories of gestures have been identified by researchers. The most familiar are the so-called emblems or quotable gestures. These are conventional, culture-specific gestures that can be used as replacement for words, such as the hand wave used in the US for "hello" and "goodbye". A single emblematic gesture can have a very different significance in different cultural contexts, ranging from complimentary to highly offensive. Another broad category of gestures comprises those gestures used spontaneously when we speak. These gestures are closely coordinated with speech. The so-called beat gestures are used in conjunction with speech and keep time with the rhythm of speech to emphasize certain worse or phrases. These types of gestures are integrally connected to speech and thought processes. Other spontaneous gestures used when we speak are more contextual and may echo or elaborate the meaning of the co-occurring speech.

Gestures can also be categorized as either speech-independent or speech-related. Speech-independent gestures are dependent upon culturally accepted interpretation and have a direct verbal translation. A wave hello or a peace sign are examples of speech-independent gestures. Speech related gestures are used in parallel with verbal speech; this form of Non-verbal communication is used to emphasize the message that is being communicated. Speech related gestures are intended to provide supplemental information to a verbal message such as pointing to an object of discussion. Gestures such as Mudra (Sanskrit) encode sophisticated information accessible to initiates that are privy to the subtlety of elements encoded in their tradition.

Haptics: Touching in communication

Haptics is the study of touching as Non-verbal communication. Touches that can be defined as communication include handshakes, holding hands, back slapping, a pat on the shoulder, and brushing an arm. Touching of oneself may include licking one's lips, picking, holding, and scratching. These behaviors are referred to as "adaptor" and may send messages that reveal the intentions or feelings of a communicator. The meaning conveyed from touch is highly dependent upon the context of the situation, the relationship between communicators, and the manner of touch.

Humans communicate interpersonal closeness through a series of non-verbal actions known as immediacy behaviors. Examples of immediacy behaviors are: smiling, touching, open body positions, and eye contact. Cultures that display these immediacy behaviors are known to be high contact cultures. Haptic communication is the means by which people and other animals communicate via touching. Touch is an extremely important sense for humans; as well as providing information about surfaces and textures it is a component of Non-verbal communication in interpersonal relationships, and vital in conveying physical intimacy.

Touch is the earliest sense to develop in the fetus. The development of an infant's haptic senses and how it relates to the development of the other senses such as vision has been the target of much research. Human bahies have been observed to have enormous difficulty surviving if they do not possess a sense of touch, even if they retain sight and hearing. Babies who can perceive through touch, even without sight and hearing, tend to fare much better. Touch can be thought of as a basic sense in that most life forms have a response to being touched, while only a subset have sight and hearing.

Striking, pushing, pulling, pinching, kicking, strangling and hand-to-hand fighting are forms of touch in the context of physical abuse. In a sentence like "I never touched him/her" nor "Don't you dare to touch him/her" the term touch may be meant as euphemism for either physical abuse or sexual touching.

Oculesics Eye gaze

The study of the role of eyes in Non-verbal communication is sometimes referred to as "oculesics". Eye contact can indicate interest, attention, and involvement. Gaze comprises the actions of looking while talking, looking while listening, amount of gaze, and frequency of glances, patterns of fixation, pupil dilation, and blink rate.

Paralanguage: vocalic non-verbal cues of the voice

Paralanguage (sometimes called vocalic) is the study of Non-verbal cues of the voice. Various acoustic properties of speech such as tone, pitch and accent, collectively known as prosody, can all give off Non-verbal cues. Paralanguage may change the meaning of words.

The linguist George L. Trager developed a classification system which consists of the voice set, voice qualities, and vocalization. The voice set is the context in which the speaker is speaking. This can include the situation, gender, mood, age and a person's culture. The voice qualities are volume, pitch, tempo, rhythm, articulation, resonance, nasality, and accent. They give each individual a unique "voice print". Vocalization consists of three subsections: characterizers, qualifiers and segregates. Characterizers are emotions expressed while speaking, such as laughing, crying, and yawning. A voice qualifier is the style of delivering a message – for example, yelling "Hey stop that!, as opposed to whispering "Hey stop that".

G.W. Porter divides non-verbal communication into four broad categories

a) Physical

This is the personal type of communication. It includes facial expressions, tone of voice, sense of touch, sense of smell, and body motions.

b) Aesthetic

This is the type of communication that takes place through creative expressions: playing instrumental music, dancing, painting and sculpturing.

c) Sings

This is the mechanical type of communication, which includes the use of signal flags, the 21-gun salute, horns, and sirens.

d) Symbolic

This is the type of communication that makes use of religious, status, or ego-building symbols.

The relation between verbal and non-verbal communication and their relative importance

In person-to-person communications our messages are sent on two levels simultaneously. If the Non-verbal cues and the spoken message are incongruous, the flow of communication is hindered. Right or wrong, the receiver of the communication tends to base the intentions of the sender on the non-verbal cues he receives.

An interesting question is: When two people are communicating face-to-face, how much of the meaning is communicated verbally, and how much is communicated non-verbally? This was investigated by Albert Mehrabian. He concluded: "It is suggested that the combined effect of simultaneous verbal, vocal, and facial attitude communications is a weighted sum of their independent effects. Clues from spoken words, from the voice tone, and from the facial expression, contribute 7%, 38%, and 55% respectively to the total meaning. Since then, other studies have analyzed the relative contribution of verbal and non-verbal signals under more naturalistic situations. Argyle using video tapes shown to the subjects, analyzed the communication of submissive/dominant attitude and found that non-verbal cues had 4.3 times the effect of verbal cues. The most important effect was that body posture communicated superior status in a very efficient way. The relative importance of spoken words and facial expressions may be very different in studies using different set-ups.

Verbal Vs. Oral communication in the context of non-verbal communication

Scholars in this field usually use a strict sense of the term "verbal", meaning "of or concerned with words", and do not use "verbal communication" as a synonym for oral or spoken communication. Thus, vocal sounds that are not considered to be words, such as a grunt, or singing a wordless note, are non-verbal. Sign languages and writing are generally understood as forms of verbal communications, as both make use of words – although like speech, both may contain paralinguistic elements and often occur alongside non-verbal messages. Non-verbal communication can occur through any sensory channel – sight, sound, smell, touch or taste.

Interaction of verbal and non-verbal communication

When communicating, non-verbal messages can interact with verbal messages in six ways: repeating, conflicting, complementing, substituting, regulating and accenting/ moderating.

i) Repeating

Repeating consists of using gestures to strengthen a verbal message, such as pointing to the object of discussion.

ii) Conflicting

Verbal and non-verbal messages within the same interaction can sometimes send opposing or conflicting messages. A person verbally expressing a statement of truth while simultaneously fidgeting or avoiding eye contact may convey a mixed message to the receiver in the interaction. Conflicting messages may occur for a variety of reasons often stemming from feelings of uncertainty, ambivalence, or frustration. When mixed messages occur, Non-verbal communication becomes the primary tool people use to attain additional information to clarify the situation; great attention is placed on bodily movements and positioning when people perceive mixed messages during interactions.

iii) Complementing

Accurate interpretation of messages is made easier when non-verbal and verbal communication complements each other. Non-verbal cues can be used to elaborate on verbal messages to reinforce the information sent when trying to achieve communicative goals; messages have been shown to be remembered well when non-verbal signals affirm the verbal exchange.

iv) Substituting

Non-verbal behavior is sometimes used as the sole channel for communication of a message. People learn to identify facial expressions, body movements, and body positioning as corresponding with specific feelings and intentions. Non-verbal signals can be used without verbal communication to convey messages; when non-verbal behavior does not effectively communicate a message; verbal methods are used to enhance understanding.

v) Regulating

Non-verbal behavior also regulates our conversations. For example, touching someone's arm can signal that you want to talk next or interrupt.

vi)Accenting/Moderating

Non-verbal signals are used to alter the interpretation of verbal messages. Touch, voice pitch, and gestures are some of the tools people use to accent or amplify the message that is sent; non-verbal behavior can also be used to moderate or tone down aspects of verbal messages as well. For example, a person who is verbally expressing anger may accent the verbal message by shaking a fist.

Dance and drama non-verbal communication

Dance and drama is a form of non-verbal communication that requires the same underlying faculty in the brain for conceptualization, creativity and memory, as doe's verbal language in speaking and writing. Means of self-expression, both forms have vocabulary (steps and gestures in dance), grammar (rules for putting the vocabulary together) and meaning. Dance, however, assembles (choreographs) these elements in a manner that more often resembles poetry with its ambiguity and multiple, symbolic and elusive meanings.

These forms of non-verbal communication have a special significance in the shape of traditional folk methods of communication used by extension agents in reaching out to the rural audience with important messages.

The need and ways to improve non-verbal communication skills in extension workers

Development experiences of the last decades have shown that human resources development is essential for food security and market integration. Achieving sustainable agricultural development is less based on material inputs (e.g., seeds and fertilizer) than on the people involved in their use. This focuses on human resources calls for increased knowledge and information sharing about agricultural production, as well as on appropriate communication methodologies, channels and tools. New agricultural technologies are generated by research institutes, universities, private companies, and by the farmers themselves. Agricultural advisory services (including traditional extension, consultancy, business development and agricultural information services) are expected to disseminate new technologies amongst the farmers.

The role of extension workers is to give highly accurate, specific and unbiased technical and management information and advice in direct response to the needs of the farmers. The adoption of new agricultural technologies by farmers is often very slow. In many countries low agricultural production has been attributed, among other factors, to poor linkages between Research-Advisory Service-Farmers and to ineffective technology delivery systems, including poor information packaging, inadequate communication systems and poor communication skills of the extension personnel. A lack of appropriate communication structures, methodologies and tools, result in poor identification of farmers' needs and priorities, inappropriate research programs, poor or irrelevant extension information and technologies and finally, low farmers' take-up of technology innovations.

Communication for development is based on the premise that successful rural development calls for the conscious and active participation of the intended beneficiaries at every stage of the development process. Rural development cannot take place without changes in attitudes and behavior among the people concerned. And for this efforts should be made to enhance the pedagogical and communication skills of development agents (at all levels) so that they may have more effective dialogue with their audience. There is a need to apply communication technology to training and extension programmes, particularly at the grassroots level, in order to improve their quality and impact.

Rural communication is an interactive process in which information, knowledge and skills, relevant for development are exchanged between farmers, extension/advisory services, information providers and research either personally or through media such as radio, print and more recently the new "Information and Communication Technologies" (ICTs).

Amongst all the rural communication channels, interpersonal communication *i.e.* face-to-face talking is perhaps the most important and effective methodology. It is here that it is not only what an extension worker says in a training/meeting that is important, but its how he/she says it that can make the difference to participating farmers in finally adopting a particular practice that can bring about significant change in their behavior for a better livelihood and in the present context, it is rightly said:

"It's not what you say; it's how you say it"

Non-verbal messages are an essential component of communication in the teaching process. Extension workers should be aware of non-verbal behavior in a farmers' training/meeting for three major reasons:

- An awareness of non-verbal behavior will allow one to become better receivers of farmers' messages.
- One will become a better sender of signals that reinforce learning.
- This mode of communication increases the degree of the perceived psychological closeness between the extension worker and farmer.

Extension workers who make eye contact with the farmers open the flow of communication and convey interest, concern, warmth and credibility. Smiling is a powerful cue that transmits:

- Happiness
- Friendliness
- Warmth
- Liking
- Affiliation

Thus, if the trainer smiles frequently he/she will be perceived as more likable, friendly, warm and approachable. Smiling is often contagious and farmers will react favorably and learn more. If an extension worker fails to gesture while speaking, he/she may be perceived as boring, stiff and unanimated. A lively and animated teaching style captures farmers' attention, makes the material more interesting, facilitates learning and provides a bit of entertainment. Head nods, a form of gestures, communicate positive reinforcement to farmers and indicate that the trainer is listening.

Standing erect, but not rigid, and leaning slightly forward communicates to farmers that the trainer is approachable, receptive and friendly. Furthermore, interpersonal closeness results when the extension worker and farmers face each other. Speaking with one's back turned or looking at the floor or ceiling should be avoided; it communicates disinterest to the training/meeting.

Cultural norms dictate a comfortable distance for interaction with farmers. The extension worker should look for signals of discomfort caused by invading farmers' space. Some of these are:

- Rocking
- Leg swinging
- Tapping
- Gaze aversion

Typically, in large gatherings space invasion is not a problem. In fact, there is usually too much distance. To counteract this, the extension worker should try and move around the meeting hall to increase interaction with the audience. Increasing proximity enables one to make better eye contact and increases the opportunities for farmers to speak.

The vocal elements of non-verbal communication include:

- Tone
- Pitch
- Rhythm
- Timbre
- Loudness
- Inflection

For maximum teaching effectiveness, the extension worker should learn to vary these six elements of his/her voice. One of the major criticisms is of instructors who speak in a monotone. Listeners perceive these instructors as boring and dull. Farmers learn less and lose interest more quickly when listening to extension agents who have not learned to modulate their voices.

Humor is often overlooked as a teaching tool, and it is too often not encouraged in formal trainings or meetings. Laughter releases stress and tension for both instructor and the trainee farmers. One should develop the ability to laugh at himself/herself and encourage farmers to do the same. It fosters a friendly classroom environment that facilitates learning.

Obviously, adequate knowledge of the subject matter is crucial to the trainer's success; however, it's not the only crucial element. Climates, which facilitate learning and retention, demand good non-verbal and verbal skills too.

Non-verbal communication is a rapidly flowing back-and-forth process. Successful non-verbal communication depends on emotional self-awareness and an understanding of the cues you're sending, along with the ability to accurately pick up on the cues the farmers are sending you. This requires your full concentration and attention. If you are planning what you're going to say next and thinking about something else, you are almost certain to miss non-verbal cues and other subtleties in the conversation. You need to stay focused on the moment-to-moment experience in order to fully understand what's going on.

Some more tips for successful non-verbal communication are:

- Take a time out if you're feeling overwhelmed by stress. Stress compromises your ability to communicate. When you're stressed out, you're more likely to misread other people, send off confusing or off-putting non-verbal signals, and lapse into unhealthy knee-jerk patterns of behavior. Take a moment to calm down before you jump back into the conversation. Once you've regained your emotional equilibrium, you'll be better equipped to deal with the situation in a positive way.
- Pay attention to inconsistencies. Non-verbal communication should reinforce what is being said. If you get the feeling that someone isn't being honest or that something is "off", you may be picking up on a mismatch between verbal and non-verbal cues. Is the person is saying one thing, and their body language something else? For example, are they telling you "yes" while shaking their head no?
- Look at non-verbal communication signals as a group. Don't read too much into a single gesture or non-verbal cue. Consider all of the non-verbal signals you are sending and receiving, from eye contact to tone of voice and body language. Are your non-verbal cues consistent – or inconsistent – with what you are trying to communicate?

Communication Based on Style and Purpose

Based on the style of communication, there can be two broad categories: formal and informal communication that has their own set of characteristic features.

Formal

It includes all the instances where communication has to occur in a set formal format. Typically this can include all sorts of business or corporate communication. The style of communication in this form is very formal and official. Official conferences, meetings and written memos and corporate letters are used for this form of communication. It can also occur between two strangers when they meet for the first time. Hence, formal communication is straightforward, official and always precise and has a stringent and rigid tone to it.

Informal Communication

The informal channel of communication is also known as 'Grapevine'. It is the result not of any official action, but of the operation of social forces at work-place. The term 'grapevine' arose during the days of U.S. Civil War. At that time, intelligence telephone lines were strung loosely from tree to tree in the manner of a grapevine, and the message thereon was often distorted; hence any rumour was said to be from the grapevine. Today the term applies to all informal communication. While formal communication exists to meet the utilitarian needs of the organization, informal communication is the method by which people carry on social, non-programmed activities within the formal boundaries of the system. It, thus, exists outside the official network. Though continuously interacting with it, this informal channel is generally multiple in nature; same person having social relationships with a number of people working in the same organization.

Informal communication includes instances of free and unrestrained conversation between people who share a casual rapport with each other. It requires two people to have a similar wavelength and hence occurs between friends and family. It does not have any rigid rules and guidelines. The conversations need not necessarily have boundaries of time, place or even subjects for that matter since we all know that friendly chats with our loved ones can simply go on and on.

Even though the whole process of communication may seem so simple, the effectiveness of each type depends to a great extent on certain internal and external environmental factors and also the communicator's ability to send, receive, decode and send a response. This is why the field of communication has been and continues to be widely studied and is given a place of importance in any walk of life.

Chapter 9
Audio-visual Aids

Audio-Visual Aids are defined as any device used to aid in the communication of an idea. From this definition, virtually anything can be used as an aid, providing it successfully communicates the idea or information for which it is designed. In this Chapter we not only use the term "Audio-Visual Aids" but "Instructional Aids", "Teaching Aids" and "Visual Aids" as well. An audiovisual product is any audiovisual (AV) item such as still photography, motion picture, audio or video tape, slide or filmstrip, that is prepared singly or in combination to communicate information or to elicit a desired audience response. Even though early aids, such as maps and drawing are still in use, advances in the audio visual field have opened up new methods of presenting these aids such as videotapes and multimedia equipment which allow more professional and entertaining presentations to be presented.

'Audio-visual aids are any device which can be used to make the learning experience more concrete, more realistic, more dynamic.'

Kinder S James (1959): 'All materials used in the classroom or other teaching situations to facilitate the understanding of the written and spoken words are audio-visual aids.'

Dale (1965): an audio-visual aid is an instructional device that can be heard and seen.

Haas and Packer (1964): Audio visual aids are those sensory objects or images which initiate or stimulate and reinforce learning.

Burton (1967): Audio visual aids are those aids which help in completing the triangular process of learning that is motivation, classification and stimulation.

Carter V Good (1963): Audio visual aids are anything by means of which learning process may be encouraged or carried on through the sense of hearing or sense of sight.

Good's dictionary of education: Audio visual aids are supplementary devices by which the teacher, through the utilization of more than one sensory channel is able to clarify, establish and correlate concepts, interpretations and appreciations.

McKean and Roberts: An audio visual aid is an instructional device in which the message can be heard as well as seen.

K P Neeraja: Literally audio refers to sound waves that can normally be heard by the human ear .The term visual implies relating to the sense of seeing. The term audio-visual aid refers to anything that an extension agent uses to help to convey the message when communicating with farmers. The spoken word is the agent's main communication tool, but, whether the agent is speaking to a large village meeting or discussing a problem in a field with a group of farmers, its impact and effectiveness can be greatly increased by the use of suitable audio-visual aids. When selected and used properly, audio-visual aids can help in the following ways:

- The interest of the audience can be maintained if the agent varies the mode of presentation. It is difficult to concentrate for long on what someone is saying; but if the agent refers to a wall chart, or illustrates a point with some slides, his audience's attention can be maintained.
- When information is presented to more than one sense (sight and touch, for example, as well as hearing), more is taken in and it is better understood and remembered.
- Processes and concepts that are difficult to express in words alone can be explained. The procedure for applying for a loan, for example, may sound confusing, but a simple chart or diagram can make the process clearer. Again, the life cycle of a crop pest can be explained by showing a series of slides or drawings.
- The effects of decisions and actions that farmers might take can be shown. Photographs of a cattle dip or a model of a cooperative store can give farmers a clear idea of just what it is they might be considering.
- Pictures can have a more immediate impact on our emotions than words. Photographs of a heavy crop, for example, are likely to arouse interest more effectively than details of yields read out by an extension agent.

Values of Audio-Visual aids

- Best motivators
- Antidote to the disease of verbal instruction

- Clear images
- Vicarious experiences
- Variety
- Freedom
- Opportunities to handle and manipulate
- Retentively
- Based on maximum of teachers
- Helpful in attracting attention
- Helpful in fixing up new learning
- Saving of energy and time
- Realism
- Vividness
- Meeting individual difference Encouragement to healthy class room interaction
- Spread on education on a mass scale
- Promotion of scientific temper
- Development of higher faculties
- Reinforcement to learners
- Positive transfer of learning and training
- Positive environment for creative discipline

Characteristics

1. They should be meaningful and purposeful
2. They should be accurate in every aspect
3. They should be simple
4. They should be cheap
5. They should be up to date
6. They should be easily portable
7. They should be according to the mental level of students
8. They should be motivate the learners
9. They should be improvised.
10. They should be large enough to be properly seen by the students for whom they are mean.

Principles in the use of Audio-Visual Aids

Principle of Selection: 1. The age level; 2. Other personality angles; 3. They should have specific educational values; 4. They should help in the realization of learning desired objectives.

Principle of Preparation: 1. As far as possible, the local material should be used in the preparation of aids. 2. The teacher must also receive training in the preparation of aids. 3. The teacher him/herself can prepare some aids or can take help of students too.

Principle of Physical Control: This is concerned with the arrangement of keeping aids safely and also to facilitate their lending to the teachers for use.

Principle of Proper Presentation: 1. Teacher should carefully visualize the use of teaching aids before their actual presentation; 2. They should be well acquainted with the use & manipulation of the aids to be shown in the class room; 3. The aids should be displayed properly. So that, all the students can see it, observe it, and can derive maximum benefits from it.

Principle of Response: This is the important principle. This tells the teacher how to guide the students to respond actively to the audio visual stimulus so that they derive the maximum benefits in learning.

Principle of Evaluation: This Principle stipulates that there should be continuous evaluation of Audio Visual Aids materials & accompanying techniques in the light of desired objective

Advantages of Audio-Visual Aids

a) **Best Motivators:** They are the best motivators. Students work with more interest and zeal. They are more attentive.

b) **Fundamental to Verbal Instructions:** They help to reduce verbalism which is a major weakness of our schools. They convey the same meaning as words mean. They provided clear understanding of the concepts and thus help to bring accuracy in learning.

c) **Clear Images:** Clear images are formed when we see, hear, touch, taste and smell as our experiences are direct, concrete and more or less permanent. Learning through the senses becomes the most natural and consequently the easiest process.

d) **Vicarious Experience:** Everyone agrees to the fact that the first hand experience is the best type of educative experience but such an experience cannot always be provided to the pupils an so in some situations certain substitutes have to be provided. For this we find a large number of inaccessible objects and phenomenon. For example all the students in India cannot possibly be shown Taj Mahal etc. In all such cases audio-visual aids provide us the best substitutes.

e) **Variety:** Audio-Visual aids provide variety and provide different tools in the hands of the teacher.

f) **Freedom:** The use of audio-visual aids provide various occasions for the pupil to move about, talk, laugh and comment upon. Under such an atmosphere the students work because they want to work and not because the teacher wants them to work.

g) **Opportunities to Handle and Manipulate:** The use of audio-visual aids provides immense opportunities to the pupils to see handle and manipulate things.

The range of Audio-Visual Aids

Extension agents often use sophisticated audio-visual aids which require electricity and complex machinery such as projectors or television sets. However there are many simple aids that the agent can make locally, and these have several advantages. They do not require a power source or heavy equipment, they do not cost much to produce and they can be made to suit the precise needs of the agent. Between these two extremes lies a wide range of more or less sophisticated aids. The distinction between a mass medium and an audio-visual aid lies only in the way it is used. All the mass media described earlier can be used as audio-visual aids. A film is a mass medium, in that it is shown to large audiences in many different places; but for an individual extension agent who uses it to increase the impact of a talk, it is an audio-visual aid. Many of the principles of media use discussed earlier also apply to audio-visual aids. The audio-visual aids available to the extension agent can now be examined.

Principles to be followed for the effective use of Audio-Visual Aids

1. Audio visual materials should function as an integral part of the educational program.
2. Audio visual aids should be centralized, under specialized direction and leadership in educational programs.
3. An advisory committee consisting of representative from all areas of curriculum should be appointed to assist in selection and coordination of Audio visual materials.
4. An education program should be flexible.
5. Audio visual materials should be carefully located to eliminate duplication, easy accessibility and convenient use.
6. Audio visual materials should be available whenever and wherever they are needed for effective utilization as an integral part of curriculum.
7. Budget appropriations should be made regularly for Audio visual education programs.
8. Periodic evaluation to be done to assess the function of, utilization and expenditure of the program.

Some criteria for selection and use of Audio-Visual Aids

What role does media play in people's lives:

How frequently and extensively is emotional or artistic expression? This would indicate the Vitality of the tradition in which they reset.

What is the usual size and composition of audience attracted in terms of age, sex, educational, economic and social groupings?

What is the importance of the functions performed in a framework of prevailing social needs and values?

a. Concern for plasticity of the medium, or capacity of a particular form to be "loaded" with specific innovative messages. Examination of religious or social functions attributed to the medium should be made so as not to distort its special role. In some cases, examination may reveal such use to be unacceptable. However, it has been found that religious themes can be interpreted in modern context, providing impetus for social change.

b. It would be determined whether the medium has an age or individuality (in which case positive identification of audience with actor can be the agent of change), or whether it is entirely non-individual. It has sometimes been found that certain messages can be best expressed without the personalities of, actors in evidence during performance, as in puppet show, mask presentations and giant figures.

c. It should be noted, whether a particular medium or any aspect of its traditions and features could be used by or extended to, mass media presentation. If so, is this desirable in relation to both a medium and the message?

d. Are the required resources, talent, materials, texts, available for extensive utilization of the medium?

e. Does the selected medium have entertainment value and / or artistic appeal?

f. Is it unique enough to prove a valuable communication channel?

g. Is the form of the proposed medium flexible enough to incorporate a sufficiently broad range of content materials, such as family planning and other developmental messages?

h. Is the medium sufficiently versatile to reach varied audiences . To properly assess this criterion, the country must be in a position to assess the range of traditional media and their suitability for audiences according to local and provincial make-up.

i. Is the selected medium, relevant to the intended audience?

Classification of Audio Visual Aids

Various classifications are given for Audio visual aids according to the type of projection by various authors.

1. According to Evolution

a. First generation media

e.g. Hand made chart, graph, exhibits, model, hand writing materials etc.

b. Second generation media

e.g. Printed or illustrated text, printed graph, work books.

c. Third generation media

e.g. Photographs, slides, filmstrips, radio, films, recordings, telecast etc.

d. Fourth generation media

e.g. Television, programmed instruction, language laboratories, electronic digital computers etc.

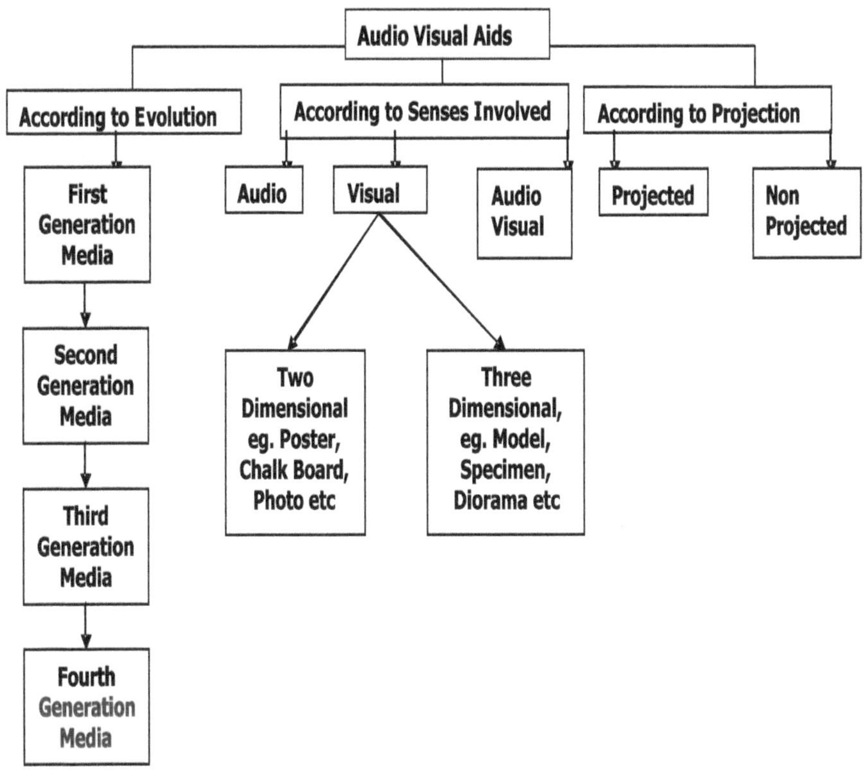

While using the foregoing methods the extension worker uses these methods independently or takes the help of certain audio-visual aids. The word audio-visual comprises three words namely, Audio refers to sense of hearing; Visual refers sense of seeing and Aid for instructional devices.

2. According to Senses involved

Audio	*Visual*	*Audio-Visual*
Tape recorder, Public address system, Telephone, Radio etc	Flash cards Black board Pictures etc	Cinema projector, telecast, Video recording, Television, Sound-synchronized slides, Drama etc

Audio Aids

Tape recorder:

A tape recorder is a portable electronic gadget to record, reproduce, erase and re record sound on a magnetic tape. This device can be used without much fuss by anybody by operating the following press buttons attached to the recorder, viz, stop, play, wind, rewind, record, pause, and eject.

k0002326 www.fotosearch.com

Uses: It is used to learn foreign languages, rhymes, and songs with clarity. GRAME PHONES Like radio gramophones are also important teaching devices. It helps to listen to famous speeches.

Using a record player for teaching

a. A record player can be used in the following ways in the actual class room situation

b. A record player can be used to supplement a lesson.

c. A record player can be used for an appreciation lesson in music.

d. A record player can be used for an appreciation lesson in literature.

e. A record player can be used for students to acquire the singing ability, deliver a speech properly, and recite a poem in the right way.

f. The player can be used to end or conclude a lesson;

g. Introduce a lesson and review a lesson.

h. A record player can be used for physical exercises accompanied with music

Public address system (PA system)

A public address system (PA system) is an electronic sound amplification and distribution system with a microphone, amplifier and loudspeakers, used to allow a person to address a large public, for example for announcements of movements at large and noisy air and rail terminals.

The term is also used for systems which may additionally have a mixing console, and amplifiers and loudspeakers suitable for music as well as speech, used to reinforce a sound source, such as recorded music or a person giving a speech or distributing the sound throughout a venue or building. Simple PA systems are often used in small venues such as school auditoriums, churches, and small bars. PA systems with many speakers are widely used to make announcements in public, institutional and commercial buildings and locations. Intercom systems, installed in many buildings, have microphones in many rooms allowing the occupants to respond to announcements. Sound reinforcement systems and PA systems may use some similar components, but with differing application, although the distinction between the two is not clear-cut. Sound reinforcement systems are for live music or performance, whereas PA systems are primarily for reproduction of speech. In Britain any PA system is sometimes colloquially referred to as a Tannoy, after the company of that name now owned by TC Electronic Group, which supplied a great many of the PA systems used in the past.

History

Edwin Jensen and Peter Pridham of Magnavox began experimenting with sound reproduction in the 1910s; working from a laboratory in Napa, California, they filed the first patent for a moving coil loudspeaker in 1911. Four years later, in 1915, they built a dynamic loudspeaker with a 1-inch (2.5 cm) voice coil, a 3-inch (7.6 cm) corrugated diaphragm and a horn measuring 34 inches (86 cm) with a 22-inch (56 cm) aperture. The electromagnet created a flux field of approximately 11,000 G.

Their first experiment used a carbon microphone. When the 12 V batteries were connected to the system, they experienced one of the first examples of acoustic feedback. They then placed the loudspeaker on the laboratory's roof, and claims say that the amplified human voice could be heard 1 mile (1.6 km) away. Jensen and Pridham refined the system and connected a phonograph to the loudspeaker to be able to broadcast recorded music. This demonstration was official presentation of the working system, and approximately 100,000 people gathered to hear Christmas music and speeches "with absolute distinctness". The first outside broadcast was made one week later, again supervised by Jensen and Pridham. On December 30, when Governor of California Hiram Johnson was too ill to give a speech in person, loudspeakers were installed at the Civic Auditorium in San Francisco, connected to Johnson's house some miles away by cable and a microphone, from where he delivered his speech. Jensen oversaw the governor using the microphone while Pridham operated the loudspeaker. The following year, Jensen and Pridham applied for a patent for what they called their "Sound Magnifying Phonograph". Over the next two years they developed their first valve amplifier. In 1919 this was standardized as a 3-stage 25 watt amplifier.

This system was used by former US president William Howard Taft at a speech in Grant Park, Chicago, and first used by a current president when Woodrow Wilson addressed 75,000 people in San Diego, California. Wilson's speech was part of his nationwide tour to promote the establishment of the League of Nations. It was held on September 9, 1919 at City Stadium. As with the San Francisco installation, Jensen supervised the microphone and Pridham the loudspeakers. Wilson spoke into two large horns mounted on his platform which channelled his voice into the microphone. Similar systems were used in the following years by Warren G. Harding and Franklin D. Roosevelt. By the early 1920s, Marconi had established a department dedicated to public address and began producing loudspeakers and amplifiers to match a growing demand. In 1925, George V used such a system at the British Empire Exhibition, addressing 90,000 via six long-range loudspeakers. This public use of loudspeakers brought attention to the possibilities of such technology. The 1925 Royal Air Force Pageant at Hendon Aerodrome used a Marconi system to allow the announcer to address the crowds, as well as amplify the band. In 1929, the Schneider Trophy race at Calshot Spit used a public address system that had 200 horns, weighing a total of 200 long tons.

Telephone paging systems

Some analog or IP private branch exchange (PBX) telephone systems use a paging facility that acts as a liaison between the telephone and a PA amplifier. In other systems, paging equipment is not built into the telephone system. Instead the system includes a separate paging controller connected to a trunk port of the telephone system. The paging controller is accessed as either a designated directory number or central office line. In many modern systems, the paging function is integrated into the telephone system, and allows announcements to be played over the phone speakers. Many retailers and offices choose to use the telephone system as the sole access point for the paging system, because the features are integrated. Many schools and other larger institutions are no longer using the large, bulky microphone PA systems and have switched to telephone system paging, as it can be accessed from many different points in the school.

Radio

Radio is the wireless transmission of signals through free space by electromagnetic radiation of a frequency significantly below that of visible light, in the radio frequency range, from a few Hz to 300 GHz. These waves are called radio waves. Electromagnetic radiation travels by means of oscillating electromagnetic fields that pass through the air and the vacuum of space. Information, such as sound, is carried by systematically changing (modulating) some property of the radiated waves, such as their amplitude, frequency, phase, or pulse width. When radio waves strike an electrical conductor, the oscillating fields induce an alternating current in the conductor. The information in the waves can be extracted and transformed back into its original form

Etymology

The etymology of "radio" or "radiotelegraphy" reveals that it was called "wireless telegraphy", which was shortened to "wireless" in Britain. The prefix radio in the sense of wireless transmission was first recorded in the word radio conductor, a description provided by the French physicist Édouard Branly in 1897. It is based on the verb to radiate (in Latin "radius" means "spoke of a wheel, beam of light, ray").

The word "radio" also appears in a 1907 article by Lee De Forest. It was adopted by the United States Navy in 1912, to distinguish radio from several other wireless communication technologies, such as the photo-phone. The term became common by the time of the first commercial broadcasts in the United States in the 1920s. (The noun "broadcasting" itself came from an agricultural term, meaning "scattering seeds widely.") The term was adopted by other languages in Europe and Asia. British Commonwealth countries continued to commonly use the term "wireless" until the mid-20th century, though the magazine of the BBC in the UK has been called Radio Times ever since it was first published in the early 1920s.

In recent years the more general term "wireless" has gained renewed popularity through the rapid growth of short-range computer networking, e.g., Wireless Local Area Network (WLAN), Wi-Fi, and Bluetooth, as well as mobile telephony, e.g., GSM and UMTS. Today, the term "radio" specifies the actual type of transceiver device or chip, whereas "wireless" refers to the lack of physical connections; one talk about radio transceivers, but another talk about wireless devices and wireless sensor networks.

Radio Communication System

A radio communication system sends signals by radio. Types of radio communication systems depend on technology, standards, regulations, radio spectrum allocation, user requirements, service positioning, and investment. The radio equipment involved in communication systems includes a transmitter and a receiver, each having an antenna and appropriate terminal equipment such as microphone at the transmitter and a loudspeaker at the receiver in the case of a voice-communication system. The power consumed in a transmitting station varies depending on the distance of communication and the transmission conditions. The power received at the receiving station is usually only a tiny fraction of the transmitter's output, since communication depends on receiving the information, not the energy that was transmitted.

Classical radio communications systems use frequency-division multiplexing (FDM) as a strategy to split up and share the available radio-frequency bandwidth for use by different parties communications concurrently. Modern radio communication systems include those that divide up a radio-frequency band by time-division multiplexing (TDM) and code-division multiplexing (CDM) as alternatives to the classical FDM strategy. These systems offer different tradeoffs in supporting multiple users, beyond the FDM strategy that was ideal for broadcast radio but less so for applications such as mobile telephony.

A radio communication system may send information only one way. For example, in broadcasting a single transmitter sends signals to many receivers. Two stations may take turns sending and receiving, using a single radio frequency; this is called "simplex." By using two radio frequencies, two stations may continuously and concurrently send and receive signals; this is called "duplex" operation.

Telephone

A telephone, or phone, is a telecommunication device that permits two or more users to conduct a conversation when they are not in the same vicinity of each other to be heard directly. A telephone converts sound, typically and most efficiently the human voice, into electronic signals suitable for transmission via cables or other transmission media over long distances, and replays such signals simultaneously in audible form to its user. The word telephone has been adapted into the vocabulary of many languages. First patented in 1876 by Alexander Graham Bell and further developed by many others, the telephone was the first device in history that enabled people to talk directly with each other across large distances. Telephones rapidly became indispensable to businesses, government, and households, and are today some of the most widely used small appliances.

The essential elements of a telephone are a microphone (transmitter) to speak into and an earphone (receiver) which reproduces the voice of the distant person. In addition, most telephones contain a ringer which produces a sound to announce an incoming telephone call, and a dial used to enter a telephone number when initiating a call to another telephone. Until approximately the 1970s most telephones used a rotary dial, which was superseded by the modern Touch-Tone push-button dial, first introduced by AT&T in 1963. The receiver and transmitter are usually built into a handset which is held up to the ear and mouth during conversation. The dial may be located either on the handset, or on a base unit to which the handset is connected by a cord containing wires. The transmitter converts the sound waves to electrical signals which are sent through the telephone network to the receiving phone. The receiving telephone converts the signals into audible sound in the receiver, or sometimes a loudspeaker. Telephones are a duplex communications medium, meaning they allow the people on both ends to talk simultaneously.

A landline telephone is connected by a pair of wires to the telephone network, while a mobile phone, such as a cellular phone, is portable and communicates with the telephone network by radio transmissions. A cordless telephone has a portable handset which communicates by radio transmission with the handset base station which is connected by wire to the telephone network.

Visual Aids

Visual communication refers to communication through visual aid and is described as the conveyance of ideas and information in forms that can be read or looked upon. Visual communication in part or whole relies on vision, and is primarily presented or expressed with two dimensional images, it includes: signs, typography, drawing, illustration, Industrial design, graphic design, advertising, animation colour and electronic resources. It also explores the idea that a visual message accompanying text has a greater power to inform, educate, or persuade a person or audience.

Visual aids are two types

1. Two-dimensional visuals (Poster, Chalk Board, Photo etc.)
2. Three dimensional visuals (Model, Specimen, Diorama etc.)

Audio-Visual Aids

Movie Projector

Movie projector is an opto-mechanical device for displaying motion picture film by projecting it onto a screen. Most of the optical and mechanical elements, except for the illumination and sound devices, are present in movie cameras.

History

The first movie projector was the Zoopraxiscope, invented by British photographer Eadweard Muybridge in 1879. The zoopraxiscope projected images from rotating glass disks in rapid succession to give the impression of motion. The stop-motion images were initially painted onto the glass, as silhouettes. A second series of discs, made in 1892–94, used outline drawings printed onto the discs photographically, then colored by hand.

A more sophisticated movie projector was invented by Frenchman Louis Le Prince while working in Leeds. In 1888 Le Prince took out a patent for a 16-lens device that combined a motion picture camera with a projector. In 1888, he used an updated version of his camera to film the first ever motion picture, the Roundhay Garden Scene. The pictures were privately exhibited in Hunslet.

The Lumière brothers invented the first successful movie projector. They made their first film, Sortie de l'usine Lumière de Lyon, in 1894, which was publicly screened at L'Eden, La Ciotat a year later. The first commercial, public screening of cinematographic films happened in Paris on 28 December 1895. The cinematograph was also exhibited at the Paris Exhibition of 1900. At the Exhibition, films made by the Lumière Brothers were projected unto a large screen measuring 16 by 21 meters.

Physiology

According to the theory of persistence of vision, the perceptual processes of the brain and the retina of the human eye retain an image for a brief moment of time. This theory is said to account for the illusion of motion which results when a series of film images is displayed in quick succession, rather than the perception of the individual frames in the series.

Persistence of vision should be compared with the related phenomena of beta movement and phi movement. A critical part of understanding these visual perception phenomena is that the eye is not a camera, i.e.: there is no "frame rate" or "scan rate" in the eye. Instead, the eye/brain system has a combination of motion detectors, detail detectors and pattern detectors, the outputs of all of which are combined to create the visual experience.

The frequency at which flicker becomes invisible is called the flicker fusion threshold, and is dependent on the level of illumination. Generally, the frame rate of 16 frames per second (frame/s) is regarded as the lowest frequency at which continuous motion is perceived by humans. (Interestingly this threshold varies across different species; a higher proportion of rod cells in the retina will create a higher threshold level). It is possible to view the black space between frames and the passing of the shutter by the following technique:

Close your eyelids, then periodically rapidly blink open and closed. If it done fast enough, you will be able to randomly "trap" the image between frames, or during shutter motion. This will not work with television due to the persistence of the phosphors nor with LCD or DLP light projectors due to the continuity of image; although certain color artifacts may appear with some digital projection technologies.

Silent films usually were not projected at constant speeds but rather were varied throughout the show at the discretion of the projectionist, often with some notes provided by the distributor. Speeds ranged from about 18 numbers frame/s on up - sometimes even faster than modern sound film speed (24 numbers frame/s). Contrary to received opinion, 16 numbers of frame/s though sometimes used as a camera shooting speed - was dangerously inadvisable for projection, due to the high risk of the nitrate-base prints catching fire in the projector. (A dramatic rendition of a nitrate print fire and its potentially devastating effects is famously found in Nuovo Cinema Paradiso, which revolves around the goings-on of a projectionist.)

Since the birth of sound film, virtually all film projectors in commercial movie theaters project at a constant speed of 24 numbers frame/s. This speed was chosen for both financial and technical reasons. When Warner Bros. and Western Electric were trying to find the proper projection speed for the new sound pictures, Western Electric went to the Warner Theater in LA and noted the average speed at which films were projected there. They set that as the sound speed at which a satisfactory reproduction and amplification of sound could be conducted. There are some specialist formats (e.g. Showscan and Maxivision) which project at higher rates 60 numbers frames/sec for Showscan and 48 for Maxivision. The Hobbit was shot at 48 numbers frames/sec and projected at the higher frame rate at specially equipped theaters.

Electronic Video Recorder

Teletext (or broadcast teletext)

It is a television information retrieval service created in the United Kingdom in the early 1970s by the Philips Lead Designer for VDUs, John Adams. Teletext is a means of sending text and diagrams to a properly equipped television screen by use of one of the vertical blanking interval lines that together form the dark band dividing pictures horizontally on the television screen. It offers a range of text-based information, typically including news, weather and TV schedules. Subtitle (or closed captioning) information is also transmitted within the television signal. The first test transmissions were made by the BBC in 1973, known as Ceefax. After adoption in the UK the standards became international as the European Teletext standards and as the World System Teletext (WST). The World Wide Web began to take over some of the functions of teletext from the late 1990s, and many broadcasters have ceased broadcast of teletext CNN in 2006 and the BBC in 2012. The decline of teletext has been hastened by the introduction of digital television, though an aspect of teletext continues in closed captioning.

Electronic Video Recording

Electronic Video Recording, or EVR, was a film-based video recording format developed by Hungarian-born engineer Peter Carl Goldmark at CBS Laboratories in the 1960s. CBS announced EVR in October 1967. The 750-foot film was stored on a 7-inch-diameter (180 mm) spool in a plastic cartridge. It used a twin-track 8.75 mm film onto which video signals was transferred by electron beam recording, one monochrome track in each direction of travel.

Some EVR films had a separate chroma track in place of the opposite-direction monochrome track for color EVR films. The images stored on an EVR film were visible frames much like motion picture film, and were read by a flying-spot scanner inside an EVR player to be converted to a video signal to be sent to a television set. EVR was also released by CBS as a professional version for television broadcasting, called BEVR (Broadcast EVR). As a professional medium, the format offered extremely high quality. It was, however, quickly superseded by professional and consumer magnetic tape formats.

Electronic Video Recorder

Television

Television, or **TV** for short, (from French télévision, meaning "television"; from Ancient Greek (tèle), meaning "far", and Latin visio, meaning "sight") is a telecommunication medium typically used in the 2000s for transmitting and receiving moving color images and sound. In the broader sense, television can also refer to images that can be monochrome (black-and-white) or with or without accompanying sound. "Television" may also refer specifically to a television set, television program, or television transmission.

Commercially available since the late 1920s (but in extremely limited amounts during those years), the television set has since become commonplace in homes, businesses and institutions, particularly as a vehicle for advertising, a source of entertainment, and news. Since the 1950s, television has been the main medium for molding public opinion. Since the mid-1960s, color televisions have been widely available. Since the 1970s, the availability of media such as video cassettes (mid 1970s), laserdiscs (1978), DVDs (1997) and high-definition Blu-ray Discs (2006) has enabled viewers to use the television set to watch recorded material, such as movies, as well as broadcast material. In recent years, Internet television has seen the rise of television available via the Internet through services such as iPlayer and Hulu.

In 2009, 78% of the world's households owned at least one television set, an increase of 5% over 2003. One major change in television technology was the replacement of the long - dominant CRT technology with various flat-panel TVs using LCD, plasma or LEDs. In 2013, 87 percent of TVs sold were color LCD screen TVs. Although other forms such as closed-circuit television (CCTV) are in use, the most common usage of the medium is for broadcast television, which was modeled on the existing radio broadcasting systems developed in the 1920s, and uses high powered radio frequency transmitters to broadcast the television signal to individual TV receivers.

The broadcast television system is typically disseminated via radio transmissions on designated channels in the 54–890 MHz frequency band. Signals are now often transmitted with stereo or surround sound in many countries. Until the 2000s, broadcast TV programs were generally transmitted as an analog television signal but during the decade several countries went almost exclusively in digital. In addition to over-the-air transmission, television signals are also distributed by cable and satellite systems. A standard television set comprises multiple internal electronic circuits, including those for receiving and decoding broadcast signals. A visual display device which lacks a tuner is properly called a video monitor, rather than a television. A television system may use different technical standards such as digital television (DTV) and high definition television (HDTV). Television systems are also used for surveillance, industrial process control, and guiding of weapons, in places where direct observation is difficult or dangerous. Some studies have found a link between infancy exposure to television and ADHD.

Drama

Drama is the specific mode of fiction represented in performance. The term comes from a Greek word meaning "action" (Classical Greek: δρᾶμα, drama), hich is derived from the verb meaning "to do" or "to act" (Classical Greek: dra). The enactment of drama in theatre, performed by actors on a stage before an audience, presupposes collaborative modes of production and a collective form of reception. The structure of dramatic texts, unlike other forms of literature, is directly influenced by this collaborative production and collective reception. The early modern tragedy Hamlet (1601) by Shakespeare and the classical Athenian tragedy Oedipus the King (c. 429 BC) by Sophocles are among the masterpieces of the art of drama. A modern example is Long Day's Journey into Night by Eugene O'Neill (1956).

The two masks associated with drama represent the traditional generic division between comedy and tragedy. They are symbols of the ancient Greek Muses, Thalia and Melpomene. Thalia was the Muse of comedy (the laughing face), while Melpomene was the Muse of tragedy (the weeping face). Considered as a genre of poetry in general, the dramatic mode has been contrasted with the epic and the lyrical modes ever since Aristotle's Poetics (c. 335 BC) the earliest work of dramatic theory.

The use of "drama" in the narrow sense to designate a specific type of play dates from the 19th century. Drama in this sense refers to a play that is neither a comedy nor a tragedy for example, Zola's Thérèse Raquin (1873) or Chekhov's Ivanov (1887). It is this narrow sense that the film and television industry and film studies adopted to describe "drama" as a genre within their respective media. "Radio drama" has been used in both senses originally transmitted in a live performance, it has also been used to describe the more high-brow and serious end of the dramatic output of radio. Drama is often combined with music and dance: the drama in opera is generally sung throughout; musicals generally include both spoken dialogue and songs; and some forms of drama have incidental music or musical accompaniment underscoring the dialogue (melodrama and Japanese N , for example). In certain periods of history (the ancient Roman and modern Romantic) some dramas have been written to be read rather than performed. In improvisation, the drama does not pre-exist the moment of performance; performers devise a dramatic script spontaneously before an audience.

3. According to projection Audio-Visual Aids can be classified as projected and non projected aids

Projected Audio Visual Aids

a. Magic Lantern
b. Slide projector
c. Over Head Projector
d. Motion pictures
e. Over Head Transparencies
f. The Opaque Projector
g. Filmstrip
h. LCD Projector
i. Home theatre Projector

Projected Audio Visual Aids

a) Magic Lantern

'The Magic Lantern' was the forerunner of the modern slide projector. However, researchers do not know who invented it. In 1676, a type of Magic Lantern called the Sturm Lantern was invented and it may have been one of the first.

The practice of projecting images from glass plates began centuries before the invention of photography. However, in the 1840s, Philadelphian daguerreotypists, William and Frederick Langenheim, began experimenting with 'The Magic Lantern' as an apparatus for displaying their photographic images. Since the opaque nature of the daguerreotype disallowed its projection, the brothers looked for a medium that could create a transparent image.

b. Slide Projector

A slide is a small piece of transparent material on which a single pictorial image or scene or graphic image has been photographed or reproduced otherwise.

Slides are a form of projected media that are easy to prepare. They are still pictures on positive film which you can process and mount individually yourself or send to a film laboratory. The standard size of the slides is 2 "X 2 "any 35mm camera will make satisfactory slides.

Slide History

Throughout history, slide film has produced excellent quality pictures, offering sharp images and clear bright colours, perfect for capturing vivid scenery and landscapes. In the mid 20th century, the invention of modern slide projectors enabled people at home to share images and create slideshows to enjoy with their friends and family.

Despite huge popularity in the 20th century, slides have gradually declined with the advent of digital photography. Now, a huge variety of slide photographs from all over the world have been reproduced digitally and can be seen and bought in exhibitions and in large databases and archives on the internet. Projectors entertained people throughout the 1950s to the 1970s and slide photography formed a large part of amateur and professional photography right throughout the 20th century capturing beautiful, sharp images and memories.

Before modern slide technology, Magic Lanterns were the forerunners of modern slide projectors. The practice of projecting images from glass plates began centuries before the invention of photography, but in the 1840s, Philadelphian daguerreotypes, William and Frederick Langenheim, were the first to begin experimenting with the Magic Lantern as an apparatus for displaying photographic images. The Langheims were able to create a transparent positive image on glass suitable for projection, and they patented this and called it the Hyalotype. The

Magic Lanterns were eventually mass produced so they could be used in schools and universities and were popular with both photographers and public speakers. The Magic Lantern projectors eventually went out of use when modern slide projectors and slide film were introduced in the mid 20th century.

Modern slides are positive transparencies, mounted in a plastic frame for projection and replaced the glass Hyalotypes. The most popular modern slide film introduced were the 35mm transparency slides which became popular as they were cheaper to make and easier to use than the heavier Hyalotype glass slides. Oscar Barnack, sometimes called the grandfather of 35mm photography, first used a 35mm film in a still camera in 1913. The camera, called the Ur-Leica, was first mass produced in 1924. The 35mm film Barnack used was especially well-suited to still and moving photography because it provided great image quality for the best price. The advent of modern slides also signaled the invention of the first modern colour film. A huge benefit of the new transparency slides were that they could be used in a projector and shown with complete colour accuracy. In 1935 Kodak introduced the new Kodaslide projector which played the first modern 'integral tripak' colour film called Kodachrome. The projector played 2x2 inch colour slides, projecting them onto a 6.5 inch translucent screen.

Initially, Kodachrome was available only as 16mm film for home movies, but in 1936 it was also introduced as 8mm home movie film and short lengths of 35mm film for still photographs. The new slides could also withstand the heat of extended projection and be produced from the newly popular 35mm cameras. People could create slideshows of their 35mm slides and share them with their families at gathering and events and the projectors were also popular in art schools to teach art history due to the colour accuracy. In 1937, Kodachrome shots by W. Robert Moore were used for the first time in National Geographic magazine. This was the first time a photographer could capture action shots in colour on 35mm film. 'Everyone just went wild over them', lab technician B. Anthony Stewart recalled in The National Geographic Society: 100 Years of Adventure and Discovery 'The iridescence of colour was just something that colour photographers had never before dreamed of'. In 1941 Kodak also made it possible for people to order prints of their Kodachrome slides. Carousel Slide projectors arrived in the 1950s, originally designed by Louis Misurace and then bought by Kodak, they could hold dozens of slides and move them forward sequentially. Kodak gradually increased the sizes of their slide projectors for a higher number of slides and also created a 'pocket' Carousel model for miniature slides. In the 1940s, Kodak produced 'Ektachrome' film which was used extensively in National Geographic Magazine for colour photographs in settings where Kodachrome was too slow. It also allowed amateurs to process their own film as the process was much less complex. Kodak's only real competitions were Fuji who launched the Velva slide in the 1990s which was hugely popular due to its sharpness and vivid colour production.

However, in the mid 1990s, scientists and engineers created a new technology that eventually led to the first multimedia projector. The newest technology of the day was digital processing and in 2004 with the advent of digital photography the production of Kodak's projectors were eventually discontinued. Many examples of slide photography have now been converted to a digital format and a huge archive of these photographs can be viewed online including on the National Library of Australia website which has digitally reproduced thousands of slide photographs, these photographs not only illustrate the history of Australia but tell the story of the development of photographic methods throughout the 20th century and beyond.

Types of slides

1. Photographic slides

a) Black and white 2″ X 2″

b) Colored 3″ X 4″

2. Handmade slides: can be made with

a. Acetate sheet

b. Cellophane

c. Etched glass

d. Plain glass

Slides can be made from photographs and pictures by teachers and pupils taking photographs and snapshots when they go on fieldtrips for historical, geographical, literacy or scientific excursions. The arrangement of slides in proper sequence, according to the topic discussed, is an important aspect of teaching with them.

Advantages

1. Requires only filming, processing and mounting by self or laboratory.
2. Results in colorful, realistic, reproduction original subject.
3. Preparation with any 35mm camera for most uses.
4. Easy to revise and up-date.
5. Easily handled, stored and re-arranged for various uses.
6. Can be combined with tape narration or can control time for discussion.
7. May be adapted to group or individual use.

c. Over head projector (OHP)

The over head projector is the most used in all audio visual aids. It projects transparencies with brilliant screen images suitable for use in a lighted room. The teacher can write or draw diagrams on the transparency while he teaches. These are simultaneously projected on the screen by the OHP.

During presentation

a. Keep the screen above the heads of the participants.

b. Keep the screen in full view of participants

c. Make sure you are not blocking any ones view when presenting.

d. Darken the room appropriately by blocking out sunshine and dimming nearby.

e. Turn the screen off between slides if you are going to talk for more than two minutes.

f. Talk to the audience, not to the screen

Purposes

a. To develop concepts and sequences in a subject matter area;

b. To make marginal notes on the transparencies for the use of the teacher that can carry without exposing them to the class;

c. To test students performances, while other classmates observe;

d. To show relationships by means of transparent overlays in contrasting color;

e. To give the illusion of motion in the transparency.

Advantages

a. It allows the teacher to stand in front of the class while using the projector, thus enabling her to point out features appearing on the screen by pointing to the materials at the projector itself and at the same time, to observe the students reactions to her discussion.

b. It gains attention of the student.

d. Motion picture

A film, also called a movie or motion picture, is a series of still images which, when shown on a screen, creates the illusion of moving images due to phi phenomenon. A film is created by photographing actual scenes with a motion picture camera; by photographing drawings or miniature models using traditional animation techniques; by means of CGI and computer animation; or by a combination of some or all of these techniques and other visual effects. Contemporary definition of cinema is the art of simulating experiences, which communicate ideas, stories, perceptions, feelings, beauty or atmosphere by the means of recorded or programmed moving images along with other sensory stimulations.

The process of filmmaking is both an art and an industry. Films were originally recorded onto plastic film which was shown through a movie projector onto a large screen; more modern techniques may use wholly digital filming and storage, such as the Red One camera which records onto hard-disk or flash cards. Films usually include an optical soundtrack, which is a graphic recording of the spoken words, music and other sounds that are to accompany the images. It runs along a portion of the film exclusively reserved for it and is not projected.

Films are cultural artefacts created by specific cultures. They reflect those cultures, and, in turn, affect them. Film is considered to be an important art form, a source of popular entertainment, and a powerful medium for educating-

or indoctrinating-citizens. The visual basis of film gives it a universal power of communication. Some films have become popular worldwide attractions by using dubbing or subtitles to translate the dialog into the language of the viewer. Some have criticized the film industry's glorification of violence.

Advantages

1. Attracts audience and arouses their interests.
2. Presents facts in a realistic and interesting way.
3. Exposes the audience to a new practice in a short time.
4. Provides experiences normally difficult to get.
5. Helps in changing attitudes.
6. Can reach illiterate people also.
7. May be shown according to convenience over and over origin.

e. Over Head Transparencies

Transparencies are popular instructional medium. They are simple and easy to prepare, uncomplicated to operate with the over head projector which is light weight. A 10x10 inches sheet with printed, written or drawn material is placed on the platform of the projector (OHP) and a large image is projected on a screen behind. The projector is used from near to the front of the room with the teacher standing or sitting beside, facing the student.

Guidelines for making effective transparencies

a. Have one main idea an each transparency.
b. Include only related figures and diagrams.
c. Use simple lettering style in writing.
d. Use diagrams in proposition to its lettering.
f. Keep the message clear and simple.
g. Emphasize the key messages.
h. Use color and lettering with discretion.

Advantages

a. Permits face to face interaction with the students.
b. Can be used in daylight conditions.
c. Can present information in systemic developmental sequences.
d. Requires limited planning and can be prepared in variety of inexpensive methods.
f. Easily available.

f. The Opaque Projector

Opaque projector is the only projector on which you can project a variety of materials ex: book pages, objects, coins, postcards, or any other similar flat material that is non-transparent.

The opaque projector will project and simultaneously enlarge, directly from the originals, printed matter, all kinds of written or pictorial matter in any sequence derived by the teacher. It requires a dark room, as the projector is large and not really movable.

Advantages

a. Stimulates attention and arouses interest;
b. Can project a wide range of materials like stamps, coins, specimen, when one copy is available;
c. Can be used for enlarging drawings, pictures and maps;
d. Does not require any written or typed materials, hand-written material can be used;
e. Helps students to retain knowledge for longer period;
f. Review instructional problems;
g. Test knowledge and ability;
h. Simple operation.

Disadvantages

i. Costly equipment;
ii. Needs to use it with care;
iii. Needs a dark room for projection

g. Filmstrip

Film strips are sequence of transparent still pictures with individual frames on 35mm film. A tap recorded narration can be synchronized with film strip. Each strip contains from 12 to 18 or more pictures. It is a fixed sequence of related stills on a roll of 35mm film or 8mm film.

Principles

a. Preview filmstrips before using them and selected carefully to meet the needs of the topic to be taught.
b. Show again any part of the filmstrip needing more specific study.
c. Use filmstrip to stimulate emotions, build attitudes and to point up problems.
d. It should be introduced appropriately and its relationship to the topic of the study brought out.
e. Use a pointer to direct attention, to specific details on the screen.

Types of filmstrip

Discussion filmstrip: it is continuous strip of film consisting of individual frames arranged in sequence usually with explanatory titles.

Sound slide film: it is similar to filmstrip but instead of explanatory titles or spoken discussion recorded explanation is audible, which is synchronized with the pictures.

Advantages

a. Are compact, easily handled and always in proper sequence?

b. Can be supplemented with recordings;
c. Are inexpensive when quantity reproduction is required;
d. Are useful for group or individual study at projection rate are controlled by instructor or user;
e. Are projected with simple light weight equipment.

h. LCD Projector

A liquid-crystal display (LCD) is a flat panel display, electronic visual display, or video display that uses the light modulating properties of liquid crystals. Liquid crystals do not emit light directly. LCDs are available to display arbitrary images (as in a general-purpose computer display) or fixed images which can be displayed or hidden, such as preset words, digits, and 7-segment displays as in a digital clock. They use the same basic technology, except that arbitrary images are made up of a large number of small pixels, while other displays have larger elements.

LCDs are used in a wide range of applications including computer monitors, televisions, instrument panels, aircraft cockpit displays, and signage. They are common in consumer devices such as video players, gaming devices, clocks, watches, calculators, and telephones, and have replaced cathode ray tube (CRT) displays in most applications. They are available in a wider range of screen sizes than CRT and plasma displays, and since they do not use phosphors, they do not suffer image. LCDs are, however, susceptible to image persistence. The LCD screen is more energy efficient and can be disposed of more safely than a CRT. Its low electrical power consumption enables it to be used in battery-powered electronic equipment. It is an electronically modulated optical device made up of any number of segments filled with liquid crystals and arrayed in front of a light source (backlight) or reflector to produce images in color or monochrome. Liquid crystals were first discovered in 1888. By 2008, annual sales of televisions with LCD screens exceeded sales of CRT units worldwide, and the CRT became obsolete for most purposes. Each pixel of an LCD typically consists of a layer of molecules aligned between two transparent electrodes, and two polarizing filters (parallel and perpendicular), the axes of transmission of which are (in most of the cases) perpendicular to each other. Without the liquid crystal between the polarizing filters, light passing through the first filter would be blocked by the second (crossed) polarizer. Before an electric field is applied, the orientation of the liquid-crystal molecules is determined by the alignment at the surfaces of electrodes. In a twisted nematic device (still the most common liquid-crystal device), the surface alignment directions at the two electrodes are perpendicular to each other, and so the molecules arrange themselves in a helical structure, or twist. This induces the rotation of the polarization of the incident light, and the device appears gray. If the applied voltage is large enough, the liquid crystal molecules in the centre of the layer are almost completely untwisted and the polarization of the incident light is not rotated as it passes through the liquid crystal layer. This light will then be mainly polarized perpendicular to the second filter, and thus be blocked and the pixel will appear black. By controlling the voltage applied across the liquid crystal layer in each pixel, light can be allowed to pass through in varying amounts thus constituting different levels of grey. LCD with top polarizer removed from device and placed on top, such that the top and bottom polarizer's are parallel.

The optical effect of a twisted nematic device in the voltage-on state is far less dependent on variations in the device thickness than that in the voltage-off state. Because of this, these devices are usually operated between crossed polarizers such that they appear bright with no voltage (the eye is much more sensitive to variations in the dark state than the bright state). These devices can also be operated between parallel polarizer's, in which case the bright and dark states are reversed. The voltage-off dark state in this configuration appears blotchy, however, because of small variations of thickness across the device. Both the liquid crystal material and the alignment layer material contain ionic compounds. If an electric field of one particular polarity is applied for a long period of time, this ionic material is attracted to the surfaces and degrades the device performance. This is avoided either by applying an alternating current or by reversing the polarity of the electric field as the device is addressed (the response of the liquid crystal layer is identical, regardless of the polarity of the applied field). Displays for a small number of individual digits and/or fixed symbols (as in digital watches and pocket calculators) can be implemented with independent electrodes for each segment. In contrast full alphanumeric and/or variable graphics displays are usually implemented with pixels arranged as a matrix consisting of electrically connected rows on one side of the LC layer and columns on the other side, which makes it possible to address each pixel at the intersections. The general method of matrix addressing consists of sequentially addressing one side of the matrix, for example by selecting the rows one-by-one and applying the picture information on the other side at the columns row-by-row.

LCDs have many characteristics that have caused their demand to increase over the past several years. They are lightweight, aesthetically appealing, energy efficient and long-term cost effective. Some of the disadvantages are that they are not suitable for dimly lit applications, can require frequent adjustments throughout the day, have limited viewing angles, and their purchase cost is relatively expensive.

	LCD Advantages
Brightness	Produces very bright images due to high peak intensity. Very suitable for environments that are brightly lit .
Emissions	Produce considerably lower electric, magnetic and electromagnetic fields than CRTs.
Geometric Distortion	No geometric distortion at the native resolution. Minor distortion can occur for other resolutions.
Power Consumption	Energy efficient. Consume less than 1/3 the power of a comparable CRT. Consume less electricity than a CRT and produce little heat.
Physical Aspects	Take up about 40% less desk space. LCDs are thin and compact.
Screen Shape	Completely flat screen.
Sharpness	At the native resolution, the image is perfectly sharp. Adjustments are required at all other resolutions which can result in measurable degradation to the image.

LCD Disadvantages

Aspect Ratio	The aspect ratio and resolution are fixed.
Black-Level	Not proficient at producing black and very dark grays. In a "standard" configuration, not appropriate for use in dimly lit and dark conditions.
Contrast	Lower contrast than CRTs due to a poor black-level.
Colour and Gray-Scale Accuracy	Colour saturation is reduced at low intensity levels due to a poor black-level. Images are satisfactory, but not accurate due to problems with black-level, gray-scale and Gamma.
Cost	Considerably more expensive purchase price than comparable CRTs . (Cheaper lifetime cost: lasts about 13,000 - 15,000 more hours than a typical CRT.)
Gray-Scale	Have an irregular intensity scale and typically produce fewer than 256 discrete intensity levels. For some LCDs portions of the gray-scale may be dithered.
Motion Artefacts	Slow response times and scan rate conversion result in severe

i. 3D Home theatre Projector

At one time, the slide projector was the presentation method of choice when visual media needed to be presented to a large group. Travellers often took photographs on slide film for the express purpose of holding slide shows of their trips when they returned, and slides could also be prepared for business and educational presentations. In an art history course, for example, the teacher might use slides of famous works of art to display examples for discussion.

Slides are actually a great presentation tool, because of the very high image quality. However, slide projectors can be difficult to work with, especially in the case of carousel projectors, which can become recalcitrant with time. It is also difficult to remember to insert the slides properly so that the images are not reversed or inverted, and to organize the slides in order. As many people who have made presentations with a slide projector know, one of the worst things which can happen is dropping the carousel on the way to the presentation podium, causing all the slides to fall out.

Because slides can be difficult to work with, other projection methods such as projectors designed to link to computers or read presentations from CDs and DVDs have become more commonplace. These projectors may be of lower quality, but they are simple enough for almost anyone to operate.

Non Projected Audio-Visual Aids

Graphic aids

It is a combination of graphic and pictorial material designed for the orderly and logical visualizing of relationships between key facts and ideas example: comparisons, relative amounts developments, processes, classification or organization. It includes the following:

1. Charts

Introduction

These visual symbols used for summarizing, comparing, contrasting or performing other services in explaining subject matter. A chart is a combination of pictorial, graphic, numerical or vertical material, which presents a clear summary.

Definition:

Chart is defined as a visual aid which depicts pictorial and written key information in systematic way to summarize, compare, example: anatomical charts and figure, diagrams etc.

Purposes

a. To visualize an item, it is otherwise difficult to explain only in words.

b. To highlight important points.

c. To provide outline for materials covered in presentation.

d. To show continuity in process.

e. For creating problems and stimulating thinking.

f. For showing development of structure.

Types of charts

a. **Narrative chart**: Arrangement of facts and ideas for expressing the events in the process or development of a significant issue to its point of resolution or we can show an improvement over a period of years;

b. **The cause and effective chart:** Arrangement of facts and ideas for expressing the relationship between rights and responsibilities or between a complex of conditions and change or conflict;

c. **The chain chart**: Arrangement of facts and ideas for expressing transitions or cycles;

d. **The evolution chart**: Facts and ideas for expressing changes in specific items from beginning data and its projections in to future;

e. **Strip tease chart:** It enables speaker to present the information step by step. It increases the interest and imagination of the audience. The information on the chart is covered with thin paper strips to which it has been applied either by wax, tape or sticky substance or pins. As the speaker wishes to visually reinforce a point with words or symbols, he removes the appropriate strip or paper. It produces interest.

f. **Pull chart:** It consists of written messages which are hidden by strips of thick paper. The message can be shown to the viewer, one after another by pulling out the concealing strips;

g. **Flow chart:** Diagrams used to show organizational elements or administrative or functional relationships. In this chart lines, rectangles, circles, are connected by lines showing the directional flow;

h. Tabulation chart: It shows the schedule of an activity or of an individual eg: time-table of a class. These are very valuable aid in the teaching situation where breakdown of a fact or a statement is to be listed. Also it is a useful aid for showing points of comparison, distinction, and contrasts between two or more things. While making the table charts the following points must be kept in mind.

i. The chart should be 50 X 75 cm or more in size.

j. The chart should be captioned in bold letters.

k. The vertical columns should be filled in short phrases rather than complete sentences;

l. Flip chart: A set of charts related to specific topic have been tagged together and hang on a supporting stand. The individual charts will carry a series of related materials or messages in sequence. The silent points of specific topic will be presented;

m. Pie chart: A circle will be drawn and divisions will be made into different sections, each section will be coded differently and code key will be given at right corner of the chart as legend. The circumference is divided into suitable sections. It is relevant for showing the component part.

2. Flash Cards

Definition

"Flash cards are a set of pictured paper cards of varying sizes that are flashed one by one in a logical sequence."

"Flash cards can be self made or commercially prepared and are made up of chart or drawing paper, plain paper using colors or ink on them for drawings."

Purposes

a. To teach the students

b. To give health education

c. Useful for small group

d. Used in group discussions

Principles

a. The messages can be brief, simple line drawing or photographs, cartoons and the content will be written in few lines at the back of the each card.

b. 10" X 12" or 22" X 28" is commonly used size.

c. 10-12 cards for one talk can be used. It should not be less than 3 and more than 20.

d. Prepare a picture for each idea which will give visual impact to the idea.

e. The height of writing on the flash card is to be approximately 5cm for better visualization.

Using the flashcards

For class room instruction, the flash card is to be properly used. The following steps should be taken while displaying flash cards.

a. Give brief introduction about the lesson to students.

b. Give instructions to students about their actions while you flash the cards;

c. Flash the card in front of the class by holding it high with both your hands so that all the students can see it.

d. Let the student respond as per instructions already given.

f. Review the lesson by selectively using flash cards.

Advantages

a. Flash cards can be used to introduce and present topics;

b. It can be used to apply information already gained by students to new situations;

c. It can be used to review a topic;

d. Can be used for drill and practice in elementary classes;

e. To develop the cognitive abilities of recognition and recall of students;

f It can work as a useful supplementary aid and can be effectively used with other material.

Disadvantages

a. Cannot be used for a large group

b. Prone to get spoiled soon

c. Preparation is time consuming.

3. Posters

Definition

"Posters are the graphic aids with short quick and typical messages with attention capturing paintings."

Purposes

a. To provide general motivation;

b. To create an aesthetic or atmospheric effect;

c. To communicate a more general idea. To thrust the message for leading to action;

d. For the class room and community.

Preparation and rules

a. To do a special job.

b. To promote one point.

c. To support local demonstration.
d. Planned for specified people.
e. Tell the message at single glance.
f. Use bold letters.
g. Use pleasing colors.
h. It should place, where people pass or gather.

Features of a good poster

a. Brevity: message should be concise
b. Simplicity: message should be easily understandable
c. Idea: should base on single idea and it should be relevant.
d. Color: suitable color and combination should be used to make the poster attractive and eye catching.
e. Display: while displaying one should be sure to find a place where there is adequate light and where the larger population will see it.

Advantages

a. It attracts attention;
b. It conveys the message very quickly;
c. It does not require a detailed study;
d. Good poster leads to action with good motivation;
f. It can stand alone and is self explanatory.

Disadvantages

a. Poster does not always give enough information;
b. When a poster is seen for longer time it may not attractive;
c. So it should be dynamic.

4. Graphs

Definition

Graphs are the visual teaching aids for presenting statistical data and contrasting the trends or changes of certain attributes.

Method of preparation

a. Before making the bar chart makes a rough sketch of it in a note book;
b. For drawing the bar graph use the chart paper of 50x 72 cm size;
c. Use two different color shades for the two contrasting groups;
d. The bars should be Spaced equally;
e. Write the key to the bar graph in a box on the right corner of the chart paper;

f. Numbers specifying the magnitude of the bars should be on the top on the bars.

Types

a. Pie graph

These are called 'circle diagrams'. The data are presented thorough the sections of portions of a circle.

a. In determining the circumference of a circle we have to take in to consideration a quantity known as pie.
b. The surface area of a circle is to cover 360 degree.
c. The total frequencies or value us equated to 360 degree and then the angles corresponding to component parts are calculated.
d. After determining their angle, the required sectors in the circle are drawn.

b. Bar graph

The graphic presentation extends the scale horizontally along the length of bars. Each bar must be of the same width, height of the bar over a period represents the corresponding time of the variable. Graphs are available in 2 forms that is vertical and horizontal

c. Line graph

To show the trends and relationships ex: single line shows the relation and the variation in the quantity. Quantitative data are plotted or when the data is continuous. The concepts are represented with the help of lines drawn either horizontally or vertically. The plotted points are connected to one another, instead of the base thus producing the curve.

d. Pictorial graph

It is an outstanding method of graphic representation. Pictures are used for the expression of ideal; they are more attractive and easily understood. Vivid pictures will be used to create rapid association with the graphic message; each visual symbol may be used to indicate quantity.

5. Map

Definition

map is a graphic aid representing the proportionately as a diagram, the surface of the earth, world or parts thereof. It conveys the message by lines, symbols, words and colors.

Types of maps

Political maps: These maps show political divisions of the world, a continent, a nation.
Physical maps: Shows the physical contour of a place, area, and region.
Relief maps: It shows the actual elevations and depressions in a place, area, and region.
Weather maps: Shows the amount of rains, temperature extremes, humidity in an area, region country.

Population maps: Shows the distribution of population in various parts of region, country.
Picture or tourist maps: Shows historical spots monumental sites.etc..
Road maps: Shows the roads of a region connecting various parts and points together.
Railway maps: Shows the railway links between various points.
Air maps: Shows the air routes between various points.
Sea root maps: Shows the sea routes between various sea ports

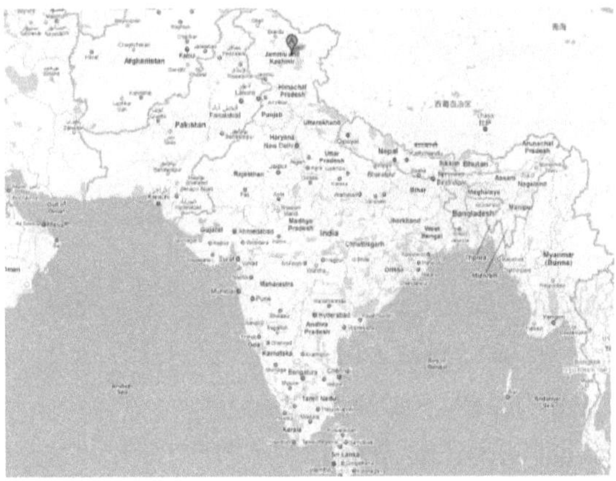

Map of India

6. Cartoons

The word cartoon has various meanings, based on several very different forms of visual art and illustration. The term has evolved over time. The original meaning was in fine art, and there cartoon meant a preparatory drawing for a piece of art such as a painting.

Definition

A cartoon is humorous caricature which gives a subtle message. In a cartoon the features of objects and people are exaggerated along with generally recognized symbols.

Principles

i. The quality of the drawing should be high primarily for visual effectiveness.
ii. The symbols used should be familiar and represent a concept or idea to which students can react intellectually.

Advantages

A cartoon can be effectively used to initiate certain lesson. It can be used for making a lesson lively and interesting.

a. Fantasy
b. Satire
c. Exaggeration

7. News Papers

It can furnish health messages in local languages which can reach to the public easily. The information will be available in low cost, easy to read and understand simple language .the people may learn to read and interpret the contents along with pictures to enhance easy grasping.

Advantages

a. Best method to reach a large group;
b. Pictures will help in easy understanding;
c. Attractive and easy to understand;
d. Lot of information can be obtained in various fields.

Disadvantages

a. Useful for literates only;
b. Detailed information cannot be produces.

8. Comic strips

Definition

A comic strip is the graphic depiction in a series of pictures or sketches of some character and events full of action. This medium of communication is found very interesting and exciting by children.

Uses

a. Comic strips fire the imagination of children.
b. It boosts the courage of children and builds up the spirit of adventure.
c. It makes communication detailed and vivid.
d. It stimulates reality and involvement.

Limitations

a. Comic strips misguide children by depicting characters with supernatural powers divorced from the hard realities of life;
b. Comic strips hamper the development of language of children;

c. Classics brought out in the form of comics develop the tendency in children to ignore or by pass the original work;

d. Comics can soon become an obsession with young children and they tend to avoid serious studies.

9. Display Board

a. Chalk Board

Definition

A **chalkboard** or **blackboard** is a reusable writing surface on which text or drawings are made with chalk or other erasable markers. Blackboards were originally made of smooth, thin sheets of black or dark grey slate stone. Modern versions are often green or brown and are thus sometimes called a green board or brown board instead.

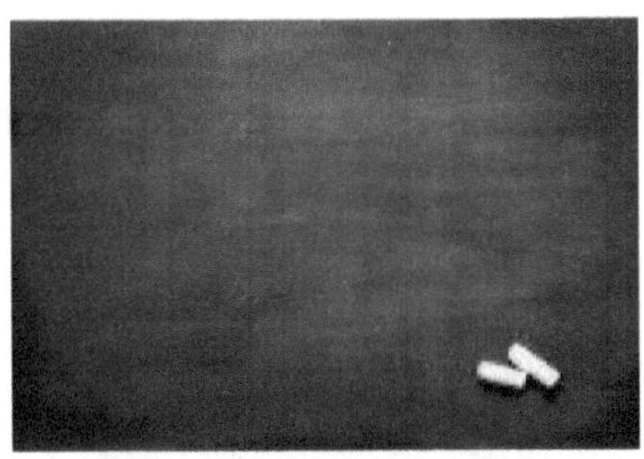

A blackboard can simply be a piece of board painted with matte dark paint (usually black or dark green). A more modern variation consists of a coiled sheet of plastic drawn across two parallel rollers, which can be scrolled to create additional writing space while saving what has been written. The highest grade chalkboards are made of rougher version porcelain enameled steel (black, green, blue or sometimes other colours). Porcelain is very hard and chalkboards made of porcelain usually last 10-20 years in intensive use.

Disadvantages

They produce a fair amount of dust, depending on the quality of chalk used. Some people find this uncomfortable or may be allergic to it, and there has been speculation about links between chalk dust and respiratory problems. The dust also precludes the use of chalk in areas shared with dust-sensitive equipment such as computers. However, these alternative methods of displaying information have drawbacks of their own. The scratching of fingernails on a blackboard is a sound that is well-known for being extremely irritating. Blackboards are also used in many establishments (typically public houses) as a form of advertising often for upcoming events and menus - as well as to keep the score in darts matches.

b. Flannel graph

Flannel graph is a storytelling system that uses a board covered with flannel fabric, usually resting on an easel. It is very similar to Fuzzy felt, although its primary use is as a story telling medium, rather than as a toy. This teaching tool is called by different names as Visual Board, Frick Board, Slap Board, Felt Board, Cohero graph and Video graph.

How to use

The principle involved is the Inter-locking of fibers of two rough or hairy surfaces, so that the pieces pressed on to a background which is hard and vertical will stay. It can be illustrated on a larger scale by pressing two tooth brushes or hair brushes together, so the bristle inter-1ook. In case of flannel graph similar principle of friction helps an object to cling to the surface of the board. The flannel board is usually painted to depict a background scene appropriate to the story being told. Paper cutouts of characters and objects in the story are then place on the board, and moved around, as the story unfolds. These cutouts are backed, either with flannel, or with some other substance that adheres lightly to the flannel background, such as coarse sandpaper.

Advantages

1. Permits numerous and varied arrangements of visua1 materials.
2. Permits the use of either chart or small pieces of material Materials can be packed and transported complete notes.
3. Permits the development of a complete story.
4. Promotes conscientious planning, which must precede the development of the material in the first place.
5. Challenges one to develop symbols to portray such things as abstractions.
6. Easier to construct materials for flannel board than to make slides or movies.

Disadvantages

1. Transportation and storing of boards and materials is a problem. Suitable tables to support boards must be available.
2. Time and cost of making material for presentation present a problem.
3. Cost of boards themselves can't be overlooked.

4. Presentation is limited to a new idea and involves a lapse of time before the new material can be added
5. Might tend to deter one from using other more effective methods and techniques when it is evident that other methods might be more appropriate.
6. To tell a complete story it often takes much board space and for that smaller design and materials cannot be seen well.

c. Bulletin Board

Definition

It is a soft board which will hold pins or tags almost suitable. Simple device placed either indoor or outdoor. Items generally displayed are photographs, publications, posters, news paper cut outs.

Advantages

1. Explains important events
2. Reports special activities

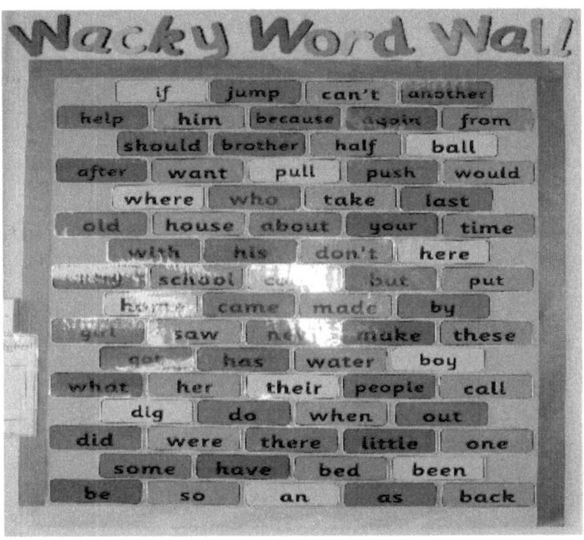

Disadvantages

1. Not effective for illiterate group.
2. Takes a lot of preplanning and preparation

A bulletin board (pin board, pin board or notice board in British English) is a place where people can leave public messages, for example, to advertise things for buying or selling, to announce events or provide information. Dormitory corridors, well-trafficked hallways, lobbies, and freestanding kiosks often have cork boards attached to facilitate the posting of notices. At some universities, lampposts, bollards, trees, and walls often become impromptu sites for putting up posters in areas where official boards are sparse in number.

d. Peg Board

It is a type of board which contains small holes to fix certain letters into the holes which is used especially in the offices to display certain items like name of the personal or faculty member.

e. Magnetic Board

It is a framed iron sheet carrying porcelain coating in some dark color generally black or green. It can be used to display pictures, cutouts and light objects with disc magnets or magnetic holders.

Advantages

Movement of visual material is easy.

10. 3-Dimentional Aids

a. Models

Definition

A model is a recognizable representation of a real thing three dimensionally, that is height, width, and depth is felt as reality.

Types of models

a. **Solid models**: It is the replica of an original thing made with some suitable material like clay, plaster of Paris, wood, iron etc. to show the external parts of the things. Example: globe, clay model of human and animal;

b. **Cutaway and x-ray models**: Are the replicas of the original things to show internal parts of a thing. Cross sectional models are difficult to make in the class room or institutions as they require expertise to construct them. Example: cross sectional model of human body;

c. **Working models**: These models are either actual working things or their miniature replicas. For illustrating an operation. Example: a motor, a generator;

d. **Sand models**: Made by using sand, clay, saw dust, example a tribal village, a forest area.

Advantages

a. Models heighten reality of things and make learning direct and meaningful as they are three dimensional;

b. Models illustrate the application side of certain principles and laws;

c. Models explain the complex and intricate operations in a simplified way and thus make comprehension easier;

d. Models are lasting and ultimately work out to be cheaper teaching aids;

e. Still models are easy to make with the help of discarded materials like empty boxes, pins, clips, nails, and clay;

f. Models are to reasonable size and convenient to handle;

g. Models involve the use of all the five senses and thus make learning effective.

Limitations

a. It requires expertise to make;

b. Time consuming;

c. Some of the models may be very expensive.

b. Objects and Specimen

Definition

A collection of real things for instructional use refers to objects.

A specimen is a sample of the real object or a material.

Using objects and specimens: while using the specimen and objects as teaching aids, a teacher must keep the following points in her mind.

a. Plan your teaching with certain simple and direct observations of the object or specimen being referred to;

b. Ask questions from the students to elicit more details of the features of the object or specimen under observation;

c. Clarify and emphasize important structural details of the object or specimen under observation;

d. Provide review and practice to make learning permanent.

Sources of objects and specimens

a. Local markets;

b. Manufacturers and factories;

c. Discarded material from the houses;

d. Specimen found in the nature can be collected by students from field trips and nature hunt;

e. Plasters casts can be purchased;

f. Wild flowers, leaves shells, stones butterflies moths, insects can also be procured.

Mounting the objects and specimens

Objects and specimens should be mounted in shallow boxes in an artistic way and the boxes should be covered with cellophane paper. Also label each object or specimen using self adhesive paper.

Advantages of objects and specimens

a. Collection of objects and specimens by students requires interaction with others leading to development of social skills and values;

b. Students when collect and display objects and specimens derive satisfaction of contributing to the school and teacher something worthwhile;

c. Student's power of observation and first hand experiences is enhanced by collection of objects and specimens;

d. Student's personal collection of objects and specimens can be good source of doing investigatory projects;

e. Collection of objects and specimens become an interesting educational pursuit of the teacher and students alike;

f. It arouse some interest among students in learning;

g. Objects and specimens involve all the five senses in the process of learning;

h. It heighten the reality in the class room;

i. It makes teaching lively.

c. Exhibition

Many times in the school, a department of the school or a class put up their work for showing it to the people outside the school, and such a show called exhibitions. The pieces of work done by the students for an exhibition are called exhibits.

Requisites for exhibition

a. The exhibition should have a central theme with a few sub themes to focus attention to a particular concept;

b. The exhibits should be clean, labeled properly;

c. The concepts of contrast in color and size should be used for lying out the exhibitions;

d. The exhibits should be so placed so the most visitors , can see them;

e. The place and exhibits should be well lighted;

f. To capture attention and interest of visitors, both motion and sound should be utilized;

g. The exhibition should have some exhibits with operative mechanism such as switches, handles, to be operated by the visitors to observe some happenings;

h. The exhibition should include lot of demonstrations as they involve deeply the students and the visitors;

i. The exhibition should be able to relate various subjects' areas to provide integrated learning.

Advantages

a. Exhibitions inspire the students to learn by doing things themselves and they get a sense of involvement;

b. Exhibitions give students a sense of accomplishment and achievement

c. Exhibitions develop social skills of communication, cooperation, coordination;

d. Exhibitions foster better school community relations and make community members conscious about the school;

f. Exhibitions couple information with pleasure;

g. Exhibitions foster creativity among students.

Disadvantages

a. Requires thorough preparation

b. Time consuming

c. Require funds or budget.

d.Museums

Definition

A museum is a building displaying a collection of historical relics, antiques, curiosities, works of arts, works of science, literature and other artifacts of general interest. Museums can be useful both for public education and specific class room instructions.

Setting up school museum

a. School should have enough space
b. Take the help of students, collect old and new objects and articles
c. Accept donations from various organizations who donates the articles
d. Students can be guided to prepare the exhibits
e. All the collected and prepared articles should be displayed and labeled
f. A detailed report book should be maintained giving a brief description of each museum pieces
g. The museum rooms should be well lighted
h. It should be cleaned and maintained timely.

e. Dioramas

Definitions

A diorama is a three dimensional arrangement of related objects, models, and cut outs to illustrate a central theme or concept. The objects and models are generally placed in a big box or show case with a glass covering and background printed with a shade or a scene. Ex: a harvest scene, a planting scene etc.

Advantages

a. Provide a good opportunity to learn
b. It gives the appearance of actual things which can not be brought to the class room
c. Interesting and enhance creativity
d. Live things also can shown in diorama ex: aquarium
f. Provides students to do project works

Disadvantages:

a. Needs expatriation for the preparation;
b. Sometimes cost effective
c. Require budget;
d. Sometimes it may misguide the student if is not the replica of actual thing.

f. Mock ups

It emphasizes the functional relationship between the device reality and its workability. Certain element of the original reality is emphasized to make it more meaningful for the purpose of instruction. In common usage, a mockup is a scale model of a structure or device, usually used for teaching, demonstration, testing a design, etc. Mockups are also used in the Consumer goods industry, as part of the product development process, when the size, impression and/or artworks have to be tested and approved. Mockup is also a frequently used term when talking about an early layout or sketch of a Web site or GUI program.

Examples: An artificial kidney to demonstrate dialysis.

e. Fiups Moulage

Moulage (French: casting/ moulding) is the art of applying mock injuries for the purpose of training Emergency Response Teams and other medical and military personnel. Moulage may be as simple as applying pre-made rubber or latex "wounds" to a healthy "patient's" limbs, chest, head, etc., or as complex as using makeup and theatre techniques to provide elements of realism (such as blood, vomits, open fractures, etc.) to the training simulation. The practice dates to at least the Renaissance, when wax figures were used for this purpose. Mould can be made up of plastic material to stimulate some life in objects. Example: body which shows evidence of trauma, infection, disease, surgical intervention.

f. Image perspectives moulage process

The basic material we use to create our soft tissue injuries is very inexpensive; it costs only pennies per simulation. If someone walks off with a simulation it can be recreated in a matter of minutes. You are not limited to mass-produced latex or plastic "one-size-fits-all injuries". You determine the type, scope and size of the injuries and create as many as you wish, in whatever size you need. Our simulations can be handled and bandaged, with care, as you would a real injury. The simulations will not shift and/or be damaged. When bandages are removed, the injury simulation will remain intact.

The simulations can be made liquid-proof, and with care - reusable!

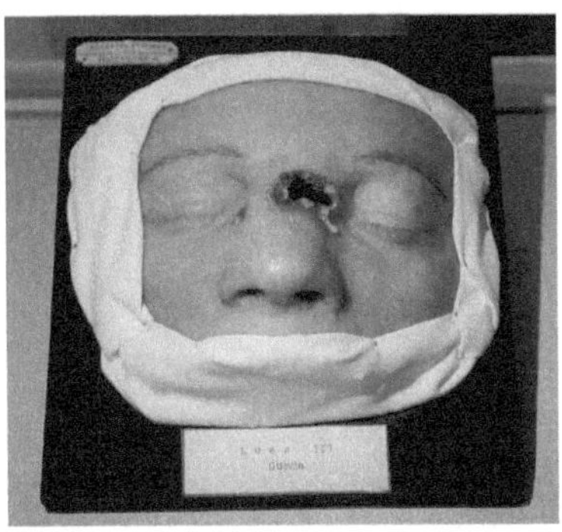

11. Folk Media

Reasons for using folk media

No communication strategy would be complete unless it included the traditional media, which must rank alongside the more commonly accepted channels of communication.

Status and function of Folk Media

a. As an intrinsically valuable form of popular entertainment or artistic expression it is worth preserving and developing for its own sake;

b. It is a means of changing values, attitudes and norms in order to provide a proper climate for social and economic progress;
c. It is a method of promoting certain behaviour acts or patterns. The aim is to get people to perform specific acts to achieve objectives of national policy (eg. visiting clinics, investing in bonds, using fertilizer);
d. It is a channel for conveying information about available techniques and facilities which people may use to solve problems.

The above purposes, it was noted, are not necessarily in compatible with each other. However, problems may arise in specific situations. Certain media may not be appropriate for certain purposes and certain purposes may not be achieved by employing particular folk media, or any folk medium. It is therefore important to evolve criteria both for selecting media and for selecting messages that pass through them.

Classification of traditional forms

Traditional forms of expression were identified. The classification covering the performing arts and the visual arts included music, song, drama, skits, puppet shows, poetry, speech, sounds, gesture, gossip, jokes, proverbs, painting, certain printed literature, sculpture, handicrafts, costuming, patterning and colouring of material and use of head-gear. Symbolic meanings may often be an important consideration in use of any one or combination of these forms of expression and therefore the thought-processes behind selection of forms assumed significance.

Advantages

Traditional media comes from the people and antecede the mass media; their appeal has historically been both functional and aesthetic. Whether of roughly hewn folk origins or of classifically rigid moulds, they have always served to entertain, educate to reinforce existing ideas or ideologies or to change existing values and attitudes.

Being close to people at the local level, these channels are potentially useful in the service of social concerns, as determined by local, provincial or national authorities themselves. They are, moreover, abundantly present in areas where mass media technology has not been fully or effectively developed to capture sustained interest at local, provincial or national levels. Folk media are personal forms of entertainment and communication. This is important because behavioral changes are most easily brought about by personal interaction. These forms of art are a part of the way of life of a community and provide acceptable means of bringing development issues into the community in its own-terms. They are capable of reaching intimate social groups, thus making use of already established communication networks in the audience.

These forms of entertainment evolved as grassroots expressions of the values and life styles of the people, dealing with values as well as information. These features are significant for development, planners in the field of population, since it is rapidly becoming apparent that something more than information is necessary to bring about the behavioral changes involved in the adoption and continued practice of health and Family Planning.

Another advantage of the traditional media is that they attract people who might not attend an educational meeting. With skill, new content might be added to the old forms which are already familiar and dear to the people. Finally, unlike mass media programmes, produced for large and often diverse audience, the folk forms can use familiar dialects for the most intimate and local communication at the village level.

Folk media rather deals with values that relate to family life, such as the status of women, conjugal relationships, education and standard of living. Folk media can be seen as part of the social fabric of the community. While they could reinforce relevant social changes, that are already occurring (or be used directly to introduce new ideas), folk forms should not be used for propaganda.

Limitations, on the other hand, included the dangers in "overloading" certain traditional channels with too many "messages" of an instrumental nature. Channels could only be loaded according, to their capacity and with respect, in particular to cultural, religious, or other sensitively integral elements. Sociologists and anthropologists could help planners define these sensitive factors affecting such limitations.

12. Puppets

One of the old and popular arts in Indian villages is puppetry. Puppetry is an education cum entertaining aid in which puppets manipulated by the performer is a person termed as a characters in a story to be depicted.

Definition

A puppet is a manipulative doll dressed as a character and the performer is a person termed as a puppeteer. A good puppeteer has to blend his art with dramatization to produce the desired effect. It is used as an effective teaching aid for languages and social sciences.

Types of puppets

1. **String or marionettes puppets:-**Marionettes consist of puppets with hinged body parts which are controlled by nine strings produces required movements in the puppet. These puppets are mainly manipulated by professional puppeteers.

2. **Stick puppets: -** stick puppet are the painted cutouts attached by sticks. The actions of these puppets are manipulated by the teacher and students by hiding behind a screen so that only puppets are visible to the audience or the class.
3. **Shadow puppets**: shadow puppets are silhouettes of cardboard which produce shadows on white screen. The motion of these silhouettes is manipulated by the teacher and students.
4. **Finger of hand puppet:** Hand puppets are round balls painted as heads with overflowing colorful costumes. These are worn on fingers which operate their movements. These are operated from below the stage.

Selection

In writing or selecting a puppet play, the age, background and tastes of the students should be taken in to consideration. A short puppet play is always preferable.

Advantages

1. Creates interest.
2. Gives the knowledge in a brief period.
3. Puppet is an effective method in teaching.
4. Motivate students.
5. Easy to carry and operate.

Disadvantages

1. Needs group cooperation and coordination.
2. Requires skills in preparation and supply.
3. Skills needed in presentation.

Activity aids

There are certain learning situations in which student participation through direct experiences can be easily incorporated, these are called activity aids. The activity teaching aids are really of great value as they put students in a role of active seekers of knowledge. There are five important activity teaching aids, which are listed below:-

1. Field trips
2. Demonstrations
3. Experiments
4. Dramatizations

a. Field Trips

Definition

According to Hedger ken Field trip may be defined as "an educational procedure by which the student studies firsthand objects and materials in their natural environment."

Types of field trip

Depending on the place of visit and its duration, field trips are mainly of the following four types, namely:-

1. Local school trips
2. Community trip
3. Educational trips
4. The natural hunt

Advantages of field trip

Field trip provides learning experience in the real life situation by direct contact with objects, process, and systems and thus has many advantages which are enumerated as follows:

1. It provides accurate information about objects, process, and systems in their real life setting.
2. It provides meaningful first-hand experience and hence results in lasting impression on the minds.
3. The students learning can be easily diverted towards effective learning.
4. Field trips are valuable aids for students who are curious about the natural and man-made process and objects.
5. Field trips can effectively supplement the classroom learning through application and reviewing the experiences of student.

Limitations of field trip

1. A field trip may be occasional activity which at best supplement some learning segments of the syllabus.
2. They can be expensive and out of reach for many disadvantaged and poor students.

3. Field trips require proper and detailed planning to make them meaningful otherwise the trip leads to confusion, and fails to fulfill the requirement

b. Demonstrations

Demonstration method is a concrete visual aid, because of its wide use in the teaching of nurses. In nursing education, it is used for this purpose and also for clinics, conferences, laboratory classes, symposia, autopsies, and teaching of health to patients. The demonstration method teaches by explanation and exhibition. In short, it is a performance to show a process or activity to others. When a teacher demonstrates, students observe and imitate and thereby learn.

Farmers are keen on seeing how a new idea works and how it might affect their crop production. Both of these can be done by a demonstration. An appropriate and practical demonstration is an important method in extension, especially among farmers who cannot read. This is because they have an opportunity to watch the differences between the new idea and the old one. The demonstration should be simple and it should illustrate concrete results to the farmers. In extension work, the agents use two different kinds of demonstrations. These are called method demonstration and result demonstration. Both of these require a great deal of thought, planning and competent implementation.

Advantages of demonstration

The following are the advantages of demonstration method.

a. It activates several senses. This increases learning, because it gives a better opportunity for observational learning.
b. It clarifies the underlying principles by demonstrating the 'why' or 'how' of the procedure.
c. It provokes interest by use of concrete illustrations.
d. It correlates theory with practice engages student's attention and concentration.
f. It encourages student's participation in learning through questions and answers as the teacher performs.

c. Experiment

An experiment is a learning activity in which students collect and interpret observations using measuring instruments to reach some conclusions. In science subjects experiments are used invariably used as instructional aid as they encourage learning by doing. While giving a lesson on an experiment, the teacher should organize the instruction so as to make the students aware of the following steps of the experiments:

a. Objectives of the experiments
b. Apparatus required
c. Procedure or methodology
d. Observations of data

e. Computation (totaling) of the observations made.

f. Results or conclusion

g. Precautions

h. Ideas for future work

The student performs the experiment and writes a report on it. Showing the cause and effect relationship.

National Demonstration

Anationwideprogramme of demonstrations known as National Demonstration on major food crops was launched in 1964-1965 with a view to obtain maximum return per unit area per unit time on the cultivators fields. It was to provide opportunity to scientists to demonstrate validity of their findings on farmer's field. NDP was implemented in selected district through State Agricultural Universities with the help of a team of scientists, training as well as demonstration and other extension activities. The rationale behind the scheme was that unless the scientists could demonstrate what they advocated, their advice might not be heeded by the farmers .It was a nationwide project with a uniform design and pattern. The scientists were expected to conduct different types of demonstrations as given below-

- Multiple cropping demonstrations
- Rain fed cropping demonstration
- Special problem soil demonstration
- Entire farming system demonstration

It differed from other demonstration programme in the following four major aspects –

- There was a specific yield target and there was no separate control plot near the demonstration. The idea behind this principle was that the entire living memory of the farmers about the yield potential of the crop, as well as the entire block in which the demonstration had been laid out, would serve as the control.
- The area of the demonstration plot was about one hectare it could be one acre if bigger plots were not available. So that the feasibility of raising a good crop could be strikingly and unquestionably demonstrated.
- The farmers in whose plots the demonstrations were laid out were the actual cultivators with small holdings, so that the high yields obtaine4d were not attributed to the effects of affluence.
- The agricultural scientists conducted these demonstrations in association with local extension agencies/workers.

Method Demonstration

Method demonstration, basically, show how to do something to the farmers. The farmers are shown step by step how to do something, for example how to sow fodder seeds or operate a specific machine. The extension agent is most likely dealing with farmers, who are already familiar with the practice being illustrated. However, now they want to learn how to do it by themselves. Major advantage of this method is that it allows the agent to communicate with the farmers and give details about simple farming methods to numerous people. In other words, the agent is growing the impact of his extension work. In addition, when the farmers have a chance to try it by themselves, the possibility of the demonstration helping them increases rather than in a situation where the farmers are given a lecture about the same subject. The key disadvantage of the method is if the demonstration is being observed by a large number of farmers, only some of the farmers get to see, hear and do.

Result Demonstration

In result demonstration, the agent shows the farmers that the new idea that is being demonstrated can work in local conditions. The difference is important in result demonstration, whether it is the difference between poor seed and selected seed, or using a fertilizer or not using it. The farmers tend not to believe the agent's words if he only says it to the farmers, that is why the old saying "seeing is believing" applies here. After the farmer sees the results, the agent builds confidence among the farmers and can encourage the farmers to try it by themselves. The major disadvantage of the method is that it takes a long time to mature; hence it is a costly method. If a demonstration fails, due to for example lack of rain, it can have devastating consequences.

Front-Line Demonstration

It is the new concept of field demonstration evolved by the Indian Council of Agricultural Research with the inception of the Technology Mission on Oilseed Crops during mid-eighties. The field demonstrations conducted under the close supervision of scientists of the National Agriculture Research System are called front-line demonstrations because the technologies are demonstrated for the first time by the scientists themselves before being fed into the main extension system of the State Department of Agriculture. The main objective of Front-Line Demonstrations is to demonstrate newly released crop production and protection technologies and its management practices in the farmers' field under different agro-climatic regions and farming situations. While demonstrating the technologies in the farmers' field, the scientist are required to study the factors contributing higher crop production, field constrains of production and thereby generate production data and feedback information. Front-Line Demonstrations are conducted in a block of two or four hectares land in order to have better impact of the demonstrated technologies on the farmers and field level extension functionaries.

Special Features

The Front Line Demonstrations are different than the normal demonstrations conducted by the extension functionaries. The special features of Front Line Demonstrations are:

- ❖ Front Line Demonstrations are conducted under the close supervision of the scientists of the National Agriculture Research System comprising of ICAR Institute, National Research Centers, Project Directorates, Krishi Vigyan Kendras, and the State Agricultural Universities and its regional Research Stations.
- ❖ Only newly released technologies or those likely to be released in near future are selected for the Front Line Demonstrations.
- ❖ Front Line Demonstrations are organized in a block of two to four hectares involving all those farmers whose plots fall in the identified demonstration block.
- ❖ Only critical inputs and training are provided from the scheme budget, remaining inputs are supplied by the farmers themselves.
- ❖ Training of the farmers associated with the Front Line Demonstrations is a prerequisite for conducting such demonstrations.
- ❖ The target audience of the Front Line Demonstration is both farmers and the extension officers. The purpose is to be convince extension functionaries and farmers together about the potentialities of the technologies for further wide scale diffusion; and
- ❖ Front Line Demonstration is used as a source of generating data on factors contributing higher crop yields and constraints of production under various farming situations.

Field Demonstration

Field demonstration is a long term educational activity conducted in a systematic manner in farmers' fields to show worth of a new practice/ technology. "Seeing is believing" is the basic philosophy of field demonstrations. Only proven technologies are therefore selected for field demonstrations. Field demonstrations educate farmers through results obtained in terms of varieties resistant to disease and pest, quality of the grains and overall higher yields. In addition, it also educates the farmers in term of input-output ratio and economic gains in terms of net returns. Basically, there are two types of field demonstrations:

i. Single practice demonstration and

ii. Composite demonstration.

Single Practice Demonstration

Single Practice Demonstration aims at proving the worth of a single practice such as effect of balance fertilizers in rice crops, higher yields from newly released varieties of hybrid maize, effect of irrigation at crown root initiation stage of wheat, effect of new pesticide on fruit borer in gram etc.

Composite Demonstration

Composite Demonstration is a combination of field based result demonstrations and a chain of skill oriented method demonstrations. A long term sequential method demonstrations (Composite Demonstration), aim at demonstrating the superiority of a package of practices in growing a field crops. Here the effect of one practice in harnessing the effect of other practices is also demonstrated and studied. For example, combined effect of irrigation and fertilizer application on grain yield and quality of a newly released mustard variety may be demonstrated as a composite demonstration. Many times, there is an over emphasis on field demonstrations in extension work. Good field demonstrations lead to higher adoption of demonstrated practices by the farmers as they developed the confidence amongst them in the practices demonstrated. Field demonstrations provide an effective learning situation as farmers "See the crops themselves", "interact with the scientists and extension workers on the fields", and "get doubts clarified then and there itself. It is, therefore, essential that whatever demonstrations are conducted, they should be well planned and execute giving no chance to fail. One bad demonstration can vanish the impact of many good demonstrations. Scientists, therefore, have to be very careful in planning and conducting the field demonstrations. Without specific purpose, demonstrations should not be organized. Demonstration should be conducted only when situation demands.

Steps in Conducting Field Demonstrations

Since field demonstration is often used as an extension method, it is sometimes laid out in a routine manner. A well conducted demonstration should help the scientists to give finishing touch to changing attitude of farmers and extension workers and improve their knowledge, understanding and skills. The following steps need to be followed in conducting field demonstration.

A. Planning Phase

1. Know the Vicinity

The scientists need to develop an understanding of the farmers, their farming systems, resources and establish rapport with them. It is essential to gather information on cropping system, present level of use of inputs and productivity of major crops of the area. There are different ways of knowing vicinity. Some are formal and some are informal. A few areas under

a. Visiting villages and farmers.

b. Collection of information using PRA tools.

c. Meeting people individually and in groups.

d. Meeting opinion leaders.

e. Exchanging information with local extension workers; and

f. Consulting office records of population and basic agriculture.

2. Select Technologies

Select only proven technologies which have higher potentialities interms of yield, disease resistance, quality, and can fit in the existing farming systems and situations of the area/farmers. Technology should be frontier ones i.e. recently released technologies or such which are at advance stage of release. Be sure that the technology selected for demonstration is much superior to the technology being already in use. At this stage scientists responsible for demonstration must have a lot of consultations with the research scientists who are responsible for released of the technology. They should ask a number of questions from the research scientists and be satisfied about the superiority of the technology.

3. Select Demonstration Site

Avoid isolated farm. Demonstration site should be easily accessible for the farmers and extension workers. As far as possible, block of demonstration site should have a good number of farmers of all categories of land holding and status. Never conduct block demonstration in a single farmer's plot. Pay also attention to farm size, layout of the field, soil type, fertility status, irrigation facilities and drainage system.

4. Select Demonstration Farmers

A group of farmers land holdings in the selected demonstration block and who are willing to cooperate in the conduct of demonstration should be selected. Demonstration farmers should be selected finally by holding a meeting in the village where the purpose of demonstration should be clearly stated and suggestion sought from the farmers. Any difference in opinion may be sorted out tactfully otherwise there is a chance of having non cooperation from those who are not selected as demonstrating farmers which may ultimately jeopardize the very purpose of demonstration.

5. Finalize Package of Practices

This is an important step in planning the field demonstrations. Collect the new technologies from the ICAR Institute/SAU's and ensure these technologies are frontier ones showing substantial increase in yields. Involve as many scientists of the research station in the discussion as possible. This will help in working out minute details of sequences of method demonstration required, identification of important tasks/practices in which presence of scientists should be necessary and critical input for demonstration. Knowledge about farming conditions will be useful at this stage. Involve also demonstrating farmers in finalizing package of practices. This will help in understanding the level of farmer's practices, resource base to sustain the technologies and their perspectives.

6. Prepare for demonstration

Arrange critical inputs for the demonstration. Critical inputs are those agricultural inputs which are vital to help the selected technologies to exhibit its production potentialities on farmer's field and not earlier being used by the farmers. Arrange such inputs viz. seeds, fertilizers, farm equipments and other inputs in time. Only critical inputs need to be supplied by the scientists. Other inputs should be arranged by the farmers themselves. Ensure that the inputs which are to be given by the farmers are available with them. The farmers should never be given an impression that the demonstration is a means of receiving free inputs. Rather they should be educated to understand the educational value of such demonstration.

B. Conducting Phase

1. Layout of Demonstration

Guide and assist the farmers in laying out the field. Special training programme may be arranged for all farmers in whose plots demonstrations are to be laid. Keep the control plot if needed; otherwise treat all other neighboring/surrounding plots as control plots. In case of block demonstration, one acre plot as a control is adequate. Sometimes a control plot is not necessary as the "entire memory of the farmer" is taken as a "control" or the neighboring plots would serve the purpose of control plots. Put a publicity board on the fringe of the demonstration plot. Mention the details of the demonstration on the board.

2. Crucial Farm Operations

Ensure your presence at the time of important operations like, seeding, fertilizer application, weeding, irrigation, plant protection measures, harvesting, threshing and weighing of produce. Two things are important at this stage: (i) using demonstration for farmers' training, and (ii) record keeping in demonstration. Each operation should be used as input of training of farmers. Encourage questions from the farmers at each of these operations. This will help in better understanding of the task/operations.

3. Field Day

Arrange a field day to project the new technologies demonstrated in front of a large manageable group of interested farmers. It is an intensive educational activity in which farm experts, extension workers and farmers are involved and learn from each other. Plan the field day when the crop is fully matured yet green. Ask the demonstration farmers to explain the story of demonstration one by one to the assembled group of farmers and extension workers. Arrange a few method demonstrations on operation of farm machines and equipments, operation of seed drill, seed treatment, fertilizer application, plant protection, etc. Farmer-scientist-extension workers discussion should be an important feature of the Field Day.

4. Harvesting

Make an eye estimate of the field. Arrange harvesting in the presence of identified groups of farmers. Ask the farmers to estimate the yield and to say in what way the demonstrated technologies are superior to the earlier ones. Are they satisfied with the performance of the technologies? What lessons they have learned from the demonstration? Will they advice other fellow farmers to adopt this practice? Will they exchange the seed materials of new variety with other fellow farmers? What are the expected profits? Will it be more than what they used to get from their own practices? What are the difficulties in following the demonstrated practices? Idea is to ascertain as to what extent farmers are satisfied with the demonstrated technology and what is the possibility of their continued adoption.

C. Follow up Phase

Some farmers may revert to old practices in the absence of follow-up. They need information reinforcement, timely supply of inputs or on the spot guidance.

Group approach in follow-up will give better results. It is better to link your follow-up programmes with the local institutions like Farmers Club, Farmers Cooperative Society, Village Panchayat etc.

D. Record Keeping

There are two types of records which one should maintain for each block demonstration.

1. Information Card

This card contains basic information about the demonstration site viz. previous crops and varieties grown, fertility status of the plots, present productivity of crops, size of holdings of each farmers in the demonstration block, extent of use of inputs etc. It should also contain detail information of demonstration like size of block, variety of crop, seed rate, sowing date, inputs applied, irrigation schedule followed, intercultural operation performed, plant protection measures taken, date of maturity, date of harvesting, incidence of disease and pests, average numbers of tillers, yields of crop etc. The card should remain with the demonstration farmers and is filled up by them or by an educated person in their family or by the KVK staff who visit the demonstration site from time to time.

2. Technical Report

The technical report should contain information on soil analysis, variety of crop, germination, plant population, pest and diseases, irrigation, fertilizer application, harvesting, final yield, extension activities undertaken etc. It should also contain information on cost-benefit ratio of the demonstration. This will help to work out the economic returns. A favorable cost-benefit ratio will fully convince the extension officers and the farmers about the profitability of the technologies demonstrated. Submission of technical reports to Zonal Coordinator/Council in time is as important as concluding good field demonstrations.

Late submission of reports may constrain the Zonal coordinator/Council not to include them in the Zonal report/national report. Remember, demonstration report helps in planning agricultural development and farm research programme of your area, state and country. Therefore send technical report of block demonstration in time. Each block demonstration should be concluded with a feature story written in simple and easily understandable language. The story should have supporting data and photographs of demonstration to convince the readers. Role of the scientists in writing of feature story is very important. Such story becomes the important learning materials in the Krishi Vigyan Kendras.

d. Dramatization

Dramatization is a very potent method of keeping the class room instruction lively and interesting. When a teacher dramatizes a lesson, the students become both the spectators and participants. This makes learning easy and permanents.

Types of dramatizations suitable for class room instruction:-

a. Role-play

b. Lets Play

c. Pageant

d. Pantomime

Role playing

Role-playing refers to the changing of one's behaviour to assume a role, either unconsciously to fill a social role, or consciously to act out an adopted role. While the Oxford English Dictionary offers a definition of role-playing as "the changing of one's behaviour to fulfil a social role", in the field of psychology, the term is used more loosely in four senses.

Lets Play

Play Let is a series of screenshots or a recorded video documenting a play through of a video game, usually including commentary by the gamer. A Lets Play differs from a walkthrough or strategy guide by focusing on an individual's subjective experience with the game, often with humorous, irreverent, or even critical commentary from the gamer, rather than being an objective source of information on how to progress through the game.

Pantomime

Pantomime (informally Panto), not to be confused with the theatrical medium of mime, is a type of musical comedy stage production, designed for family entertainment. It was developed in the United Kingdom and is generally performed during the Christmas and New Year season. Modern pantomime includes songs, slapstick comedy and dancing, employs gender-crossing actors, and combines topical humour with a story loosely based on a well-known fairy tale. It is a participatory form of theatre, in which the audience is expected to sing along with certain parts of the music and shout out phrases to the performers.

Pantomime has a long theatrical history in Western culture dating back to classical theatre, and it developed partly from the 16th-century commedia dell'arte tradition of Italy, as well as other European and British stage traditions, such as 17th-century masques. An important part of the pantomime, until the late 19th century, was the harlequinade. The pantomime is performed today throughout Britain and, to a lesser extent, in other English-speaking countries.

- To refer to the playing of roles generally such as in a theatre, or educational setting;
- To refer to taking a role of an existing character or person and acting it out with a partner taking someone else's role, often involving different genres of practice;
- To refer to a wide range of games including role-playing video game, play-by-mail games and more;
- To refer specifically to role-playing games.

Advantages of dramatization

a. Dramatization gives an added advantage of students working as both observers (spectators) and doers (participants) unlike in experiment where there are just doers and in demonstration where there are just observers;

b. Dramatization makes learning a pleasure children love to act and show off;

c. Dramatization involves students totally and they appreciate the lessons remember it better;

d. Dramatization develops the social skills required for them such as cooperation, co-ordination, punctuality, and human relations etc;

e. Dramatization makes students creative, sensitive, and alert.

The disadvantages of Audio Visual Aids in Teaching

Audiovisual aids reinforce the information being given in a presentation.

Audio-visual aids can be a great tool to use in classrooms. Often they can spice up otherwise dull topics, keep students more engaged in their lesson and enliven their imaginations. Video presentations, slide shows, PowerPoint and other media have revolutionized the way teachers reach their students. But these aids are not without their disadvantages. It is important to weigh both the pros and the cons when considering how you will use audiovisual aids in your classroom.

Technical Problems

Regardless of your equipment, there are many things that could go wrong that may derail your presentation. A bulb might burn out on your projector or it might be blurry or hard to focus. When using a PowerPoint presentation, font and colors may show up differently on screen or the music and sound might not play. CDs can skip. DVDs may be scratched or not compatible with your player.

Student Distractions

Visual aids are more of a distraction if used throughout the entire presentation versus during key points. Special effects and light from the projector can draw attention away from the speaker. Too much information can also be distracting, because the students will be trying to read and/or take notes while the speaker has moved on to the next subject. There also may be too many slides for the students to interpret or the audio and visual may not match up. An entertaining video can take the focus off the speaker and the presentation, while a dry video and low lighting can put students to sleep.

Expenses

Professionally put together audiovisual aids can be expensive. Slides, videos, and films can cost money unless borrowed from a library, and the necessary equipment can be cost-prohibitive. Due to budget constraints, some schools may not have enough audiovisual equipment for every classroom, and availability could be limited.

Time

Audiovisual presentations can take a considerable amount of time to prepare. It takes time to produce your own video or slides. Films may be difficult to obtain, or it can take time to get permission to use them. Posters and transparencies may require extensive preparation. Creating a PowerPoint presentation can also be time consuming. Time spent in preparing these materials could take away from the time it takes to prepare a clear, well-organized lesson plan.

Space

If you choose to use an audiovisual aid, the size of the room should be taken into consideration. It is critical that all students are able to see or hear your presentation. If the room is too large for everyone to see the visual aid, or if part of your audience is forced to view the presentation at odd angles, some students will struggle to keep up with your lesson.

Convenience

Certain audiovisual aids such as VCRs for older video recordings, can be bulky and difficult to transport.

Part: II
Agricultural Journalism & Electronic Media

Chapter 10
Cyber Journalism

Internet and 'world wide web' new horizon of communication

The Internet is a global system of interconnected computer networks that use the standard Internet Protocol Suite (TCP/IP) to serve billions of users worldwide. It is a network of networks that consists of millions of private, public, academic, business, and government networks, of local to global scope, that are linked by a broad array of electronic, wireless and optical networking technologies. The Internet carries a vast range of information resources and services, such as the inter-linked hypertext documents of the World Wide Web (www) and the infrastructure to support electronic mail. Most traditional communications media including telephone, music, film, and television are reshaped or redefined by the Internet, giving birth to new services such as Voice over Internet Protocol (VoIP) and IPTV. Newspaper, book and other print publishing are adapting to Web site technology, or are reshaped into blogging and web feeds. The Internet has enabled or accelerated new forms of human interactions through instant messaging, Internet forums, and social networking. Online shopping has boomed both for major retail outlets and small artisans and traders. Business-to-business and financial services on the Internet affect supply chains across entire industries.

The origins of the Internet reach back to research of the 1960s, commissioned by the United States government in collaboration with private commercial interests to build robust, fault-tolerant, and distributed computer networks. The funding of a new U.S. backbone by the National Science Foundation in the 1980s, as well as private funding for other commercial backbones, led to worldwide participation in the development of new networking technologies, and the merger of many networks. The commercialization of what was by the 1990s an international network resulted in its popularization and incorporation into virtually every aspect of modern human life. As of 2009, an estimated quarter of Earth's population used the services of the Internet. The Internet has no centralized governance in either technological implementation or policies for access and usage; each constituent network sets its own standards. Only the overreaching definitions of the two principal name spaces in the Internet, the Internet Protocol address space and the Domain Name System, are directed by a maintainer organization, the Internet Corporation for Assigned Names and Numbers (ICANN). The technical underpinning and standardization of the core protocols (IPv4 and IPv6) is an activity of the Internet Engineering Task Force (IETF), a non-profit organization of loosely affiliated international participants that anyone may associate with by contributing technical expertise.

Terminology

Internet is a short form of the technical term internet work, the result of interconnecting computer networks with special gateways or routers. The Internet is also often referred to as the 'Net.' The term the 'Internet', when referring to the entire global system of IP networks has been treated as a proper noun and written with an initial capital letter. In the media and popular culture a trend has also developed to regard it as a generic term or common noun and thus write it as "the internet", without capitalization. Some guides specify that the word should be capitalized as a noun but not capitalized as an adjective.

The terms Internet and 'World Wide Web' are often used in everyday speech without much distinction. However, the Internet and the World Wide Web are not one and the same. The Internet is a global data communications system. It is a hardware and software infrastructure that provides connectivity between computers. In contrast, the Web is one of the services communicated via the Internet. It is a collection of interconnected documents and other resources, linked by hyperlinks and URLs. In many technical illustrations when the precise location or interrelation of Internet resources is not important, extended networks such as the Internet are often depicted as a cloud. The verbal image has been formalized in the newer concept of cloud computing.

History

The USSR's launch of Sputnik spurred the United States to create the Advanced Research Projects Agency (ARPA or DARPA) in February 1958 to regain a technological lead. ARPA created the Information Processing Technology Office (IPTO) to further the research of the Semi Automatic Ground Environment (SAGE) program, which had networked country-wide radar systems together for the first time. The IPTO's purpose was to find ways to address the US military's concern about survivability of their communications networks, and as a first step interconnect their computers at the Pentagon, Cheyenne Mountain, and Strategic Air Command headquarters (SAC). J. C. R. Licklider, a promoter of universal networking, was selected to head the IPTO. Licklider moved from the Psycho-Acoustic Laboratory at Harvard University to MIT in 1950, after becoming interested in information technology. At MIT, he served on a committee that established Lincoln Laboratory and worked on the SAGE project. In 1957 he became a Vice President at BBN, where he bought the first production PDP-1 computer and conducted the first public demonstration of time-sharing.

At the IPTO, Licklider's successor Ivan Sutherland in 1965 got Lawrence Roberts to start a project to make a network, and Roberts based the technology on the work of Paul Baran, who had written an exhaustive study for the United States Air Force that recommended packet switching (opposed to circuit switching) to achieve better network robustness and disaster survivability. Roberts had worked at the MIT Lincoln Laboratory originally established to work on the design of the SAGE system. UCLA professor Leonard Kleinrock had provided the theoretical foundations for packet networks in 1962, and later, in the 1970s, for hierarchical routing, concepts which have been the underpinning of the development towards today's Internet.

Sutherland's successor Robert Taylor convinced Roberts to build on his early packet switching successes and come and be the IPTO Chief Scientist. Once there, Roberts prepared a report called Resource Sharing Computer Networks which was approved by Taylor in June 1968 and laid the foundation for the launch of the working ARPANET the following year.

After much work, the first two nodes of what would become the ARPANET were interconnected between Kleinrock's Network Measurement Center at the UCLA's School of Engineering and Applied Science and Douglas Engelbart's NLS system at SRI International (SRI) in Menlo Park, California, on 29 October 1969. The third site on the ARPANET was the Culler-Fried Interactive Mathematics center at the University of California at Santa Barbara, and the fourth was the University of Utah Graphics Department. In an early sign of future growth, there were already fifteen sites connected to the young ARPANET by the end of 1971.

The ARPANET was origin of today's Internet. In an independent development, Donald Davies at the UK National Physical Laboratory developed the concept of packet switching in the early 1960s, first giving a talk on the subject in 1965, after which the teams in the new field from two sides of the Atlantic ocean first became acquainted. It was actually Davies' coinage of the wording packet and packet switching that was adopted as the standard terminology. Davies also built a packet-switched network in the UK, called the Mark I in 1970. Bolt Beranek and Newman (BBN), the private contractors for ARPANET, set out to create a separate commercial version after establishing "value added carriers" was legalized in the U.S. The network they established was called Telenet and began operation in 1975; installing free public dial-up access in cities throughout the U.S. Telenet was the first packet-switching network open to the general public.

Following the demonstration that packet switching worked on the ARPANET, the British Post Office, Telenet, DATAPAC and TRANSPAC collaborated to create the first international packet-switched network service. In the UK, this was referred to as the International Packet Switched Service (IPSS), in 1978. The collection of X.25-based networks grew from Europe and the US to cover Canada, Hong Kong and Australia by 1981. The X.25 packet switching standard was developed in the CCITT (now called ITU-T) around 1976. X.25 was independent of the TCP/IP protocols that arose from the experimental work of DARPA on the ARPANET, Packet Radio Net and Packet Satellite Net during the same time period.

The early ARPANET ran on the Network Control Program (NCP), implementing the host to host connectivity and switching layers of the protocol stack, designed and first implemented in December 1970 by a team called the Network Working Group (NWG) led by Steve Crocker. To respond to the network's rapid growth as more and more locations connected, Vinton Cerf and Robert Kahn developed the first description of the now widely used TCP protocols during 1973 and published a paper on the subject in May 1974. Use of the term "Internet" to describe a single global TCP/IP network originated in December 1974 with the publication of RFC 675, the first full specification of TCP that was written by Vinton Cerf, Yogen Dalal and Carl Sunshine, then at Stanford University. During the next nine years, work proceeded to refine the protocols and to implement them on a wide range of operating systems. The first TCP/IP-based wide-area network was operational by 1 January 1983 when all hosts on the ARPANET were switched over from the older NCP protocols. In 1985, the United States' National Science Foundation (NSF) commissioned the construction of the NSFNET, a university 56 kilobit/second network backbone using computers called "fuzzballs" by their inventor, David L. Mills. The following year, NSF sponsored the conversion to a higher-speed 1.5 megabit/second network. A key decision to use the DARPA TCP/IP protocols was made by Dennis Jennings, then in charge of the Supercomputer program at NSF.

The opening of the NSFNET to other networks began in 1988. The US Federal Networking Council approved the interconnection of the NSFNET to the commercial MCI Mail system in that year and the link was made in the summer of 1989. Other commercial electronic mail services were soon connected, including OnTyme, Telemail and Compuserve. In that same year, three commercial Internet service providers (ISPs) began operations: UUNET, PSINet, and CERFNET. Important, separate networks that offered gateways into, then later merged with, the Internet include Usenet and BITNET. Various other commercial and educational networks, such as Telenet (by that time renamed to Sprintnet), Tymnet, Compuserve and JANET were interconnected with the growing Internet in the 1980s as the TCP/IP protocol became increasingly popular.

The adaptability of TCP/IP to existing communication networks allowed for rapid growth. The open availability of the specifications and reference code permitted commercial vendors to build interoperable network components, such as routers, making standardized network gear available from many companies. This aided in the rapid growth of the Internet and the proliferation of local-area networking. It seeded the widespread implementation and rigorous standardization of TCP/IP on UNIX and virtually every other common operating system.

Although the basic applications and guidelines that make the Internet possible had existed for almost two decades, the network did not gain a public face until the 1990s. On 6 August 1991, CERN, a pan European organization for particle research, publicized the new World Wide Web project. The Web was invented by British scientist Tim Berners-Lee in 1989. An early popular web browser was 'Viola www', patterned after HyperCard and built using the X Window System. It was eventually replaced in popularity by the Mosaic web browser. In 1993, the National Center for Supercomputing Applications at the University of Illinois released version 1.0 of Mosaic, and by late 1994 there was growing public interest in the previously academic, technical Internet. By 1996 usage of the word Internet had become commonplace, and consequently, so had its use as a synecdoche in reference to the 'World Wide Web.

Meanwhile, over the course of the decade, the Internet successfully accommodated the majority of previously existing public computer networks (although some networks, such as FidoNet, have remained separate). During the late 1990s, it was estimated that traffic on the public Internet grew by 100 percent per year, while the mean annual growth in the number of Internet users was thought to be between 20% and 50%. This growth is often attributed to the lack of central administration, which allows organic growth of the network, as well as the non-proprietary open nature of the Internet protocols, which encourages vendor interoperability and prevents any one company from exerting too much control over the network. The estimated population of Internet users is 1.97 billion as of 30 June 2010.

From 2009 onward, the Internet is expected to grow significantly in Brazil, Russia, India, China, and Indonesia (BRICI countries). These countries have large populations and moderate to high economic growth, but still low Internet penetration rates. In 2009, the BRICI countries represented about 45 percent of the world's population and had approximately 610 million Internet users, but by 2015, Internet users in BRICI countries will double to 1.2 billion, and will triple in Indonesia.

Technology

Protocols

The complex communications infrastructure of the Internet consists of its hardware components and a system of software layers that control various aspects of the architecture. While the hardware can often be used to support other software systems, it is the design and the rigorous standardization process of the software architecture that characterizes the Internet and provides the foundation for its scalability and success. The responsibility for the architectural design of the Internet software systems has been delegated to the Internet Engineering Task Force (IETF). The IETF conducts standard-setting work groups; open to any individual, about the various aspects of Internet architecture. Resulting discussions and final standards are published in a series of publications; each called a Request for Comments (RFC), freely available on the IETF web site. The principal methods of networking that enable the Internet are contained in specially designated RFCs that constitute the Internet Standards. Other less rigorous documents are simply informative, experimental, or historical, or document the best current practices (BCP) when implementing Internet technologies.

The Internet Standards describe a framework known as the Internet Protocol Suite. This is a model architecture that divides methods into a layered system of protocols (RFC 1122, RFC 1123). The layers correspond to the environment or scope in which their services operate. At the top is the 'Application Layer, the space for the application specific networking methods used in software applications, e.g., a web browser program. Below this top layer, the 'Transport Layer' connects applications on different hosts via the network (e.g., client server model) with appropriate data exchange methods. Underlying these layers are the core networking technologies, consisting of two layers. The Internet Layer enables computers to identify and locate each other via Internet Protocol (IP) addresses, and allows them to connect to one-another via intermediate (transit) networks. Lastly, at the bottom of the architecture, is a software layer, the Link Layer, that provides connectivity between hosts on the same local network link, such as a local area network (LAN) or a dial-up connection. The model, also known as TCP/IP, is designed to be independent of the underlying hardware which the model therefore does not concern itself with in any detail. Other models have been developed, such as the Open Systems Interconnection (OSI) model, but they are not compatible in the details of description, nor implementation, but many similarities exist and the TCP/IP protocols are usually included in the discussion of OSI networking.

The most prominent component of the Internet model is the Internet Protocol (IP) which provides addressing systems (IP addresses) for computers on the Internet. IP enables internetworking and essentially establishes the Internet itself. IP Version 4 (IPv4) is the initial version used on the first generation of the today's Internet and is still in dominant use. It was designed to address up to ~4.3 billion (10^9) Internet hosts. However, the explosive growth of the Internet has led to IPv4 address exhaustion which is estimated to enter its final stage in approximately 2011. A new protocol version, IPv6, was developed in the mid 1990s which provides vastly larger addressing capabilities and more efficient routing of Internet traffic. IPv6 is currently in commercial deployment phase around the world and Internet address registries (RIRs) have begun to urge all resource managers to plan rapid adoption and conversion.

IPv6 is not interoperable with IPv4. It essentially establishes a "parallel" version of the Internet not directly accessible with IPv4 software. This means software upgrades or translator facilities are necessary for every networking device that needs to communicate on the IPv6 Internet. Most modern computer operating systems are already converted to operate with both versions of the Internet Protocol. Network infrastructures, however, are still lagging in this development. Aside from the complex physical connections that make up its infrastructure, the Internet is facilitated by bi or multi lateral commercial contracts (e.g., peering agreements), and by technical specifications or protocols that describe how to exchange data over the network. Indeed, the Internet is defined by its interconnections and routing policies.

Structure

The Internet structure and its usage characteristics have been studied extensively. It has been determined that both the Internet IP routing structure and hypertext links of the World Wide Web are examples of scale-free networks. Similar to the way the commercial Internet providers connect via Internet exchange points, research networks tend to interconnect into large sub-networks such as GEANT, GLORIAD, Internet2 (successor of the Abilene Network), and the UK's national research and education network JANET. These in turn are built around smaller networks (see also the list of academic computer network organizations).

Many computer scientists describe the Internet as a "prime example of a large scale, highly engineered, yet highly complex system". The Internet is extremely heterogeneous; for instance, data transfer rates and physical characteristics of connections vary widely. The Internet exhibits "emergent phenomena" that depend on its large scale organization. For example, data transfer rates exhibit temporal self-similarity. The principles of the routing and addressing methods for traffic in the Internet reach back to their origins the 1960s when the eventual scale and popularity of the network could not be anticipated. Thus, the possibility of developing alternative structures is investigated.

Governance

The Internet is a globally distributed network comprising many voluntarily interconnected autonomous networks. It operates without a central governing body. However, to maintain interoperability, all technical and policy aspects of the underlying core infrastructure and the principal name spaces are administered by the Internet Corporation for Assigned Names and Numbers (ICANN), headquartered in Marina del Rey, California. ICANN is the authority that coordinates the assignment of unique identifiers for use on the Internet, including domain names, Internet Protocol (IP) addresses, application port numbers in the transport protocols, and many other parameters. Globally unified name spaces, in which names and numbers are uniquely assigned, are essential for the global reach of the Internet. ICANN is governed by an international board of directors drawn from across the Internet technical, business, academic, and other non commercial communities. The government of the United States continues to have the primary role in approving changes to the DNS root zone that lies at the heart of the domain name system. ICANN's role in coordinating the assignment of unique identifiers distinguishes it as perhaps the only central coordinating body on the global Internet. On 16 November 2005, the World Summit on the Information Society, held in Tunis, established the Internet Governance Forum (IGF) to discuss Internet-related issues.

Modern Uses

The Internet is allowing greater flexibility in working hours and location, especially with the spread of un metered high speed connections and web applications. The Internet can now be accessed almost anywhere by numerous means, especially through mobile Internet devices. Mobile phones, data cards, handheld game consoles and cellular routers allow users to connect to the Internet from anywhere there is a wireless network supporting that device's technology. Within the limitations imposed by small screens and other limited facilities of such pocket-sized devices, services of the Internet, including email and the web, may be available. Service providers may restrict the services offered and wireless data transmission charges may be significantly higher than other access methods.

Educational material at all levels from pre-school to post-doctoral is available from websites. Examples range from CBeebies, through school and high-school revision guides, virtual universities, to access to top end scholarly literature through the likes of Google Scholar. In distance education, help with homework and other assignments, self guided learning, whiling away spare time, or just looking up more detail on an interesting fact, it has never been easier for people to access educational information at any level from anywhere. The Internet in general and the World Wide Web in particular are important enablers of both formal and informal education.

The low cost and nearly instantaneous sharing of ideas, knowledge, and skills has made collaborative work dramatically easier, with the help of collaborative software. Not only can a group cheaply communicate and share ideas, but the wide reach of the Internet allows such groups to easily form in the first place. An example of this is the free software movement, which has produced, among other programs, *Linux, Mozilla Firefox, and OpenOffice.org*. Internet "chat", whether in the form of IRC chat rooms or channels, or via instant messaging systems, allow colleagues to stay in touch in a very convenient way when working at their computers during the day. Messages can be exchanged even more quickly and conveniently than via e-mail. Extensions to these systems may allow files to be exchanged, "whiteboard" drawings to be shared or voice and video contact between team members.

Version control systems allow collaborating teams to work on shared sets of documents without either accidentally overwriting each other's work or having members wait until they get "sent" documents to be able to make their contributions. Business and project teams can share calendars as well as documents and other information. Such collaboration occurs in a wide variety of areas including scientific research, software development, conference planning, political activism and creative writing. Social and political coll-aboration is also becoming more widespread as both Internet access and computer literacy grow. From the flash mob 'events' of the early 2000s to the use of social networking in the 2009 Iranian election protests, the Internet allows people to work together more effectively and in many more ways than was possible without it.

The Internet allows computer users to remotely access other computers and information stores easily, wherever they may be across the world. They may do this with or without the use of security, authentication and encryption technologies, depending on the requirements. This is encouraging new ways of working from home, collaboration and information sharing in many industries. An accountant sitting at home can audit the books of a company based in another country, on a server situated in a third country that is remotely maintained by IT specialists in a fourth. These accounts could have been created by home working bookkeepers, in other remote locations, based on information e-mailed to them from offices all over the world. Some of these things were possible before the widespread use of the Internet, but the cost of private leased lines would have made many of them infeasible in practice. An office worker away from their desk, perhaps on the other side of the world on a business trip or a holiday, can open a remote desktop session into his normal office PC using a secure Virtual Private Network (VPN) connection via the Internet. This gives the worker complete access to all of his or her normal files and data, including e-mail and other applications, while away from the office. This concept has been referred to among system administrators as the Virtual Private Nightmare, because it extends the secure perimeter of a corporate network into its employees' homes.

Services

Information

Many people use the terms 'Internet' and **'World Wide Web'**, or just the Web, interchangeably, but the two terms are not synonymous. The **World Wide Web** is a global set of documents, images and other resources, logically interrelated by hyperlinks and referenced with Uniform Resource Identifiers (URIs). URIs allow providers to symbolically identify services and clients to locate and address web servers, file servers, and other databases that store documents and provide resources and access them using the Hypertext Transfer Protocol (HTTP), the primary carrier protocol of the Web. HTTP is only one of the hundreds of communication protocols used on the Internet. Web services may also use HTTP to allow software systems to communicate in order to share and exchange business logic and data.

World Wide Web browser software, such as Microsoft's Internet Explorer, Mozilla Firefox, Opera, Apple's Safari, and Google Chrome, let users navigate from one web page to another via hyperlinks embedded in the documents. These documents may also contain any combination of computer data, including graphics, sounds, text, video, multimedia and interactive content including games, office applications and scientific demonstrations. Through keyword driven Internet research using search engines like Yahoo! and Google, users worldwide have easy, instant access to a vast and diverse amount of online information. Compared to printed encyclopedias and traditional libraries, the World Wide Web has enabled the decentralization of information.

The Web has also enabled individuals and organizations to publish ideas and information to a potentially large audience online at greatly reduced expense and time delay. Publishing a web page, a blog, or building a website involves little initial cost and many cost free services are available. Publishing and maintaining large, professional web sites with attractive, diverse and up to date information is still a difficult and expensive proposition, however. Many individuals and some companies and groups use web logs or blogs, which are largely used as easily updatable online diaries. Some commercial organizations encourage staff to communicate advice in their areas of specialization in the hope that visitors will be impressed by the expert knowledge and free information, and be attracted to the corporation as a result. One example of this practice is Microsoft, whose product developers publish their personal blogs in order to pique the public's interest in their work. Collections of personal web pages published by large service providers remain popular, and have become increasingly sophisticated. Whereas operations such as Angelfire and GeoCities have existed since the early days of the Web, newer offerings from, for example, Facebook and MySpace currently have large followings. These operations often brand themselves as social network services rather than simply as web page hosts.

Advertising on popular web pages can be lucrative, and e-commerce or the sale of products and services directly via the Web continues to grow. When the Web began in the 1990s, a typical web page was stored in completed form on a web server, formatted with HTML, ready to be sent to a user's browser in response to a request. Over time, the process of creating and serving web pages has become more automated and more dynamic. Websites are often created using content management or wiki software with, initially, very little content. Contributors to these systems, who may be paid staff, members of a club or other organization or members of the public, fill underlying databases with content using editing pages designed for that purpose, while casual visitors view and read this content in its final HTML form. There may or may not be editorial, approval and security systems built into the process of taking newly entered content and making it available to the target visitors.

Communication

E-mail is an important communications service available on the Internet. The concept of sending electronic text messages between parties in a way analogous to mailing letters or memos predates the creation of the Internet. Today it can be important to distinguish between internet and internal e-mail systems. Internet e-mail may travel and be stored unencrypted on many other networks and machines out of both the sender's and the recipient's control. During this time it is quite possible for the content to be read and even tampered with by third parties, if anyone considers it important enough. Purely internal or intranet mail systems, where the information never leaves the corporate or organization's network, are much more secure, although in any organization there will be IT and other personnel whose job will involve monitoring, and occasionally accessing, the e-mail of other employees not addressed to them. Pictures, documents and other files can be sent as e-mail attachments. E-mails can be cc-ed to multiple e-mail addresses.

Internet telephony is another common communications service made possible by the creation of the Internet. VoIP stands for Voice over Internet Protocol, referring to the protocol that underlies all Internet communication. The idea began in the early 1990s with walkie talkie like voice applications for personal computers. In recent years many VoIP systems have become as easy to use and as convenient as a normal telephone. The benefit is that, as the Internet carries the voice traffic, VoIP can be free or cost much less than a traditional telephone call, especially over long distances and especially for those with always on Internet connections such as cable or ADSL. VoIP is maturing into a competitive alternative to traditional telephone service. Interoperability between different providers has improved and the ability to call or receive a call from a traditional telephone is available. Simple, inexpensive VoIP network adapters are available that eliminate the need for a personal computer.

Voice quality can still vary from call to call but is often equal to and can even exceed that of traditional calls. Remaining problems for VoIP include emergency telephone number dialing and reliability. Currently, a few VoIP providers provide an emergency service, but it is not universally available. Traditional phones are line powered and operate during a power failure; VoIP does not do so without a backup power source for the phone equipment and the Internet access devices. VoIP has also become increasingly popular for gaming applications, as a form of communication between players. Popular VoIP clients for gaming include Ventrilo and Teamspeak. Wii, PlayStation 3, and Xbox 360 also offer VoIP chat features.

Data Transfer

File sharing is an example of transferring large amounts of data across the Internet. A computer file can be e-mailed to customers, colleagues and friends as an attachment. It can be uploaded to a website or FTP server for easy download by others. It can be put into a "shared location" or onto a file server for instant use by colleagues. The load of bulk downloads too many users can be eased by the use of "mirror" servers or peer to peer networks. In any of these cases, access to the file may be controlled by user authentication, the transit of the file over the Internet may be obscured by encryption, and money may change hands for access to the file. The price can be paid by the remote charging of funds from, for example, a credit card whose details are also passed usually fully encrypted across the Internet. The origin and authenticity of the file received may be checked by digital signatures or by MD5 or other message digests. These simple features of the Internet, over a worldwide basis, are changing the production, sale, and distribution of anything that can be reduced to a computer file for transmission. This includes all manner of print publications, software products, news, music, film, video, photography, graphics and the other arts. This in turn has caused seismic shifts in each of the existing industries that previously controlled the production and distribution of these products.

Streaming media refers to the act that many existing radio and television broadcasters promote Internet "feeds" of their live audio and video streams (for example, the BBC). They may also allow time shift viewing or listening such as 'Preview', 'Classic Clips' and 'Listen Again features. These providers have been joined by a range of pure Internet "broadcasters" who never had on air licenses. This means that an Internet connected device, such as a computer or something more specific, can be used to access on line media in much the same way as was previously possible only with a television or radio receiver.

The range of available types of content is much wider, from specialized technical webcasts to on demand popular multimedia services. Podcasting is a variation on this theme, where usually audio material is downloaded and played back on a computer or shifted to a portable media player to be listened to on the move. These techniques using simple equipment allow anybody, with little censorship or licensing control, to broadcast audio visual material worldwide.

'Connected' TV needs good internet connection. The image quality is depends on internet speed. As example for standard image quality needs 1 Mbps internet speed for SD 480p, requires 2.5 Mbps for HD 720p quality and the top of the line HDX quality needs 4.5 Mbps for 1080p.

Webcams can be seen as an even lower budget extension of this phenomenon. While some webcams can give full frame rate video, the picture is usually either small or updates slowly. Internet users can watch animals around an African waterhole, ships in the Panama Canal, traffic at a local roundabout or monitor their own premises, live and in real time. Video chat rooms and video conferencing are also popular with many uses being found for personal webcams, with and without two way sound. YouTube was founded on 15 February 2005 and is now the leading website for free streaming video with a vast number of users. It uses a flash-based web player to stream and show video files. Registered users may upload an unlimited amount of video and build their own personal profile. YouTube claims that its users watch hundreds of millions, and upload hundreds of thousands of videos daily.

Access

The prevalent language for communication on the Internet is English. This may be a result of the origin of the Internet, as well as English's role as a lingua franca. It may also be related to the poor capability of early computers, largely originating in the United States, to handle characters other than those in the English variant of the Latin alphabet. After English (28% of Web visitors) the most requested languages on the World Wide Web are Chinese (23%), Spanish (8%), Japanese (5%), Portuguese and German (4% each), Arabic, French and Russian (3% each), and Korean (2%). By region, 42% of the world's Internet users are based in Asia, 24% in Europe, 14% in North America, 10% in Latin America and the Caribbean taken together, 5% in Africa, 3% in the Middle East and 1% in Australia/Oceania. In Asia, South Korea has the biggest internet penetration with 81.1% users (as comparison Japan with 78.2% and USA with 77.3%). January 2011: At the end of 2010 internet users had reached 2.08 billion where last year was 1.86 billion. In year of 2000 the internet users were only 250 million. Fifty seven percent of the users are in developing countries spearheaded by China. The Internet's technologies have developed enough in recent years, especially in the use of Unicode, that good facilities are available for development and communication in the world's widely used languages. However, some glitches such as mojibake (incorrect display of some languages' characters) still remain.

Common methods of Internet access in homes include dial-up, landline broadband (over coaxial cable, fiber optic or copper wires), Wi-Fi, satellite and 3G/4G technology cell phones. Public places to use the Internet include libraries and Internet cafes, where computers with Internet connections are available. There are also Internet access points in many public places such as airport halls and coffee shops, in some cases just for brief use while standing. Various terms are used, such as "public Internet kiosk", "public access terminal", and "Web payphone". Many hotels now also have public terminals, though these are usually fee based. These terminals are widely accessed for various usage like ticket booking, bank deposit, online payment etc. Wi-Fi provides wireless access to computer networks, and therefore can do so to the Internet itself. Hotspots providing such access include Wi-Fi cafes, where would-be users need to bring their own wireless-enabled devices such as a laptop or PDA. These services may be free to all, free to customers only, or fee based. A hotspot need not be limited to a confined location. A whole campus or park, or even an entire city can be enabled. Grassroots efforts have led to wireless community networks. Commercial Wi-Fi services covering large city areas are in place in London, Vienna, Toronto, San Francisco, Philadelphia, Chicago and Pittsburgh. The Internet can then be accessed from such places as a park bench. Apart from Wi-Fi, there have been experiments with proprietary mobile wireless networks like Ricochet, various high-speed data services over cellular phone networks, and fixed wireless services. High end mobile phones such as smart phones generally come with Internet access through the phone network. Web browsers such as Opera are available on these advanced handsets, which can also run a wide variety of other Internet software. More mobile phones have Internet access than PCs, though this is not as widely used. An Internet access provider and protocol matrix differentiates the methods used to get online.

In contrast, an Internet blackout or outage can be caused by accidental local signaling interruptions. Disruptions of submarine communications cables may cause blackouts or slowdowns to large areas depending on them, such as in the 2008 submarine cable disruption. Internet blackouts of almost entire countries can be achieved by governments, such as the Internet in Egypt, which was shut down in 2011 in an attempt to stop mobilization for anti government protests.

In an American study in 2005, the percentage of men using the Internet was very slightly ahead of the percentage of women, although this difference reversed in those under 30. Men logged on more often, spend more time online, and are more likely to be broadband users, whereas women tended to make more use of opportunities to communicate (such as email). Men were more likely to use the Internet to pay bills, participate in auctions, and for recreation such as downloading music and videos. Men and women were equally likely to use the Internet for shopping and banking. More recent studies indicate that in 2008, women significantly outnumbered men on most social networking sites, such as 'Facebook' and 'Myspace', although the ratios varied with age. In addition, women watched more streaming content, whereas men downloaded more. In terms of blogs, men were more likely to blog in the first place; among those who blog, men were more likely to have a professional blog, whereas women were more likely to have a personal blog.

Social Impact

The Internet has enabled entirely new forms of social interaction, activities, and organizing, thanks to its basic features such as widespread usability and access. Social networking websites such as Face book, Twitter and My Space have created new ways to socialize and interact. Users of these sites are able to add a wide variety of information to pages, to pursue common interests, and to connect with others. It is also possible to find existing acquaintances, to allow communication among existing groups of people. Sites like LinkedIn foster commercial and business connections. YouTube and Flickr specialize in users' videos and photographs.

In the first decade of the 21st century the first generation is raised with widespread availability of Internet connectivity, bringing consequences and concerns in areas such as personal privacy and identity, and distribution of copyrighted materials. These "digital natives" face a variety of challenges that were not present for prior generations.

The Internet has achieved new relevance as a political tool, leading to Internet censorship by some states. The presidential campaign of Howard Dean in 2004 in the United States was notable for its success in soliciting donation via the Internet. Many political groups use the Internet to achieve a new method of organizing in order to carry out their mission, having given rise to Internet activism. Some governments, such as those of Iran, North Korea, Myanmar, the People's Republic of China, and Saudi Arabia, restrict what people in their countries can access on the Internet, especially political and religious content. This is accomplished through software that filters domains and content so that they may not be easily accessed or obtained without elaborate circumvention.

In Norway, Denmark, Finland and Sweden, major Internet service providers have voluntarily, possibly to avoid such an arrangement being turned into law, agreed to restrict access to sites listed by authorities. While this list of forbidden URLs is only supposed to contain addresses of known child pornography sites, the content of the list is secret. Many countries, including the United States, have enacted laws against the possession or distribution of certain material, such as child pornography, via the Internet, but do not mandate filtering software. There are many free and commercially available software programs, called content-control software, with which a user can choose to block offensive websites on individual computers or networks, in order to limit a child's access to pornographic materials or depiction of violence.

The Internet has been a major outlet for leisure activity since its inception, with entertaining social experiments such as MUDs and MOOs being conducted on university servers, and humor-related Usenet groups receiving much traffic. Today, many Internet forums have sections devoted to games and funny videos; short cartoons in the form of Flash movies are also popular. Over 6 million people use blogs or message boards as a means of communication and for the sharing of ideas. The pornography and gambling industries have taken advantage of the World Wide Web, and often provide a significant source of advertising revenue for other websites. Although many governments have attempted to restrict both industries' use of the Internet, this has generally failed to stop their widespread popularity.

One main area of leisure activity on the Internet is multiplayer gaming. This form of recreation creates communities, where people of all ages and origins enjoy the fast paced world of multiplayer games. These range from MMORPG to first-person shooters, from role playing games to online gambling. This has revolutionized the way many people interact while spending their free time on the Internet. While online gaming has been around since the 1970s, modern modes of online gaming began with subscription services such as GameSpy and MPlayer. Non-subscribers were limited to certain types of game play or certain games. Many people use the Internet to access and download music, movies and other works for their enjoyment and relaxation. Free and fee-based services exist for all of these activities, using centralized servers and distributed peer to peer technologies. Some of these sources exercise more care with respect to the original artists' copyrights than others.

Many people use the World Wide Web to access news, weather and sports reports, to plan and book vacations and to find out more about their interests. People use chat, messaging and e-mail to make and stay in touch with friends worldwide, sometimes in the same way as some previously had pen pals. The Internet has seen a growing number of Web desktops, where users can access their files and settings via the Internet.

Cyber slacking can become a drain on corporate resources; the average UK employee spent 57 minutes a day surfing the Web while at work, according to a 2003 study by Peninsula Business Services. Internet addiction disorder is excessive computer use that interferes with daily life. Some psychologists believe that Internet use has other effects on individuals for instance interfering with the deep thinking that leads to true creativity.

Internet usage has also shown a strong connection to loneliness. Lonely people tend to use the internet for an outlet for their feelings and to share their stories with other lonely people, such as in the "I am lonely will anyone speak to me" thread.

World Wide Web (www)

The World Wide Web, abbreviated as www and commonly known as the Web, is a system of interlinked hypertext documents accessed via the Internet. With a web browser, one can view web pages that may contain text, images, videos, and other multimedia and navigate between them via hyperlinks. Using concepts from earlier hypertext systems, English engineer and computer scientist Sir Tim Berners-Lee, now the Director of the World Wide Web Consortium, wrote a proposal in March 1989 for what would eventually become the World Wide Web. At CERN in Geneva, Switzerland, Berners-Lee and Belgian computer scientist Robert Cailliau proposed in 1990 to use "HyperText ... to link and access information of various kinds as a web of nodes in which the user can browse at will", and publicly introduced the project in December. "The World-Wide Web (W3) was developed to be a pool of human knowledge, and human culture, which would allow collaborators in remote sites to share their ideas and all aspects of a common project."

In the May 1970 issue of Popular Science magazine Arthur C. Clarke was reported to have predicted that satellites would one day "bring the accumulated knowledge of the world to your fingertips" using a console that would combine the functionality of the Xerox, telephone, television and a small computer, allowing data transfer and video conferencing around the globe. In March 1989, Tim Berners-Lee wrote a proposal that referenced ENQUIRE, a database and software project he had built in 1980, and described a more elaborate information management system.

With help from Robert Cailliau, he published a more formal proposal (on November 12, 1990) to build a "Hypertext project" called "World Wide Web" (one word, also "W3") as a "web" of "hypertext documents" to be viewed by "browsers" using a client server architecture. This proposal estimated that a read only web would be developed within three months and that it would take six months to achieve "the creation of new links and new material by readers, so that authorship becomes universal" as well as "the automatic notification of a reader when new material of interest to him/her has become available." See Web 2.0 and RSS/Atom, which have taken a little longer to mature. The proposal was modeled after the Dynatext SGML reader by Electronic Book Technology, a spin-off from the Institute for Research in Information and Scholarship at Brown University. The Dynatext system, licensed by CERN, was technically advanced and was a key player in the extension of SGML ISO 8879:1986 to Hypermedia within HyTime, but it was considered too expensive and had an inappropriate licensing policy for use in the general high energy physics community, namely a fee for each document and each document alteration.

A NeXT Computer was used by Berners-Lee as the world's first web server and also to write the first web browser, World Wide Web, in 1990. By Christmas 1990, Berners-Lee had built all the tools necessary for a working Web: the first web browser (which was a web editor as well); the first web server; and the first web pages, which described the project itself. On August 6, 1991, he posted a short summary of the World Wide Web project on the alt. hypertext newsgroup. This date also marked the debut of the Web as a publicly available service on the Internet. The first photo on the web was uploaded by Berners-Lee in 1992, an image of the CERN house band Les Horribles Cernettes. The first server outside Europe was set up at SLAC to host the SPIRES-HEP database. Accounts differ substantially as to the date of this event. The World Wide Web Consortium says December 1992, whereas SLAC itself claims 1991. This is supported by a W3C document entitled A Little History of the World Wide Web.

The crucial underlying concept of hypertext originated with older projects from the 1960s, such as the Hypertext Editing System (HES) at Brown University, Ted Nelson's Project Xanadu, and Douglas Engelbart's oN-Line System (NLS). Both Nelson and Engelbart were in turn inspired by Vannevar Bush's microfilm-based "memex", which was described in the 1945 essay "As We May Think". Berners-Lee's breakthrough was to marry hypertext to the Internet. In his book Weaving the Web, he explains that he had repeatedly suggested that a marriage between the two technologies was possible to members of both technical communities, but when no one took up his invitation, he finally tackled the project himself. In the process, he developed a system of globally unique identifiers for resources on the Web and elsewhere: the Universal Document Identifier (UDI), later known as Uniform Resource Locator (URL) and Uniform Resource Identifier (URI); the publishing language Hyper Text Markup Language (HTML); and the Hypertext Transfer Protocol (HTTP).

The World Wide Web had a number of differences from other hypertext systems that were then available. The Web required only unidirectional links rather than bidirectional ones. This made it possible for someone to link to another resource without action by the owner of that resource. It also significantly reduced the difficulty of implementing web servers and browsers (in comparison to earlier systems), but in turn presented the chronic problem of link rot. Unlike predecessors such as HyperCard, the World Wide Web was non-proprietary, making it possible to develop servers and clients independently and to add extensions without licensing restrictions. On April 30, 1993, CERN announced that the World Wide Web would be free to anyone, with no fees due. Coming two months after the announcement that the Gopher protocol was no longer free to use, this produced a rapid shift away from Gopher and towards the Web. An early popular web browser was Viola WWW, which was based upon HyperCard. Scholars generally agree that a turning point for the World Wide Web began with the introduction of the **Mosaic** web browser in 1993, a graphical browser developed by a team at the National Center for Supercomputing Applications at the University of Illinois at Urbana-Champaign (NCSA-UIUC), led by Marc Andreessen. Funding for Mosaic came from the U.S. High-Performance Computing and Communications Initiative, a funding program initiated by the High Performance Computing and Communication Act of 1991, one of several computing developments initiated by U.S. Senator Al Gore. Prior to the release of Mosaic, graphics were not commonly mixed with text in web pages and the Web's popularity was less than older protocols in use over the Internet, such as Gopher and Wide Area Information Servers (WAIS). Mosaic's graphical user interface allowed the Web to become, by far, the most popular Internet protocol. The World Wide Web Consortium (W3C) was founded by Tim Berners-Lee after he left the European Organization for Nuclear Research (CERN) in October, 1994. It was founded at the Massachusetts Institute of Technology Laboratory for Computer Science (MIT/LCS) with support from the Defense Advanced Research Projects Agency (DARPA), which had pioneered the Internet; a year later, a second site was founded at INRIA (a French national computer research lab) with support from the European Commission DG InfSo; and in 1996, a third continental site was created in Japan at Keio University. By the end of 1994, while the total number of websites was still minute compared to present standards, quite a number of notable websites were already active, many of which are the precursors or inspiration for today's most popular services.

Connected by the existing Internet, other websites were created around the world, adding international standards for domain names and HTML. Since then, Berners-Lee has played an active role in guiding the development of web standards (such as the markup languages in which web pages are composed), and in recent years has advocated his vision of a Semantic Web. The World Wide Web enabled the spread of information over the Internet through an easy to use and flexible format. It thus played an important role in popularizing use of the Internet. Although the two terms are sometimes conflated in popular use, World

Wide Web is not synonymous with Internet. The Web is an application built on top of the Internet.

Function

The terms Internet and World Wide Web are often used in every day speech without much distinction. However, the Internet and the World Wide Web are not one and the same. The Internet is a global system of interconnected computer networks. In contrast, the Web is one of the services that run on the Internet. It is a collection of interconnected documents and other resources, linked by hyperlinks and URLs. In short, the Web is an application running on the Internet. Viewing a web page on the World Wide Web normally begins either by typing the URL of the page into a web browser, or by following a hyperlink to that page or resource. The web browser then initiates a series of communication messages, behind the scenes, in order to fetch and display it.

First, the server name portion of the URL is resolved into an IP address using the global, distributed Internet database known as the Domain Name System (DNS). This IP address is necessary to contact the Web server. The browser then requests the resource by sending an HTTP request to the Web server at that particular address. In the case of a typical web page, the HTML text of the page is requested first and parsed immediately by the web browser, which then makes additional requests for images and any other files that complete the page image. Statistics measuring a website's popularity are usually based either on the number of page views or associated server 'hits' (file requests) that take place. While receiving these files from the web server, browsers may progressively render the page onto the screen as specified by its HTML, Cascading Style Sheets (CSS), or other page composition languages. Any images and other resources are incorporated to produce the on-screen web page that the user sees. Most web pages contain hyperlinks to other related pages and perhaps to downloadable files, source documents, definitions and other web resources. Such a collection of useful, related resources, interconnected via hypertext links is dubbed a web of information. Publication on the Internet created what Tim Berners-Lee first called the World Wide Web (in its original Camel Case, which was subsequently discarded) in November 1990.

Linking

Over time, many web resources pointed to by hyperlinks disappear, relocate, or are replaced with different content. This makes hyperlinks obsolete, a phenomenon referred to in some circles as link rot and the hyperlinks affected by it are often called dead links. The ephemeral nature of the Web has prompted many efforts to archive web sites. The Internet Archive, active since 1996, is one of the best-known efforts.

www Prefix

Many domain names used for the World Wide Web begin with www because of the long-standing practice of naming Internet hosts (servers) according to the services they provide. The hostname for a web server is often www, in the same way that it may be ftp for an FTP server, and news or nntp for a USENET news server. These host names appear as Domain Name System (DNS) subdomain names, as in www.example.com. The use of 'www' as a sub-domain name is not required by any technical or policy standard; indeed, the first ever web server was called nxoc01.cern.ch, and many web sites exist without it. Many established websites still use 'www', or they invent other sub-domain names such as 'www2', 'secure', etc. Many such web servers are set up such that both the domain root (e.g., example.com) and the www sub domain (e.g., www.example.com) refer to the same site; others require one form or the other, or they may map to different web sites.

The use of a sub-domain name is useful for load balancing incoming web traffic by creating a CNAME record that points to a cluster of web servers. Since, currently, only a sub-domain can be cname'ed the same result cannot be achieved by using the bare domain root.

When a user submits an incomplete website address to a web browser in its address bar input field, some web browsers automatically try adding the prefix "www" to the beginning of it and possibly ".com", ".org" and ".net" at the end, depending on what might be missing. For example, entering 'micro soft' may be transformed to http://www.microsoft.com/ and 'open office' to http://www.openoffice.org. This feature started appearing in early versions of Mozilla Firefox, when it still had the working title 'Firebird' in early 2003, from a much more ancient practice in browsers such as Lynx. It is reported that Microsoft was granted a US patent for the same idea in 2008, but only for mobile devices. The scheme specifiers (http:// or https://) in URIs refer to the Hypertext Transfer Protocol and to HTTP Secure respectively and so define the communication protocol to be used for the request and response. The HTTP protocol is fundamental to the operation of the World Wide Web; the added encryption layer in HTTPS is essential when confidential information such as passwords or banking information are to be exchanged over the public Internet. Web browsers usually prepend the scheme to URLs too, if omitted. In English, www is pronounced by individually pronouncing the name of characters (double-u double-u double-u). Although some technical users pronounce it dub-dub-dub this is not widespread. The English writer Douglas Adams once quipped in The Independent on Sunday (1999): "The World Wide Web is the only thing I know of whose shortened form takes three times longer to say than what it's short for," with Stephen Fry later pronouncing it in his "Podgrammes" series of podcasts as "wuh wuh wuh." In Mandarin Chinese, World Wide Web is commonly translated via a phono-semantic matching to wàn wéi w ng (万维网), which satisfies www and literally means "myriad dimensional net", a translation that very appropriately reflects the design concept and proliferation of the World Wide Web. Tim Berners-Lee's web-space states that World Wide Web is officially spelled as three separate words, each capitalized, with no intervening hyphens.

Privacy

Computer users, who save time and money, and who gain conveniences and entertainment, may or may not have surrendered the right to privacy in exchange for using a number of technologies including the Web. Worldwide, more than a half billion people have used a social network service, and of Americans who grew up with the Web, half created an online profile and are part of a generational shift that could be changing norms. Facebook progressed from U.S. college students to a 70% non-U.S. audience, and in 2009 estimated that only 20% of its members use privacy settings. In 2010 (six years after co founding the company), Mark Zuckerberg wrote, "We will add privacy controls that are much simpler to use".

Privacy representatives from 60 countries have resolved to ask for laws to complement industry self-regulation, for education for children and other minors who use the Web, and for default protections for users of social networks. They also believe data protection for personally identifiable information benefits business more than the sale of that information. Users can opt-in to features in browsers to clear their personal histories locally and block some cookies and advertising networks but they are still tracked in websites' server logs, and particularly web beacons. Berners-Lee and colleagues see hope in accountability and appropriate use achieved by extending the Web's architecture to policy awareness, perhaps with audit logging, reasoners and appliances. In exchange for providing free content, vendors hire advertisers who spy on Web users and base their business model on tracking them. Since 2009, they buy and sell consumer data on exchanges (lacking a few details that could make it possible to de-anonymize, or identify an individual). Several times a day the, Lotame Solutions captures what users are typing in real time, and sends that text to Open Amplify who then tries to determine, to quote a writer at The Wall Street Journal, "what topics are being discussed, how the author feels about those topics, and what the person is going to do about them". Microsoft backed away in 2008 from its plans for strong privacy features in Internet Explorer, leaving its users (50% of the world's Web users) open to advertisers who may make assumptions about them based on only one click when they visit a website. Among services paid for by advertising, Yahoo! could collect the most data about users of commercial websites, about 2,500 bits of information per month about each typical user of its site and its affiliated advertising network sites. Yahoo! was followed by MySpace with about half that potential and then by AOL–Time Warner, Google, Facebook, Microsoft, and eBay.

Security

The Web has become criminals' preferred pathway for spreading malware. Cybercrime carried out on the Web can include identity theft, fraud, espionage and intelligence gathering. Web-based vulnerabilities now outnumber traditional computer security concerns, and as measured by Google, about one in ten web pages may contain malicious code. Most Web based attacks take place on legitimate websites, and most, as measured by Sophos, are hosted in the United States, China and Russia. The most common of all malware threats is SQL injection attacks against websites. Through HTML and URIs the Web was vulnerable to attacks like cross-site scripting (XSS) that came with the introduction of JavaScript and were exacerbated to some degree by Web 2.0 and Ajax web design that favors the use of scripts. Today by one estimate, 70% of all websites are open to XSS attacks on their users.

Proposed solutions vary to extremes. Large security vendors like McAfee already design governance and compliance suites to meet post-9/11 regulations, and some, like Finjan have recommended active real time inspection of code and all content regardless of its source. Some have argued that for enterprise to see security as a business opportunity rather than a cost center, "ubiquitous, always-on digital rights management" enforced in the infrastructure by a handful of organizations must replace the hundreds of companies that today secure data and networks. Jonathan Zittrain has said users sharing responsibility for computing safety is far preferable to locking down the Internet.

Standards

Many formal standards and other technical specifications and software define the operation of different aspects of the World Wide Web, the Internet, and computer information exchange. Many of the documents are the work of the World Wide Web Consortium (W3C), headed by Berners-Lee, but some are produced by the Internet Engineering Task Force (IETF) and other organizations.

Usually, when web standards are discussed, the following publications are seen as foundational:

a. Recommendations for mark up languages, especially HTML and XHTML, from the W3C. These define the structure and interpretation of hypertext documents.

b. Recommendations for stylesheets, especially CSS, from the W3C.

c. Standards for ECMAScript (usually in the form of JavaScript), from Ecma International.

d. Recommendations for the Document Object Model, from W3C.

Additional publications provide definitions of other essential technologies for the World Wide Web, including, but not limited to, the following:

a. Uniform Resource Identifier (URI), which is a universal system for referencing resources on the Internet, such as hypertext documents and images. URIs, often called URLs, are defined by the IETF's RFC 3986 / STD 66: Uniform Resource Identifier (URI): Generic Syntax, as well as its predecessors and numerous URI scheme-defining RFCs;

b. HyperText Transfer Protocol (HTTP), especially as defined by RFC 2616: HTTP/1.1 and RFC 2617: HTTP Authentication, which specify how the browser and server authenticate each other.

Accessibility

Access to the Web is for everyone regardless of disability including visual, auditory, physical, speech, cognitive, or neurological. Accessibility features also help others with temporary disabilities like a broken arm or the aging population as their abilities change. The Web is used for receiving information as well as providing information and interacting with society, making it essential that the Web be accessible in order to provide equal access and equal opportunity to people with disabilities. Tim Berners-Lee once noted, "The power of the Web is in its universality. Access by everyone regardless of disability is an essential aspect." Many countries regulate web accessibility as a requirement for websites. International cooperation in the W3C Web Accessibility Initiative led to simple guidelines that web content authors as well as software developers can use to make the Web accessible to persons who may or may not be using assistive technology.

Internalization

The W3C Internationalization Activity assures that web technology will work in all languages, scripts, and cultures. Beginning in 2004 or 2005, Unicode gained ground and eventually in December 2007 surpassed both ASCII and Western European as the Web's most frequently used character encoding. Originally RFC 3986 allowed resources to be identified by URI in a subset of US ASCII. RFC 3987 allows more characters any character in the Universal Character Set and now a resource can be identified by IRI in any language

Statistics

According to a 2001 study, there were over 550 billion documents on the Web, mostly in the invisible Web, or deep Web. A 2002 survey of 2,024 million Web pages determined that by far the most Web content was in English (56.4%) next were pages in German (7.7%), French (5.6%), and Japanese (4.9%). A more recent study, which used Web searches in 75 different languages to sample the Web, determined that there were over 11.5 billion Web pages in the publicly indexable Web as of the end of January 2005. As of March 2009, the indexable web contains at least 25.21 billion pages. On July 25, 2008, Google software engineers Jesse Alpert and Nissan Hajaj announced that Google Search had discovered one trillion unique URLs. As of May 2009, over 109.5 million websites operated. Of these 74% were commercial or other sites operating in the .com generic top-level domain.

Caching

If a user revisits a Web page after only a short interval, the page data may not need to be re-obtained from the source Web server. Almost all web browsers cache recently obtained data, usually on the local hard drive. HTTP requests sent by a browser will usually only ask for data that has changed since the last download. If the locally cached data are still current, it will be reused. Caching helps reduce the amount of Web traffic on the Internet. The decision about expiration is made independently for each downloaded file, whether image, style sheet, JavaScript, HTML, or whatever other content the site may provide. Thus even on sites with highly dynamic content, many of the basic resources only need to be refreshed occasionally. Web site designers find it worthwhile to collate resources such as CSS data and JavaScript into a few site wide files so that they can be cached efficiently. This helps reduce page download times and lowers demands on the Web server.

There are other components of the Internet that can cache Web content. Corporate and academic firewalls often cache Web resources requested by one user for the benefit of all. (See also caching proxy server.) Some search engines also store cached content from websites. Apart from the facilities built into Web servers that can determine when files have been updated and so need to be re-sent, designers of dynamically generated Web pages can control the HTTP headers sent back to requesting users, so that transient or sensitive pages are not cached. Internet banking and news sites frequently use this facility. Data requested with an HTTP 'GET' is likely to be cached if other conditions are met; data obtained in response to a 'POST' is assumed to depend on the data that was POSTed and so is not cached

Weblog

A blog (a blend of the term web log) is a type of website or part of a website. Blogs are usually maintained by an individual with regular entries of commentary, descriptions of events, or other material such as graphics or video. Entries are commonly displayed in reverse-chronological order. Blog can also be used as a verb, meaning to maintain or add content to a blog.

Most blogs are interactive, allowing visitors to leave comments and even message each other via widgets on the blogs and it is this interactivity that distinguishes them from other static websites.

Many blogs provide commentary or news on a particular subject; others function as more personal online diaries. A typical blog combines text, images, and links to other blogs, Web pages, and other media related to its topic. The ability of readers to leave comments in an interactive format is an important part of many blogs. Most blogs are primarily textual, although some focus on art (Art blog), photographs (photoblog), videos (video blogging), music (MP3 blog), and audio (podcasting). Microblogging is another type of blogging, featuring very short posts. As of December 2007, blog search engine Technorati was tracking more than 112 million blogs.

Origins

Before blogging became popular, digital communities took many forms, including Usenet, commercial online services such as GEnie, BiX and the early CompuServe, e-mail lists and Bulletin Board Systems (BBS). In the 1990s, Internet forum software, created running conversations with "threads." Threads are topical connections between messages on a virtual "corkboard."

The modern blog evolved from the online diary, where people would keep a running account of their personal lives. Most such writers called themselves diarists, journalists, or journalers. Justin Hall, who began personal blogging in 1994 while a student at Swarthmore College, is generally recognized as one of the earliest bloggers like Jerry Pournelle. Dave Winer's 'Scripting News' is also credited with being one of the oldest and longest running weblogs. Another early blog was Wearable Wireless Webcam, an online shared diary of a person's personal life combining text, video, and pictures transmitted live from a wearable computer and Eye Tap device to a web site in 1994. This practice of semi-automated blogging with live video together with text was referred to as surveillance, and such journals were also used as evidence in legal matters.

Early blogs were simply manually updated components of common Web sites. However, the evolution of tools to facilitate the production and maintenance of Web articles posted in reverse chronological order made the publishing process feasible to a much larger, less technical, population. Ultimately, this resulted in the distinct class of online publishing that produces blogs that we recognize today. For instance, the use of some sort of browser-based software is now a typical aspect of "blogging". Blogs can be hosted by dedicated blog hosting services, or they can run using blog software, or on regular web hosting services.

Some early bloggers, such as 'The Misanthropic Bitch', who began in 1997, actually referred to their online presence as a zine, before the term blog entered common usage.

Chapter 11
Photojournalism

Photojournalism is a particular form of journalism, which includes the collecting, editing, and presenting of news material for publication or broadcast that creates images in order to tell a news story. It is now usually understood to refer only to still images, but in some cases the term also refers to video used in broadcast journalism. Photojournalism is distinguished from other close branches of photography such as documentary photography, social documentary photography, street photography or celebrity photography by the qualities of;

a. **Timeliness** — the images have meaning in the context of a recently published record of events.

b. **Objectivity** — the situation implied by the images is a fair and accurate representation of the events they depict in both content and tone.

c. **Narrative** — the images combine with other news elements to make facts relatable to the viewer or reader on a cultural level.

Like a writer, a photojournalist is a reporter but he or she must often make decisions instantly and carry photographic equipment, often while exposed to significant obstacles like physical danger, weather, crowds etc.

Photojournalism can be defined as photography intended, in conjunction with text, to convey information about a topical event or events. This demarcates it, albeit loosely, from documentary photography, which trades immediacy for time spent examining an ongoing situation or process; for example, Train accident in Kalka Mail, AIDS in the Congo etc.

Photojournalism as a *"medium can be used in what are interpretively neutral ways to document life as it is. It can also be a vehicle for political, social and personal statements. In many cases, the lines between objectivity and subjectivity, fact and attitude, are impossible to disentangle"*.

As William Stapp has pointed out, several conditions had to be fulfilled before photojournalism became practicable: the photographic process had to be capable of capturing newsworthy events or their aftermath with sufficient realism and clarity; transmission had to be fast enough to bring pictures home while the subject matter was still topical; and a reproduction process was needed that could multiply images rapidly, on a large scale and preferably together with text.

There are two types of photojournalism. The first type is where an image is used to illustrate a story. Many feature journalists work closely with photographers and commission them to produce images that will be published with their articles. There is no limit to how many images used. This is usually the photo editor decision.

The second is where an image is used to tell a story without any words. One single image may be used or as many as ten images are often used in magazines. If you can write - do so; it will be an advantage when you submit some images.

History

The practice of illustrating news stories with photographs was made possible by printing and photography innovations that occurred between 1880 and 1897. While newsworthy events were photographed as early as the 1850s, printing presses could only publish from engravings until the 1880s. Early news photographs required that photos be re-interpreted by an engraver before they could be published.

The first photojournalist was Carol Szathmari (Romanian painter, lithographer, and photographer) who did pictures in the Crimean War (between Russia and Ottoman Empire, 1853 to 1856). His albums were sent to European royals houses. Just a few of his photographs survived. William Simpson of the Illustrated London News and Roger Fenton were published as engravings. Similarly, the American Civil War photographs of Mathew Brady were engraved before publication in Harper's Weekly. Because the public craved more realistic representations of news stories, it was common for newsworthy photographs to be exhibited in galleries or to be copied photographically in limited numbers.

On March 4, 1880, The Daily Graphic (New York) published the first halftone (rather than engraved) reproduction of a news photograph. Further innovations followed. In 1887, flash powder was invented, enabling journalists such as Jacob Riis to photograph informal subjects indoors, which led to the landmark work How the Other Half Lives. By 1897, it became possible to reproduce halftone photographs on printing presses running at full speed.

Despite these innovations, limitations remained, and many of the sensational newspaper and magazine stories in the period from 1897 to 1927, were illustrated with engravings. In 1904 the London Daily Mirror had become the world's first daily newspaper to be illustrated entirely with photographs. In 1921, the wire photo made it possible to transmit pictures almost as quickly as news itself could travel. However, it was not until development of the commercial 35mm Leica camera in 1925, and the first flash bulbs between 1927 and 1930 that all the elements were in place for a "golden age" of photojournalism.

From 1935 to 1942, the Farm Security Administration and its predecessor the Resettlement Administration were part of Franklin Roosevelt's New Deal, and were designed to address agricultural problems and rural poverty associated with the Great Depression. A special photographic section, headed by Roy Stryker, was intended merely to provide public relations for its programs, but instead produced what some consider one of the greatest collections of documentary photographs ever created in the U.S. Whether this effort can be called "photojournalism" is debatable, since the FSA photographers had more time and resources to create their work than most photojournalists usually have.

In the "golden age" of photojournalism (1930s - 1950s), some magazines (Picture Post (London), Paris Match (Paris), Arbeiter-Illustrierte-Zeitung (Berlin), Berliner Illustrierte Zeitung (Berlin), Life (USA), Look (USA), Sports Illustrated (USA)) and newspapers (The Daily Mirror (London), The New York Daily News (New York) built their huge readerships and reputations largely on their use of photography, and photographers such as Robert Capa, Alfred Eisenstaedt, Margaret Bourke-White and W. Eugene Smith became well known names.

Henri Cartier-Bresson is held by some to be the father of modern photojournalism, although this appellation has been applied to various other photographers, such as Erich Salomon, whose candid pictures of political figures were novel in the 1930s.

Soldier Tony Vaccaro is also recognized as one of the pre-eminent photographers of World War II. His images taken with the modest Argus C3 captured horrific moments in war, similar to Capa's soldier being shot. Capa himself was on Omaha Beach on D-Day and captured pivotal images of the conflict on that occasion. Vaccaro is also known for having developed his own images in soldier's helmets, and using chemicals found in the ruins of a camera store in 1944.

Until the 1980s, largest newspapers were printed with turn of the century "letterpress" technology using easily smudged oil-based ink, off-white, low quality "newsprint" paper, and coarse engraving screens. While letterpresses produced legible text, the photoengraving dots that formed pictures often bled or smeared and became fuzzy and indistinct. In this way, even when newspapers used photographs well a good crop, a respectable size murky reproduction often left readers re reading the caption to see what the photo was all about. The Wall Street Journal adopted stippled hedcuts in 1979 to publish portraits and avoid the limitations of letterpress printing. Not until the 1980s had a majority of newspapers switched to "offset" presses that reproduce photos with fidelity on better, whiter paper.

By contrast, '***Life***', one of America's most popular weekly magazines from 1936 through the early 1970s, was filled with photographs reproduced beautifully on oversize 11 14 inch pages, using fine engraving screens, high-quality inks, and glossy paper. *Life* often published a United Press International (UPI) or Associated Press (AP) photo that had been first reproduced in newspapers, but the quality magazine version appeared to be a different photo altogether. In large part because their pictures were clear enough to be appreciated, and because their name always appeared with their work, magazine photographers achieved near-celebrity status. *Life* became a standard by which the public judged photography, and many of today's photo books celebrate "photojournalism" as if it had been the exclusive province of near-celebrity magazine photographers.

The Best of Life (1973), for example, opens with a two-page (1960) group shot of 39 justly famous *Life* photographers. However, 300 pages later, photo credits reveal that scores of the photos among *Life's* "best" were taken by anonymous UPI and AP photographers. Thus even during the golden age, because of printing limitations and the UPI and AP syndication systems, many newspaper photographers labored in relative obscurity.

During the sixties there was a growing realization that the view that photography would cause living conditions to improve was naive. The menace of the Cold War had shown that in the Korean War (1950-1953) fighting could end in a stalemate but there was still a chance that good would overcome. The sense that America could change the world in a positive way was tarnished but still there. In the sixties and early seventies the Vietnam War (1961-1975) shook that ideal and the civil rights issues threw the nation into upheaval the sense that there was going to be a better world was thrown into question and the picture magazines found it difficult to exist as the world grew more cynical. Obtaining advertising revenue for the magazines became increasingly difficult as the trend moved to the more positive lifestyle magazines here pictures of famous people and domestic abundance could attract ad revenue.

In 1947 a few famous photographers founded the international photographic cooperative Magnum Photos. In 1989 Corbis Corporation and in 1993 Getty Images were founded. These powerful image libraries sell the rights to photographs and other still images. Since the late 1970s, photojournalism and documentary photography have increasingly been accorded a place in art galleries alongside fine art photography.

The Danish Union of Press Photographers (Pressefotografforbundet) was the first national organization for newspaper photographers in the world. It was founded in 1912 in Copenhagen, Denmark by six press photographers. Today it has over 800 members. The National Press Photographers Association (NPPA) was founded in 1946 in the U.S., and has about 10,000 members. Others around the world include the British Press Photographers Association (BPPA) founded in 1984, then re-launched in 2003, and now have around 450 members. Some other famous photographers' association are Hong Kong Press Photographers Association (1989), Northern Ireland Press Photographers Association (2000), Pressfotografernas Klubb (Sweden, 1930), and PK Pressefotografenes Klubb (Norway).

News organizations and journalism schools run many different awards for photojournalists. Since 1968, Pulitzer Prizes have been awarded for the following categories of photojournalism: 'Feature Photography', 'Spot News Photography'. Other awards are World Press Photo, Best of Photojournalism, and Pictures of the Year as well as the UK based The Press Photographer's Year.

Ethical and Legal Considerations

Photojournalism works within the same ethical approaches to objectivity that is applied by other journalists. What to shoot, how to frame and how to edit are constant considerations. Often, ethical conflicts can be mitigated or enhanced by the actions of a sub-editor or picture editor, who takes control of the images once they have been delivered to the news organization. The photojournalist often has no control as to how images are ultimately used. The emergence of digital photography offers whole new realms of opportunity for the manipulation, reproduction, and transmission of images. It has inevitably complicated many of the ethical issues involved.

Issues include photo manipulation what degree is acceptable. As for example, staged photos, photos of war, and false or misleading captioning may be cited. The 2006 Lebanon War photographs controversies is a notable example of some of these issue, and see photo manipulation: use in journalism for other examples. The U.S. National Press Photographers Association, and other professional organizations, maintains codes of ethics to specify approaches to these issues.

Major ethical issues are often inscribed with more or less success into law. Laws regarding photography can vary significantly from nation to nation. The legal situation is further complicated when one considers that photojournalism made in one country will often be published in many other countries.

The Impact of New Technologies

Smaller, lighter cameras greatly enhanced the role of the photojournalist. Since the 1960s, motor drives, electronic flash, auto-focus, better lenses and other camera enhancements have made picture taking easier. New digital cameras free photojournalists from the limitation of film roll length, as thousands of images can be stored on a single memory card.

Content remains the most important element of photojournalism, but the ability to extend deadlines with rapid gathering and editing of images has brought significant changes. As recently as 15 years ago, nearly 30 minutes were needed to scan and transmit a single color photograph from a remote location to a news office for printing. Now, equipped with a digital camera, a mobile phone and a laptop computer, a photojournalist can send a high-quality image in minutes, even seconds after an event occurs. Camera phones and portable satellite links increasingly allow for the mobile transmission of images from almost any point on the earth.

There is some concern by news photographers that the profession of photojournalism as it is known today could change to such a degree that it is unrecognizable as image-capturing technology naturally progresses. Citizen journalism and the increase in user contribution and submission of amateur photos to news sites are becoming more widespread. As early as the Crimean War in the mid-19th century, photographers were using the novel technology of the box camera to record images of British soldiers in the field. However, the widespread use of cameras as a way of reporting news did not come until the advent of smaller, more portable cameras that used the enlargeable film negative to record images. The introduction of the 35 mm Leica camera in the 1930s made it possible for photographers to move with the action, taking shots of events as they were unfolding.

The age of the citizen journalist and the attainment of news photos from amateur bystanders have contributed to the art of photojournalism. Paul Levinson attributes this shift to the Kodak camera, one of the first cheap and accessible photo technologies that "put a piece of visual reality into every person's potential grasp." The empowered news audience with the advent of the Internet sparked the creation of blogs, podcasts and online news, independent of the traditional outlets, and "for the first time in our history, the news increasingly is produced by companies outside journalism".

Terminology Related to Photojournalism and Photography

Photojournalism

Photojournalism is the craft of employing photographic storytelling to document life: it is universal and transcends cultural and language bounds.

Juxtaposition

Juxtaposition is the act of placing or positioning items in the image area of a photograph side by side or next to one another to illustrate some comparison.

Composition

Composition is the arranging of informational and/or artistic elements with a viewfinder so as to form a unified whole or image that imparts information or an idea.

Format

Format generally refers to the size of the negative produced by the camera. There are three basic formats: small (35mm), medium 2 1/4 or 4 5), and large (8 10 or Sheet Film). Photographers often use the term when referring to the shape of the image as in vertical, horizontal and square images. Of late, many photography department folk will use the term "file format", referring to the type of computer format in which the electronic image has been saved, i.e. TIFF, EPS, JPEG, etc.

Long Glass

Long glass refers to a Telephoto lens that magnifies the subjects in an image and covers a narrow angle of view of a scene. At a very basic level, these lenses create the enlarging effect by extending the lens-to-film distance within the camera.

Depth of Field

It is the apparent sharpness of a photograph considering the distance between the nearest and farthest elements that appear to maintain acceptable sharpness and clarity. It is depended on aperture, focal length and focused distance. It is always the smaller the aperture, the wider the lens and further the focused distance, giving a greater depth of field and vice versa.

Transmit or Uplink

Transmit or Uplink is the process of sending a photograph from one location to another via telephone lines, cellular circuitry or satellite delivery.

Throw

The throw is the distance between the subject and the camera position with a particular lens. Usually, photographers use this term when using very long lenses with focal lengths of 300 mm or higher.

EPD (Electronic Picture Desk)

Electronic Picture Desk is the most popular picture desk in the industry today is The Associated Press' computer processing unit (CPU) known as the Leaf Picture desk. It serves primarily as a receiving station for pictures transmitted from photo services, agencies, other publications, and photographers in remote locations. It also has a dual function as an electronic darkroom where a photograph can be cropped, processed, and output for usage in the final product.

Scanner

Scanner is a front end capture system, or device that scans the film or transparency and creates a digital version of the photograph. The electronic file can be saved in several computer formats.

Zoning

Zoning refers to presetting one focus to photograph action so that the area where the action is anticipated to take place will be in focus.

Wild Art

Stand-Alone, Feature Shot, Enterprise art are found moments that are intended to be realistic slice of everyday life.

Available

Available refers to the natural light or existing lighting conditions under which an image is made.

Bracketing

Bracketing is a common photographic approach whereby the photographer will take several frames of the same scene at different exposure settings. This is usually done with a difficult lighting situation.

Dodging and Burning

Dodging and Burning simply refers to the lightening and darkening of areas of the photograph by altering the amount of light hitting the print.

Pixel

Pixel is the fundamental picture element of a digital photograph. A single digital or electronic photograph is made up of thousands of pixels.

Strobe

Strobe is an electronic flash device that provides artificial, supplementary lighting.

Body

Many photographers refer to the camera this way. Basically, they are talking about the light-proof device with a diaphramed lens through which the image of an object can be focused and recorded on a photosensitive medium.

Dynamic Range

Dynamic Range refers to the varying gray values, or the spread of gray values, in an image that are possible reproduction. Am image with a high dynamic range suggests a wide number of gray levels and is normally associated with good contrast levels.

Fast Lens and Film

Every photographer wants speed, which refers to lenses that open to a large f-stop (aperture setting). The larger f-stops allow greater amounts of light into the lens. Also, faster film bases are very sensitive to light and allow the photographer to take pictures in very low light settings.

Bounce Card

Bounce Card is a small white card used to redirect the stream of light and soften the quality of light being emitted from a strobe or flash unit.

RIP (Raster Image Processor)

A Raster Image Processor is a production device where a digital file is converted into a raster image. An electronic beam draws a file, one line at a time, to a paint or display device, film recorder, or printer.

Aspect Ratio

Aspect Ratio is the proportion of an image size given in terms of the horizontal length vs. the vertical height. A 5 7 image has a ratio of 7:5.

Glitch or Artifact

Glitch or Artifact is something unnatural in the scanning process that impairs the image quality or causes unwanted distortion.

F-Stops

F-stops sometimes called f-numbers are nothing more than the numerical designations indicating the size of the aperture that governs the amount of light entering the lens. (F/2, f/5.6, f/16, etc.). The numbers indicate the mount of light allowed to pass through the lens barrel to the sensor. It is always the lower the number, the larger the aperture.

Compression

Compression is the reduction of the picture information to reduce the file size. It is accomplished through the sampling and averaging of the file's data. By doing so, you reduce the amount of transfer time during transmission as well as the amount of space needed for storing or archiving the files.

Seeing Light

Seeing Light refers to the ability of the photographer to see and capture the effects of light and shadow in the world around us.

Moments

Photographers seek to capture the natural and candid events of history. Their quest to capture "moments" comes from the famous French documentary photographer Henri Cartier-Bresson and his consistent ability to capture decisive moments in time without disrupting a particular situation.

Shooter

Shooter is the term which has long been associated with newspaper photographers. Today the term is rather controversial. Many journalists feel that the term does not acknowledge the valuable contributions of the photojournalist or news photographer.

Digital Camera

A digital camera records an image electronically as opposed to film. To truly get the most out of a digital camera one should have a computer with a sharp monitor. Direct printing is an option but is impractical and defeats much of the purpose of a digital camera.

Digital Photograph

It is simply a photograph that has been captured electronically on a hard drive (storage device) or converted from film to digital format through an electronic scanning process.

In-Camera Manipulation

In Camera Manipulation refers to certain measures and steps taken by the photographer to produce a desired effect, such as filers, lens choice, angles, and/ or multiple exposures.

Computer Enhancement

Computer enhancement refers to contrast adjustments, toning, color correcting dodging and burning to prepare an image or photograph for production.

Photo Illustration

Photo Illustration is a set-up photograph (usually in a studio with no digital alteration) that is illustrative in nature and is clearly out of the realm of reality. Traditionally, it is an approach used for fashion, food and product photographs.

Computer Image

Computer Image is a pure digital illustration, created on a computer to achieve a particular effect, which either uses photographs as its base material or is so photorealistic that it may be perceived as being real. The images are not real events that have occurred in history but rather are manifestations of ones imagination.

Lens

Lens is one or more the one pieces of optical glass or similar material designed to collect and transfer rays of light to form a sharp image on film, paper or a projection screen. In practical photography, compound lenses made of a number of elements of different types of glass are used. This enables the manufacturer to correct most of the faults (aberrations) found in simple lenses and provide images that are sharp across the whole picture.

Lens Speed

This typically is referring to the aperture size. A lens with a large maximum aperture such as f2.8 would be considered fast. A large aperture allows more light to reach the sensor of your digital camera, allowing for a faster shutter speed. The largest aperture (smallest F stop) at which a lens can be set. Lens speed is relative: a 400 mm lens with a maximum aperture of F/3.5 is considered extremely fast, while a 28mm F/3.5 lens is considered to be quite slow.

Aperture

This refers to the hole through the lens that allows light to travel to the camera sensor. The aperture is the opening formed by the blades of the iris or diaphragm in the lens, through which light passes to expose the film. Aperture size is usually given in f-numbers, the larger the number; the smaller the opening.Typically the aperture is controllable by stopping down the lens. This means that one can turn/ dial down the aperture ring on a lens to allow less light to reach the sensor.

Telephoto

A telephoto lens is a lens that brings objects that are physically distant in closer. A telephoto lens works much like a telescope or binoculars. A decent telephoto lens is practically a necessity for wildlife photography. Typically anything above 50 mm is considered a telephoto but most commonly lenses in the 100mm to 300mm range fit this description best.

Super Telephoto

The super telephoto lens range typically is considered 400mm and above and most commonly used in professional sports and wildlife photography.

Prime

A prime lens is any lens that only consists of a single focal length, e.g. 100 mm, 300 mm etc. Prime lenses used to be regarded as far sharper than a zoom lens. Modern zooms however have closed the gap somewhat.

Zoom Lens

Zoom Lens is any lens which can be adjusted to a wide range of focal lengths without a change in focus, thus an alternative for a number of individual lenses of various focal lengths. It is a difficult type of lens to design and manufacture, but very useful for the photographer who likes to travel light.

Macro/Micro

Macro lenses enable one to photograph small objects at a close range. Typically a macro lens will focus to 1:2 life sizes. This means the object will appear half the size it actually was when you complete your image. Most professional macro lenses will focus to 1:1, or full life size. Macro lenses are highly used in medical, botany and insect photography. The most common macro range is 50mm to 105mm.

Wide Angle

Wide Angle lenses are most commonly used for landscape and architecture photography. Wide angle lenses allow for a wider field of view. Any focal length below 50mm used to be considered a wide angle however modern wide angle photography typically begins at 28mm and wider.

Fisheye

A fisheye lens is lens with an extremely wide angle of view, typically with large amounts of distortion. This is a specialized tool with limited use.

Sharpness

Sharpness refers to the resolving power of a lens, the amount of fine detail the lens is able to clearly expose. In nature photography sharpness is often considered the true measure of a lens.

Sensor

A sensor is the electronic devise that actually records the image, as opposed to film, eg. CCD- Charged coupled device, a type of sensor, CMOS- complementary metal oxide semiconductor, another type of sensor. CMOS is often considered cheaper to produce and handles noise slightly better. Full Frame refers to a digital sensor that is equal in size to a 35mm film plane.

Noise

This refers to the digital interference one will see in an image at maximum resolution. It is similar to grain in film. The less noise in an image the higher quality the image will appear. Noise can be added as an effect as well. Higher ISO setting typically include more noise in an image.

ISO

This refers to the sensors sensitivity to light. A low number ISO such as 50-100 ISO will produce a higher quality image. A higher ISO setting such as 400-800 ISO will produce a faster shutter speed but at the cost of more noise in the image.

Shutter

This is the piece of material that opens and closes to allow the light to reach the sensor. A fast shutter speed is best for sharp images but a slow shutter speed can emphasize speed or movement. Shutter speed literally refers to how fast the shutter actuates.

White Balance

This typically is a setting that is manipulated per the type of lighting one is shooting in. Sun, cloudy, flash and florescent lighting are common white balance settings. Many professional digital cameras allow for fine tuning and custom white balance settings. The automatic white balance on new cameras is typically fairly accurate. Many photographers prefer the effects of a certain white balance setting regardless of the light they are shooting in.

Flash

This is artificial lighting. Usually built into a camera or attached via a hotshoe. Speed light, the term used for on or off camera flash units and hotshoe is the electronic coupling on top of the camera to connect flash or flash cords.

APO

Apochromatic glass is typically considered to be better at displaying accurate colors. APO glass is also typically more expensive than standard glass lenses.

AF

Auto focus lenses are able to focus via an in camera drive or a separate lens drive.

MF

Manual focus lenses much be focused by hand, usually by turning part of the lens barrel.

IF

Internal Focus means the lens barrel will not rotate while focusing. IF is especially important when using filters such as a polarizer.

Aberration

Aberration is the inability of a lens to produce a true image, particularly at the edge of a photograph. It is usually, the more expensive the lens, the better its optical quality and the fewer aberrations.

Angle of View

Angle of View is the area of a scene that a lens can cover. The focal length of the lens determines the angle of view. A wide-angle (short focal length) lens includes more of a scene than a standard (normal focal length) lens or telephoto (long focal length) lens. Angle of view is basically the angle at which light rays can pass through the lens to produce an image on the film.

Aspherical Lens

Aspherical Lens is a lens with a curved, non-spherical surface. Used to reduce aberrations and achieve a more compact lens size. With a spherical lens, rays traveling from the lens periphery create the image before the ideal focal point and give a blurred image centre. With an aspherical lens, even the rays traveling from the lens periphery converge at the ideal focal point, thus producing a sharp image.

Chromatic Aberration

Chromatic Aberration is the inability of a lens to bring all light wavelengths (particularly red & blue) into the same plane of focus, thus causing overall blur. Usually found in regular large-aperture telephoto and super-telephoto lenses. Not improved by reducing aperture size. It can be corrected with low dispersion (ED, LD SD) glass.

Colour Temperature

Colour Temperature is a method of expressing the colour content and quality of light and measured in Kelvin (K). "Photographic daylight" has a colour temperature of about 5500K. Photographic tungsten lights have colour temperatures of 3200K to 3400K depending on their construction.

Electronic flash

Electronic Flash is designed to provide light where the lighting on the scene is insufficient. Electronic flash requires high voltage, usually obtained through batteries and a voltage multiplying circuit which discharge a brief, intensive burst. It is generally considered to have the same photographic effect as daylight. Modern flash units have multiple TTL exposure control functions and auto focus control.

Flash Sync Speed

Exposure time with a focal-plane shutter is measured from the moment the first curtain is released until the moment the second curtain is released. The instant the first curtain closes, the electrical contacts for X sync close and instantly fire the flash.

Focal Length

Focal length is the distance from the film to the optical centre of the lens when the lens is focused on infinity. Focal length on most adjustable cameras is marked in millimeters on the lens mount. On 35mm-format cameras, lenses with a focal length of 50mm are called normal or standard lenses. Lenses of 35mm or less are called wide angle lenses and lenses of 85mm or more are called telephoto lenses. Lenses which allow varying focal lengths without changing focus are called zoom lenses.

Perspective

Perspective is a two-dimensional representation of a three-dimensional scene. In photography this can be achieved by viewing 3-D objects from an angle rather than head-on. A photograph is also given perspective if there are objects in the foreground, middle distance and background, giving the whole scene "depth".

Single-Lens-Reflex (SLR) Camera

In Single-Lens-Reflex (SLR) Camera, light enters the camera through the lens is reflected up by a mirror behind the lens onto a ground glass screen above. This screen is viewed through the viewfinder and a glass pentaprism which turns the image the correct way up. Other camera functions such as light metering and flash control also operate through the camera lens.

Principles for Photojournalism

Ross Collins of North Dakota State University has given ten principles of photojournalism, which should be followed by photojournalists. They are as follows;

1. Get in Close

The photos are often published on low-quality newsprint for a jaded audience, which are easily distracted and detail gets lost. Photos strong enough to break through the media competition focus on a single, frame-filling centre of interest. It should not be of more than two or three people and no more than a few objects and it should be a clean and having contrasting background. Sometimes photojournalists call this the "poster effect." This is why telephoto or zoom lenses are the workhorses of photojournalism. Many photographers don't even use a normal (50 mm) lens. "If you cannot get in close, crop ruthlessly and if your pictures are not good enough then you are not close enough.", which is stated by Robert Capa, one of the top war photographers, who was killed by land mine while trying to get in close.

2. Find Unusual Angles

If the photojournalist is sitting or standing like everyone else, then he is probably going to take back some blah photos. Good photojournalists are always moving. They kneel, climb on a chair, crouch, lie down and shoot down a stairway.

Corollary to this rule: If forced to take photos of a speaker at a podium, never sit in the chairs with the rest of the audience. Do not think you will be too noticeable. Everyone already knows you're taking pictures. If you're too shy to move around in front of an audience, become a copy editor.

3. Get Identities

All subjects easily identifiable must be named. Journalism is about people, not about abstractions. Most editors will reject photos without identities, unless they are used as generic illustrations. Beginners are intimidated by the thought of walking up to strangers, but it is not really too hard.

4. Burn Pixels

Move in, move out, move high, move low, and keep on clicking. You will have more to choose from, and you will more likely have a strong image. Unlike studio photographers, photojournalists have to nab action as it is unfolding, so at least to some extent are forced to trust on luck to bring back good pictures. The old general rule used to be to shoot one roll (24 ex.) minimum at any event you're asked to cover, even a speaker at a podium. You can interpret this in the digital age. Half a football game should bag you 75 photos at least. How are you going to have any choice at all if you return to the office with a mere four shots? Film is pretty cheap and pixels are free!

5. Go Beyond the Cliché

It is always better for photojournalists to avoid the cliché. The same kind of common photographs related to any topic do not create interests among the clients. Try to capture some special moments and uncommon photographs, which have higher demand and greater acceptability to the viewers.

6. Avoid Obvious Posing

Sometimes you have to ask people to stand at someplace or do something, but that still does not mean they have to grin into the camera. Try to make the scene appear natural, as if the subject were involved in something and the photographer just happened to come by. Photojournalism aims to cover people doing things, not people posing. That is for the studio photographers.

7. Add Light, but Don't Make It Obvious

Available light is ideal, but usually photojournalists face some situations, where light conditions are insufficient. Especially when you are shooting in colour, you need to control the colour balance of grayish green florescent or difficult combinations of artificial and natural light. In some cases, light comes from ugly angles (such as nearly every standard classroom or office), or is so weak you cannot even get close to stopping the action.

The normal approach is to carry a portable electronic flash, but never to merely attach it to the hot shoe and blast away. That looks too artificial. Aim your electronic flash straight up. Tape a piece of white cardboard at an angle so the light bounces up and off the cardboard. Adjust for your available light exposure, perhaps underexposing a stop. The added flash will fill in the ugly shadows, but still maintain a feeling of natural light.

This is harder if your light is primarily florescent, as your bluish flash will clash with the greenish ambient light, but you should be able to adjust most colour cast problems fairly easily in Photoshop. When all else fails, at least take your flash off your camera, hold it a few inches away, and shoot directly. As an added note, good photojournalists carefully analyze the light falling on their subject, and find an angle which best enhances the scene. For instance, if a subject is bending over, the face may be in shadow. Wait until he stands. If a subject is seated next to a window, wait until she turns toward this attractive light source.

8. Focus Faithfully, Stay Steady

You just cannot use a blurry photo in a publication: any slight fuzziness will be enhanced by poor quality paper and fast reproduction. Most fuzzy photos can be blamed on camera movement. If you are naturally jittery, get help: buy a mini tripod and lean it on your chest, or buy a monopod and lean it on the ground. In a pinch, lean against a tree or wall. Rule of thumb is that any shutter speed more than one stops lower than the size of your lens will produce a fuzzy picture, unless you use a mechanical aid. This means if you are shooting at normal 50 mm or equivalent ("normal perspective"), you cannot hand-hold anything slower than 1/30 second. If you are shooting with a zoom set at 200 mm, 1/125 is the limit. Photoshop's blur filter can sharpen pixels, but cannot do so much with pictures that were originally blurry.

9. Dump Poorly Exposed Photographs

You have got a great shot, but the guy's face is in shadow? Got a great expression on that cute kid, but a telephone pole is growing out of her head? Do not submit poorly exposed photographs for publication, no matter how much you like the image. On the other hand, be careful to expose correctly. Automatic cameras are usually smarter than we are, but not always. Study the principles of proper exposure. It is still hard to fix up poorly exposed images, even with Photoshop software.

10. Have Guts

Walk up to ten people on the street, snap their picture, ask for their name and write it down. Get a floor pass to a rock concert. Walk as close to the sports action as you dare get a press pass if you need one. Look for interesting expressions after the posed shots are done. Get out there and shoot. Photojournalists are to a person not to be counted among the timid. But, at the same time, it is essential to know the limit of intimidation and where to stop.

Qualities of a Photojournalists

Photojournalism can be a very competitive career field. Job opportunities with newspapers and magazines are highly sought after and freelancing careers can be difficult to establish. Many photojournalists become extremely successful in the field and ten top qualities that set them apart.

Ambition

Great photojournalists are very ambitious and seek out opportunities for work, following the industry closely and pursuing opportunities as they arise.

An Eye for Detail

Great photojournalists pay close attention to detail and capture what is important. Conversely, they are also able to edit out small details that might take away from a subject in a photograph.

Business Aptitude

Great photojournalists have keen business skills and can manage business aspects of a freelancing career, including finances and marketing.

Creativity

The creative "eye" is an integral part of photography. Great photojournalists are able to intuitively judge what makes for a great photograph and what should be done with lighting, composition, and focus to make that vision a reality.

Familiarity with Legal Documents

Great photojournalists are familiar with contracts, copyright laws, and other legal issues that may present themselves.

Hand-Eye Coordination

Great photojournalists have excellent hand-eye coordination to manipulate the equipment involved in capturing top-notch photos. No shaking, quick lens replacement etc.

Networking Skills

Great photojournalists have good networking skills and can identify those who are important to know in the industry. They are able to approach important players and keep them informed of the work they do.

Rapport Building Ability

Great photojournalists have excellent rapport building ability and can work well with clients and editors. The ability to put people at ease while they are being photographed is also the key for portrait photographers.

Self-Marketing Skills

Successful photojournalists have portfolios that demonstrate the quality and variety of their work. They are also able to adequately promote their services to find new customers.

Technical Skills

Great photojournalists have excellent technical skills and can easily handle photography equipment, such as digital cameras and lighting, as well as software programs to edit photographs.

Eminent Photojournalists

Lucien Aigner

Lucien Aigner was born at Nove Zamky, Slovakia in 1901 and died at Great Barrington, Massachusetts in 1999. Lucien Aigner studied law at the University of Budapest, having previously studied theatre in Berlin and worked as an assistant cameraman to Stefan Lorant (a lifelong friend, who later became one of the most influential picture editors of the early 20th century). In 1924 he became a reporter for the Hungarian newspaper group Az Est. A keen photographer (he was given a box Brownie camera for his 9th birthday), Aigner started to illustrate his articles with his own photographs. In 1925 he moved to Paris to act as manager for a photographer, a short lived association and continued to work as a correspondent for Az Est. Due to his poor French, Aigner bought a Leica and concentrated more on photography. He soon became known for his informal 'grabbed' shots of political figures and published picture stories in the leading illustrated magazines in France, Germany and England. After a trip to America in 1936, Aigner became the Life magazine correspondent in Paris. He moved to America in 1939 and settled in New York. In 1947 he got a job in radio broadcasting, in the Hungarian section of Voice of America, but was forced out of this post in the 1950s during the McCarthy era. He moved to New England and set up a portrait studio. The discovery of the negatives of his European photographs in an old suitcase in 1970 prompted renewed interest in Aigner's life and work.

Zinn Arthur

Zinn Arthur was born in Ukraine in 1912 and died in Los Angeles in 2003. Zinn Arthur was born as Abrasha Choosidman, but when his family arrived as immigrants at Ellis Island, New York, in 1921, they changed their surname to Zinberg. Arthur's first career was in music. He led a big band in high school, where he shortened his name to 'Arty Zinn' and later changing it to Zinn Arthur. During the Second World War he toured US army bases around the world with Irving Berlin's show 'This is the Army'. After the war, Arthur turned to photography, taking his first pictures of musicians such as Nat King Cole, Duke Ellington and Louis Armstrong. He became a celebrity photographer, specializing in TV personalities and then in film. He worked on a total of 66 film sets and photographed actors including Elizabeth Taylor, Audrey Hepburn, Humphrey Bogart and Burt Lancaster for the covers of magazines such as ***Life, Look, Esquire*** and ***Harper's Bazaar***. Finally, Arthur had a third career as a restaurateur, with establishments in Long Island, New York, and south Florida. The restaurant walls were lined with pictures he had taken during his years as a photographer.

Brian Brake

Brian Brake was born in Wellington, New Zealand in 1927 and died in Auckland in 1988. Brian Brake started his career in 1945 as an assistant in a Wellington portrait studio. He joined the New Zealand National Film Unit as a cameraman in 1949 and was also involved in directing films, including the prize-winning Snows of Aorangi (1950). In 1953 he moved to Europe to pursue a career in photojournalism. Through the invitation of Henri Cartier-Bresson and Ernst Haas, Brake joined Magnum in 1954. Specializing in Asia, he became one of the most famous documentary photographers of the late 1950s and the 1960s. In 1957 Brake was one of three Magnum photographers permitted to work in China and in 1959 he became the only western photographer to cover the tenth anniversary celebrations of Chinese Communism. Brake started to publish colour picture stories, the most influential of which was 'Monsoon', first appeared in ***Life*** magazine in September 1960. His use of imaginative sequencing helped to establish Brake as one of the masters of colour photojournalism. Having worked mainly for *Life* magazine during the 1960s, Brake settled briefly in Hong Kong, where he worked in documentary film making, before returning to New Zealand in 1976. In the 1980s he helped establish the New Zealand Centre for Photography. The Brian Brake collection is held at the Museum of New Zealand Te Papa Tongarewa.

Ernest Cole

Ernest Cole was born at Eersterust, Pretoria in 1940, died in New York in 1990. Ernest Cole was the first photographer to expose the conditions of life in South Africa under the apartheid regime. As a black man he was defined as an 'unskilled labourer', so his entry into photojournalism was as a sweeper and messenger at *Zonk* magazine. In 1958 he joined the magazine Drum as a design and production assistant and registered for a correspondence course with the New York Institute of Photography. Encouraged by his tutors, he started a project to record apartheid in South Africa and finally got a job as a photographer for the newspaper Bantu World, before becoming South Africa's first freelance black photojournalist in the early 1960s. In 1966 Cole got himself reclassified as 'coloured', which meant that he was able to leave country. He travelled to France and England before arriving in New York with the prints and layout sheets of his apartheid project in September 1966. Magnum Photos organized to publish the series in book form. *House of Bondage* – the book, came out the following year but was banned in South Africa, forcing Cole into exile. After a period of working for Magnum, he moved to Sweden to take up film making, though his pictures were still published extensively, often in anti-apartheid publications.

Gisèle Freund

Gisèle Freund was born in Berlin in 1908 and died in Paris in 2000. Gisèle Freund studied sociology and history of art at the University of Freiburg before moving to Frankfurt in 1931 to study with the famous sociologists Karl Mannheim and Norbert Elias. Her father had given her a Leica and Elias noting the ever-present camera suggested that she study 19th-century photography. Freund was also active in socialist circles and in 1933 she was forced to flee to Paris. There she continued her studies, completing what was to be the first PhD thesis ever written on photography, and started to take photographs for a living. She became known for her portraits of writers (from 1938 often taken with colour film) and published reportage in *Life, Picture Post, Vu* and *Paris Match* magazines. On the eve of the German occupation of Paris, Freund a Jew and an anti fascist was again forced into exile, first to the south of France and then to Argentina. She became a member of Magnum in 1947 and covered Latin America for the agency until 1954. She returned to Paris and continued her career as a writer, photographic reporter and photographer of literary figures. In 1974 her PhD thesis was translated into English and published as the influential book *Photography and Society*.

Ernst Haas

Ernst Haas was born in Vienna in 1921 and died in New York in 1986. Ernst Haas bought his first camera in 1946 on the Viennese black market, exchanging the 10 kilograms of margarine he had been given for his 25th birthday for a 35 mm Rolleiflex. His medical studies had been interrupted during the war, so he had been working on and off in a photographers' studio. His first published photo-story, *Homecoming Prisoners of War,* marked a turning point in his career. It was taken with his new camera in 1946 and published in Heute magazine in 1949. When this piece was taken up by Life magazine, Haas became the first photographer to be asked to join Magnum, an invitation that he accepted. The following year Haas started to experiment with colour film. He played a pivotal role in the development of colour photography and in 1953 *Life* magazine devoted an unprecedented 24 pages to his first published colour essay, 'Shots of a Magic City', taken in New York. Haas went on to produce colour essays for magazines during the 1950s and 1960s, including features on Paris, Venice and England. From the mid 1950s he also started to investigate the semi-abstract depiction of movement. In 1958 readers of Popular Photography voted Haas one of the 'World's Ten Greatest Photographers'.

Philippe Halsman

Philippe Halsman was born at Riga, Latvia in 1906 and died in New York in 1979. Philippe Halsman studied engineering in Dresden before moving to Paris to open a photographic portrait studio in 1932. There he photographed public figures including politicians, intellectuals and actors. When Paris fell to the Nazi forces in 1940, Halsman fled to America. He received his first commission for *Life* magazine in 1942 and went on to produce over one hundred magazine covers between 1942 and 1971. Halsman's sitters included film stars and celebrities such as Elizabeth Taylor and Marilyn Monroe and artists such as Henri Matisse, Salvador Dali and Georgia O'Keefe. He stated that the 'fascination with the human face has never left me every face I see seems to hide and sometimes, fleetingly, to reveal the mystery of another human being capturing this revelation became the goal and passion of my life'. His work is, at times, saturated with humour. In the 1950s he produced a series of 'jump' photographs, capturing his sitters including figures such Richard Nixon, vice-president at the time mid air. Halsman claimed that these photographs, usually taken at the end of a portrait sitting, revealed the real character of the celebrity.

Frank Horvat

Frank Horvat was born at Opatija, Croatia in 1928. When he was 15, Frank Horvat swapped his stamp collection for a 35 mm Retinamat camera. After studying art at the Accademia di Brera in Milan and working in an advertising firm, he started his career as a photographer in the late 1940s, working freelance for Italian magazines. He then had a brief stay in London, working for Life and Picture Post, before moving to Paris in 1955, where he worked as both a fashion photographer and photojournalist (he was an associate member of Magnum from 1958 to 1961). In the 1950s and 1960s his photographs were published in Jardin des modes, Elle, Glamour, Vogue and Harper's Bazaar. Horvat has travelled extensively throughout his career. In 1962, for example, the German magazine Revue commissioned him to produce a series of picture essays from cities across the globe. In recent years he has focused more on book projects and personal work. In a recent piece, entitled "A Daily Report", 1999, Horvat took at least one photograph every day for the last year of the millennium.

Mary Ellen Mark

Mary Ellen Mark was born in Philadelphia in 1940. After studying painting and art history and gaining an MA in photojournalism from the University of

Pennsylvania, Mary Ellen Mark started her career as a freelance photojournalist in the mid 1960s. In 1965 she was awarded a Fulbright Scholarship to photograph in Turkey. There Mark established her signature style of combining a documentary approach with a fascination with the bizarre. She also established one of the key themes in her work: an interest in children acting like adults. Returning to America in 1967, Mark moved to New York and worked as a photojournalist, publishing photo-stories in magazines such as the *New York Times, Evergreen* and *Life*. Mark also worked on film sets, taking production stills on films including *Catch 22* and *One Flew over the Cuckoo's Nest*. Mark's first solo exhibition, Ward 81, was a result of her work on the latter, when she returned to the high security women's mental hospital where it had been filmed. An interest in people, especially women, on the edge of society is apparent throughout her work. Her projects include a series on prostitutes in Bombay's Falkland Road (1978), street kids in Seattle (1983) and a study of Indian travelling circuses (1989). Mark continues to publish and exhibit her work to great acclaim and in 2001 was awarded the Cornell Capa Infinity Award from the International Center for Photography, New York.

David Seymour

David Seymour was born in Warsaw in 1911 and died at Suez in 1956. With the intention of entering his family's printing and publishing business, David Seymour studied at the Academy of Graphic and Book Arts in Leipzig, Germany, from 1929 to 1931. Due to worsening political situation in Germany, he moved to Paris to study chemistry in 1932. Soon economic difficulties in Poland meant that he had to supplement his allowance, so he took up photography to earn a living. Adopting the nickname 'Chim' (a French phonetic abbreviation of his Polish-Jewish surname Szymin), Seymour worked as a freelance photographer, publishing picture stories in Regards, the Popular Front illustrated magazine, from 1934. From 1936 to 1938 he covered the Spanish Civil War, before moving on assignment to Mexico for the newly created Paris Match magazine. When the Second World War broke out, he changed his surname to Seymour and moved to the USA, where he worked in photo-reconnaissance for the US army. In 1947 Seymour co-founded the Magnum photographer's agency with Henri Cartier-Bresson and Robert Capa (whom he had met in Paris in the 1930s), George Rodger and William Vandivert. He returned to Europe in 1948, commissioned by UNICEF to take pictures of displaced children, a project that helped to establish his reputation. Seymour worked as Magnum's representative in Europe and Israel during the 1950s but was killed while reporting the Suez crisis of 1956.

Erich Salomon

Erich Salomon was born in Berlin in 1886 and died at Auschwitz in 1944. Erich Salomon was a key figure in the development of modern photojournalism. He graduated in law in 1913 before being called up for military service. When war ended he returned to Berlin and in 1925 he got a job in the publicity department of the publishers Ullstein. He bought a large-format press camera before acquiring a more compact Ermanox in 1927. After publishing his first press photograph in the Berliner Illustrierte Zeitung in 1928, Salomon set himself up as a freelance photographer and journalist. He travelled widely in Europe and America and soon became known for his off-guard pictures of politicians, sporting events and celebrities. Salomon went to such great lengths to disguise his camera that the editor of Graphic coined the term 'candid photography' to describe his clandestine technique. In 1931 he published a book of his work entitled Famous Contemporaries in Unguarded Moments and in 1935 held an exhibition at the Royal Photographic Society in London. However, the rise of the Nazis meant that from 1933 Salomon, who was Jewish, was no longer able to publish in German magazines, so he moved with his family to The Hague in the Netherlands. He was discovered by the Nazi authorities in 1944 and, with his wife and son Dirk, deported to Theresienstadt in Czechoslovakia and then on to Auschwitz.

Nicolas Tikhomiroff

Nicolas Tikhomiroff was born in Paris in 1927. After serving in the French army, which he joined in 1944 after the liberation of Paris, Nicolas Tikhomiroff began his photographic career working in the darkroom of a fashion photographer. He soon began publishing his own work in magazines such as Marie France and from the 1950s started working with Magnum Photos. Assignments took him to the Soviet Union, South East Asia and Africa, where he spent long periods covering the wars in Vietnam, Cambodia, Laos and Algeria. Tikhomiroff later worked in the film industry, documenting the behind-the-scenes activities of directors such as Federico Fellini, Lucchino Visconti and Orson Welles. He was also active in fashion and advertising photography. Tikhomiroff retired from professional activities in 1987 and now spends his time working on personal projects in France. Although he was never a full member of the agency, his earlier work is still distributed by Magnum Photos.

Henri Cartier-Bresson (1908 – 2004)

Henri Cartier Bresson was born in France, was both an artist and a journalist. He abandoned his studies of literature and painting after he saw the early sports photographs in illustrated magazines. When he began his career in photography, the bulky press photographer's camera that used large, single sheets of film, one sheet per image, was being replaced by the new lightweight 35 mm Leica camera. Later he spent three years in a German prisoner of war camp during World War II. Cartier-Bresson is the father of modern photojournalism. He was influenced by French surrealism, an art movement that found significance in the meanings that lie beneath the surface of everyday life. The context of a gesture, a meeting or a setting in a photograph could convey great beauty or emotion. From this he developed the style of candid "street photography," which is still in use. Cartier Bresson met and photographed the world's leading artists, writers, and politicians. Many familiar faces appear among his works: the movie director John Huston, the writer Truman Capote, novelist William Faulkner, fashion designer Coco Chanel, sculptor Alexander Calder, and writer Carson McCullers. Pictured is philosopher Jean-Paul Sartre, left, with architect Jean Pouillon on the Pont des Arts in 1946. He received international acclaim for his coverage of Gandhi's funeral in India in 1948. He is most famous for his concept of the decisive moment. "There is nothing in this world that does not have a decisive moment," he said. "Photography is the simultaneous recognition, in a fraction of a second, of the significance of an event as well as of a precise organization of forms that give that event its proper expression. In photography, the smallest thing can be a great subject. The little human detail can become a leitmotif."

Eddie Adams

Eddie Adams is the recipient of the Pulitzer award and over 500 prizes to his credit, it was the Vietnam War that saw him get recognition for his best-known photograph which shocked people around the globe (clicked on Saigon street). Adams was a known figure for his portraits of celebrities, politicians and wars. He believed the particular scene could always be manipulated to change the actual truth as seen by the photographer.

Felice Beato

Although his origins have created confusion, Felice Beato was one of the first famous war photographers. His inclination to travel saw him capture many moods and moments of the people from different lands. He went from India to Japan and even China. Felice captured the Indian rebellion of 1857 and the second Opium War on film. His powerful and long-lasting images influenced other photographers as well.

Famous Indian Photographers and Photojournalists

Photography arrived in India almost within two years of its discovery. As India was being ruled by the British, British photographers started taking pictures of the country, its scenery and monuments. When the first war of Independence happened in 1857, it was one of the first incidents of war photography in the world and you may have seen some photographs in magazines. Later on when the camera became smaller, photojournalism gained greater popularity. Many political developments were also taking place in India especially around the freedom movement. All this gave a lot of chance to photojournalists to take pictures.

Raja Deen Dayal

Raja Deen Dayal was one of the first notable Indian photojournalists. He was a court photographer in India during the rule of the sixth Nizam of Hyderabad. As he was the only native photographer, he has left behind a very impressive record of British India.

Sunil Janah

Among the famous photojournalists who worked then was Sunil Janah. A political activist and journalist, Sunil Janah began to photograph while writing assignments for his newspaper. At the time of India's independence, Janah photographed the significant events in the country and made a record of the transition from British rule to independent India. His photographs of India's partition, its people; specially the tribals as well as pictures of industries and temple structures are very famous. Photographs of Nehru and Gandhi by Sunil Janah are now seen by us everywhere.

Homai Vyarawalla

There is another name which needs a special mention here, also because in a profession dominated by men, she was the first woman photojournalist. She is Homai Vyarawalla. Her work was first published in 1938 in the Bombay Chronicle, and later in other major publications of those times. She also worked for the Illustrated Weekly of India and during World War II covered every aspect of wartime activities in India. Her documentation of the events of the freedom movement are significant. She remained a freelance photographer until 1970 and was highly respected amongst all photojournalists.

Raghubir Singh

Boasting of international standards, Raghubir Singh is one of the finest documentary photographers of the twentieth century. However, the beginning of this ace artist was a very humble one. Born in 1942, in a Thakur family, he received education from the Hindu college in Delhi. It was in his college years that Raghubir took to photography. Later, this talented photographer clicked for *Life, The New York Times* and some British publications. He was awarded Padma Shri in 1983 and was also the proud recipient of Maharaja Sawai Ram Singh Award.

Raghu Rai

Amongst the photojournalists after Independence, the name of Raghu Rai is most famous. Rai's photographs are still seen as he continues to do photography. Raghu Rai started his career in photojournalism in the 1960s in Delhi and worked for important national newspapers like the *Hindustan Times* and *Statesman*. Later he became the Chief Photographer for 'India Today' where he worked for a long time. Now Mr. Rai works as an independent photographer and his work is truly respected all over the world. His photographs of famous personalities such as Indira Gandhi and Mother Teresa are very well known. Apart from this Raghu Rai has covered all important events that have taken place in the second half of the last century such as the Bhopal gas tragedy and the Bangladesh war. He has brought out books on various subjects such as Delhi, Taj Mahal, Sikhs, Benaras etc. His early pictures are mostly in black and white but later he has photographed in colour. They all have a beautiful quality about them.

Anish Bhaseen

An ace Indian photographer, excelling in almost all the genres of photography, Anish Bhaseen is a name to reckon with, in the world of photography. His photographs have not only been admired in India, but have gained world recognition and are widely acclaimed in UK, USA and Canada. This gifted photographer is a consultant to the Photography and Digital Imaging Industry and also actively takes part in many photographic workshops and seminars in India.

Prashant Panjiar

Prashant Panjiar is another successful photojournalist of the present times. Born in Kolkata, he is a self taught photographer who has worked for many magazines in Delhi. His most successful career was with the 'Outlook' magazine as its chief photographer and associate editor. He is a founding member of this magazine and through his photographs, he gave it a popular appeal among people and now Outlook has become a leading news magazine in India. Mentioned above are just a few names of important photojournalists. Apart from them there have been so many who have made important contributions to the field of photojournalism by their work in the print media all over the world. With the improvement in printing technology, newspapers have started using more photos and in colour. This has given rise to a greater demand for trained people in the field of photojournalism.

Anand Sharan

One of the leading photographers in India, Anand Sharan made his presence felt strongly in the world forum in the year 2006, when his work was commended by the famous Saatchi Gallery in London. The Gallery today exhibits the photographs of this great artist. Based in Bangalore, Anand Sharan is one of the best known fashion photographers of India.

Vikram Bawa

Fondly known as the 'Master of Gimmicks', Vikram Bawa is an internationally acclaimed, renowned Indian fashion and landscape photographer. He is the first photographer from the Asian subcontinent to venture into (3) three dimensional photography. His photographs speak a story of their own and flaunt an astounding quality, clarity and crispness.

Camera and its type

Pinhole Camera

A pinhole camera is a simple camera without a lens and with a single small aperture, effectively a light-proof box with a small hole in one side. Light from a scene passes through this single point and projects an inverted image on the opposite side of the box. The human eye in bright light acts similarly, as do cameras using small apertures. Up to a certain point, the smaller the hole, the sharper the image, but the dimmer the projected image. Optimally, the size of the aperture should be 1/100 or less of the distance between it and the projected image. Because a pinhole camera requires a lengthy exposure, its shutter may be manually operated, as with a flap of light-proof material to cover and uncover the pinhole. Typical exposures range from 5 seconds to several hours. A common use of the pinhole camera is to capture the movement of the sun over a long period of time. This type of photography is called Solargraphy. The image may be projected onto a translucent screen for real-time viewing (popular for observing solar eclipses; see also camera obscura), or can expose photographic film or a charge coupled device (CCD). Pinhole cameras with CCDs are often used for surveillance because they are difficult to detect.

History

As far back as the 4th century BC, Greeks such as Aristotle and Euclid wrote on naturally-occurring rudimentary pinhole cameras. For example, light may travel through the slits of wicker baskets or the crossing of tree leaves. (The circular dapples on a forest floor, actually pinhole images of the sun, can be seen to have a bite taken out of them during partial solar eclipses opposite to the position of the moon's actual occultation of the sun because of the inverting effect of pinhole lenses.)

The 10th-century Ibn al-Haytham (Alhazen) published this idea in the Book of Optics in 1021 AD. He improved on the camera after realizing that the smaller the pinhole, the sharper the image (though the less light). He provides the first clear description for construction of a camera obscura (Lat. dark chamber). As a side benefit of his invention, he was credited with being the first man to shift physics from a philosophical to an experimental basis.

In the 5th century BC, the Mohist philosopher Mo Jing in ancient China mentioned the effect of an inverted image forming through a pinhole. The image of an inverted Chinese pagoda is mentioned in Duan Chengshi's (d. 863) book Miscellaneous Morsels from Youyang written during the Tang Dynasty (618-907). Along with experimenting with the pinhole camera and the burning mirror of the ancient Mohists, the Song Dynasty (960-1279 AD) Chinese scientist Shen Kuo (1031-1095) experimented with camera obscura and was the first to establish geometrical and quantitative attributes for it.

In the 13th century AD, Robert Grosseteste and Roger Bacon commented on the pinhole camera. Between 1000 and 1600, men such as Ibn al-Haytham, Gemma Frisius, and Giambattista della Porta wrote on the pinhole camera, explaining why the images are upside down. Pinhole devices provide safety for the eyes when viewing solar eclipses because the event is observed indirectly, the diminished intensity of the pinhole image being harmless compared with the full glare of the Sun itself.

Around 1600 AD, Giambattista della Porta added a lens to the pinhole camera. It was not until 1850 AD that a Scottish scientist by the name of Sir David Brewster actually took the first photograph with a pinhole camera. Up until recently it was believed that Brewster himself coined the term "Pinhole" in "The Stereoscope". The earliest reference to the term "Pinhole" has been traced back to almost a century before Brewster to James Ferguson's Lectures on select Subjects. Sir William Crookes and William de Wiveleslie Abney were other early photographers to try the pinhole technique.

Selection of Pinhole Size

Within limits, a smaller pinhole (with a thinner surface that the hole goes through) will result in sharper image resolution because the projected circle of confusion at the image plane is practically the same size as the pinhole. An extremely small hole, however, can produce significant diffraction effects and a less clear image due to the wave properties of light. Additionally, vignetting occurs as the diameter of the hole approaches the thickness of the material in which it is punched, because the sides of the hole obstruct the light entering at anything other than 90 degrees.

The best pinhole is perfectly round (since irregularities cause higher-order diffraction effects), and in an extremely thin piece of material. Industrially produced pinholes benefit from laser etching, but a hobbyist can still produce pinholes of sufficiently high quality for photographic work. One method is to start with a sheet of brass shim or metal reclaimed from an aluminum drinks can or tin foil/aluminum foil, use fine sand paper to reduce the thickness of the centre of the material to the minimum, before carefully creating a pinhole with a suitably sized needle.

A method of calculating the optimal pinhole diameter was first attempted by Jozef Petzval. The formula used today was evolved by Lord Rayleigh:

$$d = 1.9\sqrt{f\lambda}$$

Where, d is pinhole diameter, f is focal length (distance from pinhole to image plane) and λ is the wavelength of light.

For standard black-and-white film, a wavelength of light corresponding to yellow-green (550 nm) should yield optimum results. *For a pinhole-to-film distance of 1 inch (25 mm), this works out to a pinhole 0.22 mm in diameter (??).* For 5 cm, the appropriate diameter is 0.32 mm. The depth of field is basically infinite, but this does not mean that no optical blurring occurs. The infinite depth of field means that image blur depends not on object distance, but on other factors, such as the distance from the aperture to the film plane, the aperture size, and the wavelength(s) of the light source.

Pinhole cameras can be handmade by the photographer for a particular purpose. In its simplest form, the photographic pinhole camera can consist of a light-tight box with a pinhole in one end, and a piece of film or photographic paper wedged or taped into the other end. A flap of cardboard with a tape hinge can be used as a shutter. The pinhole may be punched or drilled using a sewing needle or small diameter bit through a piece of tinfoil or thin aluminum or brass sheet. This piece is then taped to the inside of the light tight box behind a hole cut through the box. A cylindrical oatmeal container may be made into a pinhole camera.

Pinhole cameras can be constructed with a sliding film holder or back so the distance between the film and the pinhole can be adjusted. This allows the angle of view of the camera to be changed and also the effective f-stop ratio of the camera. Moving the film closer to the pinhole will result in a wide angle field of view and a shorter exposure time. Moving the film farther away from the pinhole will result in a telephoto or narrow angle view and a longer exposure time.

Pinhole cameras can also be constructed by replacing the lens assembly in a conventional camera with a pinhole. In particular, compact 35 mm cameras whose lens and focusing assembly has been damaged can be reused as pinhole cameras maintaining the use of the shutter and film winding mechanisms. As a result of the enormous increase in f-number while maintaining the same exposure time, one must use a fast film in direct sunshine. Pinholes (homemade or commercial) can be used in place of the lens on an SLR. Use with a digital SLR allows metering and composition by trial and error, and is effectively free, so is a popular way to try pinhole photography. Unusual materials have been used to construct pinhole cameras, e.g., a Chinese roast duck.

The f-number of the camera may be calculated by dividing the distance from the pinhole to the imaging plane (the focal length) by the diameter of the pinhole. For example, a camera with a 0.02 inch (0.5 mm) diameter pinhole, and a 2 inch (50 mm) focal length would have an f-number of 2/0.02 (50/0.5), or 100 (f/100 in conventional notation). Due to the large f-number of a pinhole camera, exposures will often encounter reciprocity failure. Once exposure time has exceeded about 1 second for film or 30 seconds for paper, one must compensate for the breakdown in linear response of the film/paper to intensity of illumination by using longer exposures. Other special features can be built into pinhole cameras such as the ability to take double images, by using multiple pinholes, or the ability to take pictures in cylindrical or spherical perspective by curving the film plane.

Digital Camera

A digital camera (or digicam) is a camera that takes video or still photographs, or both, digitally by recording images via an electronic image sensor. It is the main device used in the field of digital photography. Most 21st century cameras are digital. Digital cameras can do things film cameras cannot: displaying images on a screen immediately after they are recorded, storing thousands of images on a single small memory device, and deleting images to free storage space. The majority, including most compact cameras, can record moving video with sound as well as still photographs. The optical system works the same as in film cameras, typically using a lens with a variable diaphragm to focus light onto an image pickup device. The diaphragm and shutter admit the correct amount of light to the imager, just as with film but the image pickup device is electronic rather than chemical. Most digicams, apart from camera phones and a few specialized types, have a standard tripod screw.

Types of Digital Camera

Digital cameras are made in a wide range of sizes, prices and capabilities. The majority are camera phones, operated as a mobile application through the cellphone menu. Professional photographers and many amateurs use larger, more expensive digital single-lens reflex cameras (DSLR) for their greater versatility. Between these extremes lie digital compact cameras and bridge digital cameras that "bridge" the gap between amateur and professional cameras. Specialized cameras including multispectral imaging equipment and astrographs continue to serve the scientific, military, medical and other special purposes for which digital photography was invented.

Compact Digital Cameras

Compact cameras are designed to be tiny and portable and are particularly suitable for casual and "snapshot" use. Hence, they are also called point and shoot cameras. The smallest, generally less than 20 mm thick, are described as subcompacts or "ultra compacts" and some are nearly credit card size. Most, apart from ruggedized or water resistant models, incorporate a retractable lens assembly allowing a thin camera to have a moderately long focal length and thus fully exploit an image sensor larger than that on a camera phone, and a mechanized lens cap to cover the lens when retracted. The retracted and capped lens is protected from keys, coins and other hard objects, thus making it a thin, pocketable package. Subcompacts commonly have one lug and a short wrist strap which aids extraction from a pocket, while thicker compacts may have two lugs for attaching a neck strap.

Compact cameras are usually designed to be easy to use, sacrificing advanced features and picture quality for compactness and simplicity; images can usually only be stored using lossy compression (JPEG). Most have a built-in flash usually of low power, sufficient for nearby subjects. Live preview is almost always used to frame the photo. Most have limited motion picture capability. Compacts often have macro capability and zoom lenses but the zoom range is usually less than for bridge and DSLR cameras. Generally a contrast-detect auto focus system, using the image data from the live preview feed of the main imager, focuses the lens. Typically, these cameras incorporate a nearly silent leaf shutter into their lenses. For lower cost and smaller size, these cameras typically use image sensors with a diagonal of approximately 6 mm, corresponding to a crop factor around 6. This gives them weaker low light performance, greater depth of field, generally closer focusing ability, and smaller components than cameras using larger sensors.

Starting in 2011, some compact digital cameras can take 3D still photos. These 3D compact stereo cameras can capture 3D panoramic photos for play back on a 3D TV. Some of these are rugged and waterproof, and some have GPS, compass, barometer and altimeter.

Bridge Cameras

Bridge are higher end digital cameras that physically and ergonomically resemble DSLRs and share with them some advanced features, but share with compacts the use of a fixed lens and a small sensor. Like compacts, most use live preview to frame the image. Their auto focus uses the same contrast detect mechanism, but many bridge cameras have a manual focus mode, in some cases using a separate focus ring, for greater control. They originally "bridged" the gap between affordable point-and-shoot cameras and the then unaffordable earlier digital SLRs.

Due to the combination of big physical size but a small sensor, many of these cameras have very highly specified lenses with large zoom range and fast aperture, partially compensating for the inability to change lenses. On some, the lens qualifies as super zoom. To compensate for the lesser sensitivity of their small sensors, these cameras almost always include an image stabilization system to enable longer handheld exposures.

These cameras are sometimes marketed as and confused with digital SLR cameras since the appearance is similar. Bridge cameras lack the reflex viewing system of DSLRs, are usually fitted with fixed (non-interchangeable) lenses (although some have a lens thread to attach accessory wide-angle or telephoto converters), and can usually take movies with sound. The scene is composed by viewing either the liquid crystal display or the electronic viewfinder (EVF). Most have a longer shutter lag than a true DSLR, but they are capable of good image quality (with sufficient light) while being more compact and lighter than DSLRs.

High-end models of this type have comparable resolutions to low and mid-range DSLRs. Many of these cameras can store images in a Raw image format, or processed and JPEG compressed, or both. The majority have a built in flash similar to those found in DSLRs.

Mirror less Interchangeable lens Camera

In late 2008, a new type of camera emerged, combining the larger sensors and interchangeable lenses of DSLRs with the live-preview viewing system of compact cameras, either through an electronic viewfinder or on the rear LCD. These are simpler and more compact than DSLRs due to the removal of the mirror box, and typically emulate the handling and ergonomics of either DSLRs or compacts. The system is used by Micro Four Thirds, borrowing components from the Four Thirds DSLR system. The Ricoh GXR of 2009 puts the sensor and other electronic components in the interchangeable sensor lens assembly or "camera unit," rather than in the camera body.

Digital Single Lens Reflex Cameras

Digital single lens reflex cameras (DSLRs) are digital cameras based on film single lens reflex cameras (SLRs). They take their name from their unique viewing system, in which a mirror reflects light from the lens through a separate optical viewfinder. In order to capture an image the mirror is flipped out of the way, allowing light to fall on the imager. Since no light reaches the imager during framing, auto focus is accomplished using specialized sensors in the mirror box itself. Most 21st century DSLRs also have a "live view" mode that emulates the live preview system of compact cameras, when selected. These cameras have much larger sensors than the other types, typically 18 mm to 36 mm on the diagonal (crop factor 2, 1.6, or 1). This gives them superior low-light performance, less depth of field at a given aperture, and a larger size. They make use of interchangeable lenses; each major DSLR manufacturer also sells a line of lenses specifically intended to be used on their cameras. This allows the user to select a lens designed for the application at hand: wide angle, telephoto, low light, etc. So each lens does not require its own shutter, DSLRs use a focal-plane shutter in front of the imager, behind the mirror. The mirror flipping out of the way at the moment of exposure makes a distinctive "clack" sound.

Line Scan Camera Systems

A line-scan camera is a camera device containing a line-scan image sensor chip, and a focusing mechanism. These cameras are almost solely used in industrial settings to capture an image of a constant stream of moving material. Unlike video cameras, line-scan cameras use a single array of pixel sensors, instead of a matrix of them. Data coming from the line-scan camera has a frequency, where the camera scans a line, waits, and repeats. The data coming from the line-scan camera is commonly processed by a computer, to collect the one-dimensional line data and to create a two-dimensional image. The collected two-dimensional image data is then processed by image-processing methods for industrial purposes. Line-scan technology is capable of capturing data extremely fast, and at very high image resolutions. Usually under these conditions, resulting collected image data can quickly exceed 100 MB in a fraction of a second. Line-scan-camera based integrated systems, therefore are usually designed to streamline the camera's output in order to meet the system's objective, using computer technology which is also affordable.

Line scan cameras intended for the parcel handling industry can integrate adaptive focusing mechanisms to scan six sides of any rectangular parcel in focus, regardless of angle, and size. The resulting 2-D captured images could contain, but are not limited to 1D and 2D barcodes, address information, and any pattern that can be processed via image processing methods. Since the images are 2-D, they are also human readable and can be viewable on a computer screen. Advanced integrated systems include video coding, optical character recognition (OCR) and finish-line cameras for high speed sports.

History

Steven Sasson as an engineer at Eastman Kodak invented and built the first digital camera using a charge-coupled device image sensor in 1975. The resolution of a digital camera is often limited by the image sensor (typically a CCD or CMOS sensor chip) that turns light into discrete signals, replacing the job of film in traditional photography. The sensor is made up of millions of "buckets" that essentially count the number of photons that strike the sensor. This means that the brighter the image at a given point on the sensor, the larger the value that is read for that pixel. Depending on the physical structure of the sensor, a color filter array may be used which requires a demosicing/interpolation algorithm. The number of resulting pixels in the image determines its "pixel count". For example, a 640x480 image would have 307,200 pixels, or approximately 307 kilopixels; a 3872x2592 image would have 10,036,224 pixels, or approximately 10 megapixels. Demanding high quality and resolution (e.g. for use in professional photography), this count is an object of manufacturer completion. As of August 2011, the highest resolution available on the market is 80.1 MP.

As the technology has improved, costs have decreased dramatically. Counting the "pixels per dollar" as a basic measure of value for a digital camera, there has been a continuous and steady increase in the number of pixels each dollar buys in a new camera, in accord with the principles of Moore's Law. This predictability of camera prices was first presented in 1998 at the Australian PMA DIMA conference by Barry Hendy and since referred to as "Hendy's Law".

Since only a few aspect ratios are commonly used (mainly 4:3 and 3:2), the number of sensor sizes that are useful is limited. Furthermore, sensor manufacturers do not produce every possible sensor size, but take incremental steps in sizes. For example, in 2007 the three largest sensors (in terms of pixel count) used by Canon were the 21.1, 17.9, and 16.6 megapixel CMOS sensors.

Transferring Photos

Many digital cameras can connect directly to a computer to transfer data:

a. Early cameras used the PC serial port. USB is now the most widely used method (most cameras are viewable as USB mass storage), though some have a FireWire port. Some cameras use USB PTP mode for connection instead of USB MSC; some offer both modes.

b. Other cameras use wireless connections, via Bluetooth or IEEE 802.11 Wi-Fi, such as the Kodak EasyShare One.

c. Cameraphones and some high-end stand-alone digital cameras also use cellular networks to connect for sharing images. The most common standard on cellular networks is the MMS Multimedia Messaging Service, commonly called "picture messaging". The second method with smartphones is to send a picture as an email attachment. Many cameraphones do not support email, so this is less common.

d. A common alternative is the use of a card reader which may be capable of reading several types of storage media, as well as high speed transfer of data to the computer. Use of a card reader also avoids draining the camera battery during the download process, as the device takes power from the USB port. An external card reader allows convenient direct access to the images on a collection of storage media. But if only one storage card is in use, moving it back and forth between the camera and the reader can be inconvenient. Many computers have a card reader built in, at least for SD cards.

Printing Photos

Many modern cameras support the PictBridge standard, which allows them to send data directly to a PictBridge-capable computer printer without the need for a computer. Wireless connectivity can also provide for printing photos without a cable connection. Polaroid has introduced a printer integrated into its digital camera which creates a small, printed copy of a photo. This is reminiscent of the original instant camera, popularized by Polaroid in 1975.

Displaying Photos

Many digital cameras include a video output port. Usually sVideo, it sends a standard definition video signal to a television, allowing the user to show one picture at a time. Buttons or menus on the camera allow the user to select the photo, advance from one to another, or automatically send a "slide show" to the TV. HDMI has been adopted by many high end digital camera makers, to show photos in their high resolution quality on an HDTV.

In January 2008, Silicon Image announced a new technology for sending video from mobile devices to a television in digital form. MHL sends pictures as a video stream, up to 1080p resolution, and is compatible with HDMI. Some DVD recorders and television sets can read memory cards used in cameras; alternatively several types of flash card readers have TV output capability.

File Formats

The Joint Photography Experts Group standard (JPEG) is the most common file format for storing image data. Other file types include Tagged Image File Format (TIFF) and various raw image formats. Many cameras, especially professional or DSLR cameras support a raw image format. A raw image is the unprocessed set of pixel data directly from the camera's sensor. They are often saved in formats proprietary to each manufacturer, such as NEF for Nikon, CRW or CR2 for Canon, and MRW for Minolta. Adobe Systems has released the DNG format, a royalty free raw image format which has been adopted by at least 10 camera manufacturers.

Raw files initially had to be processed in specialized image editing programs, but over time many mainstream editing programs, such as Google's Picasa, have added support for raw images. Editing raw format images allows more flexibility in settings such as white balance, exposure compensation, color temperature, and so on. In essence raw format allows the photographer to make major adjustments without losing image quality that would otherwise require retaking the picture.

Formats for movies are AVI, DV, MPEG, MOV (often containing motion JPEG), WMV, and ASF (basically the same as WMV). Recent formats include MP4, which is based on the QuickTime format and uses newer compression algorithms to allow longer recording times in the same space.

Other formats that are used in cameras but not for pictures are the Design Rule for Camera Format (DCF), an ISO specification for the camera's internal file structure and naming, and Digital Print Order Format (DPOF), which dictates what order images are to be printed in and how many copies.

Batteries

Off the Shelf

The first type of battery for digital cameras conforms to an established off-the shelf form factor, most commonly AA, CR2, or CR-V3 batteries, with AAA batteries in a handful of cameras. The CR2 and CR-V3 batteries are lithium based, and intended for single use. They are also commonly seen in camcorders. AA batteries are the most common; however, the non-rechargeable alkaline batteries supplied with low-end cameras are capable of providing enough power for only a very short time in most cameras. They may serve satisfactorily in cameras that are only occasionally used.

Consumers with more than an occasional need use AA Nickel metal hydride batteries (NiMH) instead, which provide an adequate amount of power and are rechargeable. NIMH batteries do not provide as much power per volume as lithium ion batteries, and they also tend to discharge when not used. To get same power, NiMH Rechargeable battery takes up to two times in volume compare to Li-on Rechargeable Battery, by weight NiMH Rechargeable Battery is three to five times heavier, but by price NiMH Rechargeable Battery is only a half compare to Li-on Rechargeable Battery. Please see Wikipedia: Table of rechargeable battery technologies in Rechargeable battery. They are available in various ampere hour (Ah) or milli-ampere-hour (mAh) rating, which affects how long they last in use. Typically mid-range consumer models and some low end cameras use off-the-shelf batteries; only a very few DSLR cameras accept them (for example, Sigma SD10). Rechargeable RCR-V3 lithium-ion batteries are also available as an alternative to non-rechargeable CR-V3 batteries. Cameras, especially earlier ones made for AA-size batteries assumed that these would be of the non-rechargeable, preferably alkaline manganese type delivering 1.5 volts per cell. Rechargeable NiCd or NiMH cells only deliver 1.2 volts, which means that many such cameras will only operate for a short time or not at all even with new and newly charged 1.2 volt units. A portable ultra-high-endurance external power-supply for the shoulder bag to operate older 6 volt cameras can be made up of five 1.2 volt C-size cells which can be either NiCd or NiMH, with a cable and 4mm DC-plug.

Proprietary

The second type of battery for digital cameras is proprietary battery formats. These are built to a manufacturer's custom specifications, and can be either aftermarket replacement parts or OEM. Almost all proprietary batteries are lithium ion. While they only accept a certain number of recharges before the battery life begins degrading (typically up to 500 cycles), they provide considerable performance for their size. A result is that at the two ends of the spectrum both high end professional cameras and low end consumer models tend to use lithium ion batteries.

Single Lens Reflex Camera (SLR)

A single-lens reflex (SLR) camera is a camera that typically uses a semi-automatic moving mirror system (hence the name "reflex", from the mirror's reflection) that permits the photographer to see exactly what will be captured by the film or digital imaging system, as opposed to pre-SLR cameras where the view through the viewfinder could be significantly different from what was captured on film.

Prior to the development of SLR, all cameras with viewfinders had two optical light paths: one path through the lens to the film, and another path positioned above (TLR or twin-lens reflex) or to the side (rangefinder). Because the viewfinder and the film lens cannot share the same optical path, the viewing lens is aimed to intersect with the film lens at a fixed point somewhere in front of the camera. This is not problematic for pictures taken at a middle or longer distance, but parallax causes framing errors in close up shots. Moreover, focusing the lens of a fast reflex camera when it is opened to wider apertures (such as in low light or while using low speed film) is not easy.

Most SLR cameras permit upright and laterally correct viewing through use of a roof pentaprism situated in the optical path between the reflex mirror and viewfinder. Light, which comes both horizontally and vertically inverted after passing through the lens, is reflected upwards by the reflex mirror, into the pentaprism where it is reflected several times to correct the inversions caused by the lens, and align the image with the viewfinder. When the shutter is released, the mirror moves out of the light path, and the light shines directly onto the film (or in the case of a DSLR, the CCD or CMOS imaging sensor). The Canon Pellix film camera was an exception to the moving mirror system, wherein the mirror was a fixed beamsplitting pellicle.

Focus can be adjusted manually by the photographer or automatically by an autofocus system. The viewfinder can include a matte focusing screen located just above the mirror system to diffuse the light. This permits accurate viewing, composing and focusing, especially useful with interchangeable lenses.

Up until the 1990s, SLR was the most advanced photographic preview system available, but the recent development and refinement of digital imaging technology with an on-camera live LCD preview screen has overshadowed SLR's popularity. Nearly all inexpensive compact digital cameras now include an LCD preview screen allowing the photographer to see what the CCD is capturing. However, SLR is still popular in high-end and professional cameras because they are system cameras with interchangeable parts, allowing customization. They also have far less shutter lag, allowing photographs to be timed more precisely. Also the pixel resolution, contrast ratio, refresh rate, and color gamut of an LCD preview screen cannot compete with the clarity and shadow detail of a direct-viewed optical SLR viewfinder.

A cross-section (or 'side-view') of the optical components of a typical SLR camera shows how the light passes through the lens assembly is reflected by the mirror placed at a 45 degree angle, and is projected on the matte focusing screen. Via a condensing lens and internal reflections in the roof pentaprism the image appears in the eyepiece. When an image is taken, the mirror moves upwards from its resting position in the direction of the arrow, the focal plane shutter opens, and the image is projected onto the film or sensor in exactly the same manner as on the focusing screen.

Most 35 mm SLRs use a roof pentaprism or penta-mirror to direct the light to the eyepiece, first used on the 1949 Contax S from East Germany. The first Japanese pentaprism SLR was the 1955 Miranda T, followed by the Asahi Pentax, Zunow, Nikon F, and the Yashica Pentamatic. Some SLRs offered removable pentaprisms with optional viewfinder capabilities, such as the waist-level finder, the interchangeable sports finders used on the Canon F1, F1n and F1; the Nikon F, F2, F3, F4 and F5; and the Pentax LX. Another prism design was the porro prism system used in the Olympus Pen F, the Pen FT, the Pen FV half-frame 35 mm SLR cameras. This was later used on the Olympus EVOLT E-3x0 series, the Leica Digilux 3 and the Panasonic DMC-L1. A right-angle finder is available that slips onto the eyepiece of most SLRs and D-SLRs and allows viewing through a waist-level viewfinder. There is also a finder that provides EVF remote capability.

History

Large format SLR cameras were probably first marketed with the introduction of C.R. Smith's Monocular Duplex (USA, 1884). SLRs for smaller exposure formats were launched in the 1920s by several camera makers. The first 35 mm SLR available to the mass market, Leica's PLOOT reflex housing along with a 200mm f4.5 lens paired to a 35mm rangefinder camera body, debuted in 1935. The Soviet ("Sport"), also a 24mm by 36mm image size, was prototyped in 1934 and went to market in 1937. K. Nüchterlein's Ihagee Kine-Exakta (Germany, 1936) was the first integrated 35mm SLR to enter the market. Additional Exakta models, all with waist-level finders, were produced up to and during World War II. Another ancestor of the modern SLR camera was the Swiss-made Alpa, which was innovative, and influenced the later Japanese cameras. The first eye-level SLR viewfinder was patented in Hungary on August 23, 1943 by Jen Dulovits, who then designed the first 35 mm camera with one, the Duflex, which used a system of mirrors to provide a laterally correct, upright image in the eye-level viewfinder. The Duflex, which went into serial production in 1948, was also the world's first SLR with an instant-return (a.k.a. auto return) mirror.

The first commercially produced SLR that employed a roof pentaprism was the Zeiss Ikon VEB Contax S, announced on May 20, 1949.

The Japanese adopted and further developed the SLR. In 1952, Asahi developed the Asahiflex and in 1954, the Asahiflex IIB. In 1957, the Asahi Pentax combined the fixed pentaprism and the right-hand thumb wind lever. Nikon, Canon and Yashica introduced their first SLRs in 1959 (the F, Canonflex, and Pentamatic, respectively).

As a small matter of history, the first 35 mm camera (non-SLR) to feature through the lens light metering may have been Nikon, with a prototype rangefinder camera, the SPX. According to the below website, the camera used Nikon 'S' type rangefinder lenses.

Pentax was the first manufacturer to show an early prototype 35 mm behind-the-lens metering SLR camera, which was named the Pentax Spotmatic. The camera was shown at the 1960 Photokina show. However, the first Through-the-lens (TTL) light metering SLR on the market was the 1963 Topcon RE Super, which had the CdS metering cell placed behind the reflex mirror. The mirror had narrow slits cut into the surface to let the light reach the cell providing average metering. Late in the following year, a production model of the Pentax Spotmatic was shown whose CdS light meter cells were on the pentaprism, reading the light off the focusing screen providing average reading, yet keeping the Spotmatic name, but now written in one word.

While auto-exposure was commonly used in the early 1960s with various 35 mm fixed lens rangefinder cameras such as the Konica Auto 'S', and other cameras such as the Polaroid Land cameras whose early models used selenium cell meters, auto-exposure for interchangeable lens SLRs was a feature that was largely absent, except for a few early leaf-shutter SLRs such as the Kowa SE-R and Topcon Auto 100. The first autofocus 35 mm SLR was the Pentax ME-F released in 1981.

The Minolta Maxxum 7000, released in 1985, was the first 35 mm SLR with integrated autofocus and motorized film-advance winder, which became the standard configuration for SLR cameras from then on. This development had significant impact on the photographic industry.

Canon, Nikon and Pentax have all developed digital SLR cameras using the same lens mounts as on their respective film SLR cameras. Konica Minolta did the same, but in 2006 sold their camera technology to Sony, who now builds DSLRs based on the Minolta lens mount. Samsung builds DSLRs based on the Pentax lens mount. Olympus, on the other hand, chose to create a new digital-only Four Thirds System SLR standard, adopted later by Panasonic and Leica.

Contax came out with a DSLR model, the Contax N-Digital. Unfortunately this model was too late and too expensive for competition with other camera manufacturers. The Contax N-digital was the last Contax to use that maker's lens system, and the camera, while having impressive features such as a full-frame sensor, was expensive and lacked sufficient write-speed to the memory card for it to be seriously considered by some professional photographers.

Early SLRs were built for large format photography, but this film format has largely lost favor among professional photographers. SLR film-based cameras have been produced for most film formats as well as for digital formats. These film-based SLRs use the 35 mm format as, this film format offers a variety of emulsions and film sensitivity speeds, usable image quality and a good market cost. 35 mm film comes in a variety of exposure lengths: 20 exposure, 24 exposure and 36 exposure rolls. Medium format SLRs provide a higher-quality image with a negative that can be more easily retouched than the smaller 35 mm negative, when this capability is required.

A small number of SLRs were built for APS such as the Canon IX series and the Nikon Pronea cameras. SLRs were also introduced for film formats as small as Kodak's 110, such as the Pentax Auto 110, which had interchangeable lenses.

Advantages

Many of the advantages of SLR cameras derive from viewing and focusing the image through the attached lens. Most other types of cameras do not have this function; subjects are seen through a viewfinder that is near the lens, making the photographer's view different from that of the lens. SLR cameras provide photographers with precision; they provide a viewing image that will be exposed onto the negative exactly as it is seen through the lens. There is no parallax error, and exact focus can be confirmed by eye, especially in macro photography and when photographing using long focus lenses. The depth of field may be seen by stopping down to the attached lens aperture, which is possible on most SLR cameras except for the least expensive models. Because of the SLR's versatility, most manufacturers have a vast range of lenses and accessories available for them.

Compared to most fixed-lens compact cameras, the most commonly used and inexpensive SLR lenses offer a wider aperture range and larger maximum aperture (typically f/1.4 to f/1.8 for a 50 mm lens). This allows photographs to be taken in lower light conditions without flash, and allows a narrower depth of field, which is useful for blurring the background behind the subject, making the subject more prominent. 'Fast' lenses are commonly used in theater photography, portrait photography, surveillance photography, and all other photography requiring a large maximum aperture.

The variety of lenses also allows for the camera to be used and adapted in many different situations. This provides the photographer with considerably more control (i.e., how the image is viewed and framed) than would be the case with a view camera. In addition, some SLR lenses are manufactured with extremely long focal lengths, allowing a photographer to be a considerable distance away from the subject and yet still expose a sharp, focused image. This is particularly useful if the subject includes dangerous animals (e.g., wildlife); the subject prefers anonymity to being photographed; or else, the photographer's presence is unwanted (e.g., celebrity photography or surveillance photography). Practically all SLR and DSLR camera bodies can also be attached to telescopes and microscopes via an adapter tube to further enhance their imaging capabilities.

In most cases, single-lens reflex cameras cannot be made as small or as light as other camera designs, such as rangefinder cameras, autofocus compact cameras and digital cameras with electronic viewfinders (EVF) — owing to the mirror box and pentaprism/pentamirror. The mirror box also prevents lenses with deeply recessed rear elements from being mounted close to the film or sensor unless the camera has a mirror lockup feature; this means that simple designs for wide angle lenses cannot be used. Instead, larger and more complex retrofocus designs are required.

The SLR mirror 'blacks-out' the viewfinder image during the exposure. In addition, the movement of the reflex mirror takes time, limiting the maximum shooting speed. The mirror system can also cause noise and vibration. Partially reflective (pellicle) fixed mirrors avoid these problems and have been used in a very few designs including the Canon Pellix and the Canon EOS-1N RS, but these designs introduce their own problems. These pellicle mirrors reduce the amount of light traveling to the film plane or sensor and also can distort the light passing through them, resulting in a less-sharp image. To avoid the noise and vibration, many professional cameras offer a mirror lock-up feature; however, this feature totally disables the SLR's automatic focusing ability. Electronic viewfinders have the potential to give the 'viewing experience' of a DSLR (through-the-lens viewing) without many of the disadvantages.

Reliability of SLRs

SLRs vary widely in their construction and typically have bodies made of plastic or magnesium. Most manufacturers don't cite durability specifications, but some report shutter life expectancies for professional models. For instance, the Canon EOS 1Ds MkII is rated for 200,000 shutter cycles and the newer Nikon D3 is rated for 300,000 with its exotic carbon fiber/kevlar shutter. Because many SLRs have interchangeable lenses, there is a tendency for dust, sand and dirt to get into the main body of the camera through the mirror box when the lens is removed, thus dirtying or even jamming the mirror movement mechanism or the shutter curtain mechanism itself. In addition, these particles can also jam or otherwise hinder the focusing feature of a lens if they enter into the focusing helicoids. The problem of sensor cleaning has been somewhat reduced in DSLRs as some cameras have a built-in sensor cleaning unit.

Chapter 12

Development Journalism in Print Media

Development Journalism

The idea of "development journalism" was conceived in the 1960s at the Press Foundation of Asia. This approach to journalism emerged out of dissatisfaction with the dominance of Western news and communication ideals in developing countries, these being inaccurately covering socio-economic development. There was a real need of reflection on new type of journalism specifically designed to function in the cultural and political structures. Since then, the approach of media promotion on development issues, the question on how journalists report on development in parallel to electronic technologies that take root or to so called "citizen journalism" emerging in wealthy countries placed journalism as a powerful tool that empowers individuals, builds stronger local communities and elevates global awareness on development. Journalists have to adapt to new ways of covering development issues as being the ears and eyes of society and the voices of the voiceless through bottom-up reporting. Indeed, this acclaimed "watchdog" role, reportages on development issues that bring attention to issues that are overlooked or under represented often only raises awareness of problems. But do they really contribute to solutions.

The National Foundation for India (NFI) is an independent, Indian grant making and fundraising body, helping people improve their own lives. It make a difference by supporting partners who look for solutions to complex problems in poor, remote and challenging parts of the country. NFI's distinct nature comes from being completely independent of corporate, religious or governmental control and our focus is nationwide. NFI has partnered with a number of Indian and international donor agencies and is a professionally managed and financially accountable foundation. NFI believes that India can develop meaningfully only when social equity and justice are guaranteed and a sense of respect for our fellow humans is achieved. The belief that a vibrant voluntary sector was needed to create a more just and equitable society led a group of eminent citizens led by the Dr. M.S. Swaminathan, late Shri C. Subramaniam and late Dr. Kamla Chowdhury to set up NFI in 1992. Through the sharing of knowledge, experiences and best practices, NFI today addresses some of the greatest challenges facing the country. NFI uses two instruments development action and development communication to work across seven thematic areas. These include: Community Health, Elementary Education, Local Governance, Livelihood Security; Peace & Justice, Citizens and Society, Development Journalism. The aim is to create lasting change by building healthy communities and promoting sustainable development.

Philosophy and Strategy

Development Journalism portfolio aims to draw public attention to the conditions in which marginalized communities live, situations requiring social change and the effectiveness of state and non-state action. This thematic area believes that using media and communication to effect awareness and change is essential to any meaningful dialogue about empowerment and social issues.

The programme area supports work that

- Uses the media in creative ways for promoting the concerns of the disadvantaged
- Developing communication initiatives that go hand in hand with the development interventions of National Foundation of India (NFI) in other thematic areas
- Orient journalists in the regional media on the effective coverage of development issues
- Encourages community based media initiatives.

Media feature services

NFI supports networks and NGOs that supply development articles by grass root practitioners and activists to the print media.

Community media projects

NFI assists NGOs in field projects that enable community participation in media forms that serve local development needs, and help build local development perspectives. This could include the medium of print, radio, cable, internet, and traditional and contemporary performing arts.

Research

NFI supports research on the manner in which the media covers development issues, the impact of such coverage, and safe online casino on measures to enhance the quality of development journalism. It also plans to support preparation of training material for use by journalism schools, and by NGOs engaged in this field.

Forums for dialogue

NFI brings together journalists, NGOs, and academicians to deliberate on the role of the media in development. This is done through lectures, workshops, seminars, and conferences, organized at the local, regional and national level.

Future Outlook

The future plans for the programme include expanding our community radio interventions to smaller NGOs in states such as Orissa, Chhatisgarh and the Northeast. While there are a number of agencies providing technical assistance to these small NGOs, there are very few with the knowledge and experience that NFI brings in helping at the level of programming and community contribution. The Media Fellowship Programme continues to grow from strength to strength and we hope to be able to use the programme to enhance the quality and quantity of development journalism at the national level, while continuing to train rural stringers and journalists at the local and regional levels. We would also like to strengthen the development action and communication links across all programmes of NFI.

The Development Journalism programme area is not only a thematic area of its own, but it also lends itself to deepen development communication across all the development action themes at NFI. This entails facilitating uses of communication through community media like community radio and TV, wallpapers and grassroots newspapers, and various community based and folk media. As the programme expands, we hope to integrate it more fully into the other thematic areas at NFI, thereby making communication an essential component of all NFI programming.

Print Journalism

1. **Print Journalism** is for those who wish to specialize in newspaper and magazine journalism. Print students learn both newspaper and magazine journalism and often publish institute's newspaper and magazine.

2. **Broadcast Journalism** is for those who wish to specialize in television and radio journalism. Broadcast Journalism students learn both television and radio journalism.

3. **Online/ Multimedia Journalism** is for those who wish to combine print, web and broadcast journalism in an on-line environment. Online/Multimedia students learn all media.

Course requirements for each of the above concentrations vary. In brief, all students are required to complete certain core courses, appropriate courses in the chosen media concentration, and one course in the area of specialization. Additionally all students are required to complete required number of electives, media workshop and master's thesis.

Print students are required to cover both areas, with special focus on newspaper or magazine journalism, respectively. Television and Radio students produce both television and radio programs as part of their Broadcast curriculum, with special focus on their area/areas of interest in both TV and radio. Online/Multimedia students are required to produce a multimedia website that incorporates all aspects of their study involving print, broadcast, and web audio, video, slideshows, etc., and also using cell phones and PDAs to upload material.

Students concentrating in newspaper journalism must take the workshop that requires producing an in-house newspaper. Magazine journalism students produce a monthly magazine. Broadcast students might produce both TV and radio news segments and documentaries, depending on their particular area(s) of interest. Students concentrating in online/multimedia journalism must take the workshop that requires producing an in-house multimedia website.

1) Print

Newspaper Journalism (Print Journalism)

More than ever, newspaper professionals must consistently demonstrate the qualities of precision, accuracy, speed, fairness, understanding and public responsibility. Students will gain these skills through a blend of theoretical and practical courses and workshops. The Institute seeks to prepare them to step into the nation's newsrooms. They will learn reporting, editing, page design and newsroom management. Teaching methods include lectures, case studies, seminars with guest speakers, individual assignments and workshops. Students specializing in Newspaper journalism must take the newspaper workshop.

Magazine Journalism (Print Journalism)

Magazine journalism students' course work will prepare them to participate in writing, editing and production. They will use the latest technology to create graphics, choose typefaces and prepare layouts. While most students select magazine work because of an interest in writing feature articles, the program seeks to instill additional interest and expertise in layout and design, and other areas. Students specializing in Magazine journalism must take the magazine workshop.

2) Broadcast

Television Journalism

Television curriculum requires acquiring skills in reporting, writing, and production for TV broadcast. Students work with both print and television faculty, and acquire the necessary background in each of the subject areas and specializations. Television workshops train students for the real world of broadcast journalism, offering best practices in the industry. Students learn to shoot and edit videotape, and to produce hard news and feature stories. IIJNM maintains close association with several TV stations in the Bangalore area with whom students get the opportunity to interact. Students specializing in Television must take the Television workshop.

Radio Journalism

Radio Journalism students learn to work on news bulletins and documentaries, as well as on in depth and longer form radio reports. Advanced courses, later in the year, stress on air production and hosting skills. They also learn editing and mixing on state of the art digital audio workstations. Theoretical sessions will be further strengthened through library research and seminar paper presentations. Students specializing in Radio journalism must take the Radio workshop.

3) Online/Multimedia

Online/Multimedia Journalism

This stream/curriculum teaches how to deliver news in any and all media the skills needed to be a converged journalist. Students learn solid reporting and writing for online, airwaves, magazines, PDAs, cell phones and electronic paper. Convergent journalism is all about what makes a news story effective today and how to recognize the best medium for a particular story. That medium may be the web, broadcast, radio, or a newspaper or magazine or, more likely, all of the above combined. Students take courses from all areas print, TV, radio, and web/new media and learn to combine the individual media skills to present a single story through multimedia. They will create a news website. The Softcopy does everything from gathering information and writing content to publishing their work. Utilizing major web-publishing software tools, students design, create and operate web publications as part of their coursework and workshops. The Online/Multimedia Journalism course teaches you to be an aspiring journalist, a news writer, an editor, a reporter, and a producer. These students are taught, how to tailor a story to meet the needs of various media and present it in today's cutting edge format and media technologies.

A Short History of Indian Journalism

"The over 2000 years history of the Indian Press, from the time of Hicky to the present day, is the history of a struggle for freedom, which has not yet ended. There have been alternating periods of freedom and of restrictions on freedom amounting to repression. The pioneering works on the Indian Press, like that of Margarita Barns, were stories of arbitrariness and despotism, of reforms and relaxation. The story of the Indian Press is a story of steady expansion but also one of Press laws."

The first newspaper meant for publication was 'announced' in 1776 by William Bolts. He asked those interested to come to his residence to read the news. This 'newspaper' had the twin function of informing the British community of news from 'home', and of ventilating grievances against the colonial administration.

Hicky's Gazette

It was not until James Augustus Hicky dared to start his Bengal Gazette (also called Hicky's Gazette) in 1780 that the age of Journalism dawned in the country. England had already had a taste of the Spectator papers of Addison and Steele, and of lesser known periodicals as well. They had learnt about the power of the periodical essayists who could laugh to scorn the manners and mores of society, and of those in high places.

Political and social corruption was rife among the British who were sent to rule the country when Hicky, a printer by profession, launched his Gazette 'in order to purchase freedom for my mind and soul'. He described the Bengal Gazette (later called Hicky's Gazette) as a 'weekly political and commercial paper open to all parties but influenced by none'. His venom was aimed at individuals like Mrs. Warren Hastings and their private affairs. He published announcements of marriages and engagements, and of 'likely' engagements.[8] The Gazette was, in essence, no better than a scandal sheet. Barely a year later, Sir Warren Hastings denied all postal facilities to Hicky who hit back with these ringing words: 'Mr. Hicky considers the Liberty of the Press to be essential to the very existence of an Englishman and a free Government. The subject should have full liberty to declare his principles and opinions, and every act which tends to coerce that liberty is tyrannical and injurious to the community'.

In June the following year (1781), Hicky was arrested and thrust into jail, from where he continued writing for the Gazette. He was stopped from 'bringing out his weekly only when the types used for printing were seized'.

Five newspapers made their appearance in Bengal in six years' time and all were started by Englishmen. Some of these newspapers received Government patronage. "The Madras Courier" and the "Bombay Herald" (which later merged with the Bombay Courier) were then launched in the two cities. They were subservient to the Government, and therefore flourished. The total circulation of all these weeklies was not more than 2,000; yet, the Government issued Press Regulations (1799) making the publication of the name of the printer, editor and proprietor obligatory. The regulation also ordered these to declare themselves to the secretary of the Government ; and to submit all was established with the aid of Government grant and in the North West Provinces, a Hindu and an Urdu periodicals started of under the Government patronage. The Bengali Press with as many as nine Newspaper.

Indian language Press

The pioneers of Indian language journalism were the Serampore Missionaries with Samachar Darpan and other Bengali periodicals, and Raja Ram Mohan Roy with his Persian newspaper Miraltool Akbar. The object of Ram Mohan Roy, the social reformer, in starting the paper was 'to lay before the public such articles of intelligence as may increase their experience, and tend to their social improvement', and to 'indicate to the rulers a knowledge of the real situation of their subjects, and make the subjects acquainted with the established laws and customs of their rules'. Roy ceased publishing his paper later in protest against the Government's Press Regulations.

The Bombay Samachar, a Gujarati newspaper, appeared in 1822. It was almost a decade before daily vernacular papers like "Mumbai Vartaman" (1830), The "Jan-e-Jamshed (1831)", and the "Bombay Darpan" (1850) began publication In 1839_Tamil, and a Telugu newspaper or had a circulation of around 200 copies each, even as the British Press with 26 newspapers (six of them dailies) grew in strength and power, under the liberal rule of Lord Metcalfe, and later of Lord Auckland.

Censorship and the mutiny

The year of what the British historians term 'the Sepoy Mutiny', however, brought back the Press restriction in the form of the Gagging Act, 1857 Lord Canning argued for them, stating that 'there are times in the existence of every state in which something of the liberties and rights, which it jealously cherishes and scrupulously guards in ordinary Seacons, must be sacrificed for the public welfare. Such is the State of India at this moment. Such a time has come upon us. The liberty of the Press is no exception.'

The mutiny brought the rule of the East India Company to a close, with the Crown taking over the colony', with the promise of religious toleration and Press freedom. The main topics of discussion in the English and vernacular Press before and after the Mutiny were sati, caste, widow remarriage, polygamy, crimes, and opposition to the teaching of English in schools and colleges. Bombay's Gujarati Press in particular, excelled in the defense of the Indian way of life. In 1876 the Vernacular Press Act was promulgated.

During the next two decades The Times of India, the Pioneer, the Madras Mail, and The Amrit Bazar Patrika came into existence all except the last edited by English men and serving the interest of English educated readers. The English Press played down the inaugural meeting of the Indian national congress on December 28, 1885 in Bombay, but it was reported at length by the vernacular papers such as Kesari (founded by Lokmanya Tilak). The Amrit Bazar Patrika and Kesari soon gained a reputation for opposing Government attempts to suppress nationalist aspirations. The Amrit Bazaar Patrika, for instance, denounced the deposition of the maharaja of Kashmir, and Kesari was foremost in attacking the Age of consent Bill of 1891, which sought to prohibit the consummation of marriage before a bride completed the age of 12.The Kesari's stand was endorsed by the Amrit Bazar Patrika and Bangabasi of Calcutta on the ground that the Government had no right to interfere with traditional Hindu customs. Tilak charged the Government with disrespect for the liberty and privacy of the Indian people and with negligence in providing relief during the countrywide famine in 1896-97, which resulted in the death of over a million people.

Such savage anti- Government sentiments could not be allowed free play and so Lord Elgin added sections to the Indian Penal Code to enable the Government to deal with promotion of 'disaffection' against the Crown; or of enmity and hatred between different classes. Also prohibited was the circulation of any reports with intent to cause mutiny among British troops, intent to cause such fear or alarm among the public as to cause any person to commit an offence against the state, or intent to incite any class or to commit an offence against the state, or intent to incite any class or community. The penalties for offences ranged from life imprisonment to short imprisonment or fines.

The man who became the most notes worthy victim of these new laws was none other than Bal Gangadhar Tilak, editor of *Kesari* and its English companion, *Mahratha*. He was arrested, convicted and jailed for six years, but Kesari continued to build up its reputation and influence as a national daily, as India woke to the 20th century. Other campaigns of Press freedom who were prosecuted at about the same time wear Aurobindo Ghose of Bande Mataram, B.B. Upadhayaya of Sandhaya of Sandhay and B.N. Dutt of Jugantar.

In 1910, the Indian Press Act clamped further controls on newspaper in the wake of the partition of Bengal and violent attacks by terrorists in Ahmedabad, Ambala and elsewhere. The Act required owners of printing presses to deposit securities of Rs.500 to Rs. 2,000, which were forfeited if 'objectionable matters' were printed. The threats of seizure of the printing press, and confiscation of copies sent by post were also included in the Act. The vernacular press suffered rigorous suppression during this period (1910-1914). The Government banned 50 works in English and 272 in the vernacular, which included 114 in Marathi, 52 in Urdu and 51 in Bengali."

World War I introduced still more severe press laws but there was no let-up in nationalist agitations. Annie Besant's New India became the mouthpiece of Home Rule advocates, ably supported by the Bombay Chronicle (edited by Benjamin Horniman), Maratha (edited by N.C.Kelkar) and other publications. The Government reacted swiftly by exiling Annie Besant , deporting Horniman and imposing new securities on offending publications. The Rowlatt Act of 1919 infuriated Indian opinion, which now came under the leadership of Mahatma Gandhi. His Non Co-operation Movement took the Press by storm. Gandhi was to remain front-page news for years to come. His arrests and imprisonments were covered with relish by the English and the vernacular Press, whose readership now rose dramatically. The Swaraj Party led by C.R Das, Vallabhbahi Patel and motilal Nehru, launched its own publications the Banglar Katha in Calcutta. The Swadesh Mitram in the South, and Hindustan Times, Pratap and Basumati in the North.

The Indian Press Ordinance (1930), like the Press Act of 1910, and five other Ordinances gave added to the Government in dealing with acts of terrorism, and inflammatory literature. The Swadeshi Movement, covered prominently by the Press, as in The Hindu (Madras) led to the imprisonment of leaders like Gandhi and Nehru, and of editors like S.A. Brelvi of Bombay Chronicle and Ganesh Shankar Vidyarthi of Pratap. The Indian Press (Emergency Powers) Act of 1931 raised deposit securities and fines, and gave Magistrates the power to issue summary actions. Several other Acts were made law during the thirties, forcing the closure of many Presses and publications.

Meanwhile, The Free Press of India, which began as a news agency, started The Indian Express and Dhenamani in Madras, The Free Press Journal in Bombay, and Gujarati and Marathi Journals. The News agency collapsed after it forfeited Rs. 20,000 security under the Indian Press (Emergency Powers) Act, but its publications continued under different owners, and The Free Press editors stated a new agency called the United Press of India (U.P.I.)

Then came the Quit India Movement, and World War II, and the Press in India, including the English language Press and that in the Indian Native States played a commendable role in reporting the struggle for freedom fairly. It opposed communal riots and the partition of the Country, and when partition did take place in the glorious year of independence, lamented it. Indeed, it could be said that that the Press played no small part in India's victory to freedom of speech and expression upheld the freedom of the Press. While the obnoxious Press Acts were repealed or amended, the Official Secrets Act and sections of the Indian code dealing with disaffection, communal hatred and incitement of armed forces to disloyalty, were retained. The Nehru Government Passed in October 1951 the Press (Objectionable Matters) Act which was reminiscent of earlier Press laws enacted by the colonial rulers. The 'objectionable matters' were quite comprehensive. So fierce was the opposition to it that in 1956, it was allowed to lapse, and the First Press Co the Nehru Government Passed in October 1951 the Press (Objectionable Matters) Act which was reminiscent of earlier Press laws enacted by the colonial rulers. The 'objectionable matters' were quite comprehensive.

The national and regional Press covered the campaigns of the first national elections of 1951-1952 with professional skill. So were the other events of the Nehru era, like the formation of the linguistic States, the second and third general elections as well as the Chinese attack, and the take-over of Goa. Unlike her father, Mrs. Indira Gandhi had never been at ease with the Press. How much freedom can the Press have in a country like India fighting poverty, backwardness, ignorance, disease and superstitions?' asked she in the first year of her regime The national dailies grew strident in their attacks on her Government, especially on the question of nationalization of banks, privy purses, the Congress split, but joined forces with her during the Bangla Desh war of liberation. The attacks reached their climax in the period prior to the emergency, with open accusations of rampant corruption, and demands for her resignation, followed by the Allahabad High Court's verdict of her being guilty of corrupt election practices.

Early Years of the Printing Media

Acta Diuna (Journal of the day) was the earliest known journal. It is a handwritten news bulletin distributed in the Forum Romanum. This was in the 1st century B.C during the time of Julius Caesar.

The first printed newspaper appeared in Peiking (Beijing) in the 8th century A.D. The Chinese did the printing using separate wooden block for type, which could be used over &over again. The Koreans also followed the Chinese. Printing ink & paper were developed in china & Egypt. But the whole process of printing had a stunted growth in Asia

Europeans, on the other hand, used the new process on large scale. They benefited from the popularization of printing, which led to the advent of affordable books & popular newspaper. This also led to the democratization of communication.

Printing led to the third major revolution in communication, the first having been the development of human speech some 35,000 years ago & second the art of representing sound in written form using an alphabet. With the spread of printing, libraries & schools sprang up. Books, libraries & higher education- all these led to major political, socioeconomic & cultural changes in Europe. The new modes of agricultural and mechanical production, migration of people from rural areas to industrial towns, rapid changes in social and cultural life and scientific and technological led to new methods of communication among people. Democratization of communication meant strengthening of democratic idea. Political democracy led to economic democracy & social reorganization. In all this, the newsletter produced in various sizes & at different frequencies played an influential role in Germany, Holland, France & England during 16th, 17th & 18th centuries.

By the early 18th century, political leaders realized how powerful an instrument the periodicals were for spreading ideas. Many of them began to produce their own papers to propagate their ideas & influence people such as the Whigs & Tories in England. A new force was detected in society, namely, the force of public opinion. Consequently, the journalism of the period was largely political in nature & thus the impression, justifiable to a great extent, was created that journalism was an adjunct of politics.

Avisa Relation Oder zeitung was the first regular printed newspaper of Europe which commenced publication in 1615. Weekly News was the first newspaper of England that lasted from 1622 to 1641. The next paper was a fortnight, the Daily Courant that started publication in 1702.

The first printing press in the Americas was established in South America by Juan Pablos, a Spanish printer, in 1539. Boston, capital of the Massachusetts colony became an active centre of printing in the 17th century. The Puritans imported the first printing press to New England in 1638 to supply printed materials to Harvard College.

The very first newspaper in the US was *Public Occurrences Foreign & Domestic*, Published by Benjamin Harris, a bookseller in Boston, in 1690. Harris could not continue publication as he was imprisoned by the British authorities for printing without 'prior consent'. Then came another news paper, the *Boston Newsletter* in 1704. James Franklin, brother of Benjamin Franklin, published the New England Courant in 1721. Famous English essayists such as Joseph Addison, Richard Steele, Dr Samuel Johnson & others either contributed essays to their own publications or to periodicals published by other. They wrote contemporary issues besides literary topics. Across the Atlantic, Benjamin Franklin, Thomas Pain, Thomas Jefferson, Alexander Hamilton & other had contributed. The Glorious Revolution in England (1688), The American Revolution (1775-76) & The French Revolution (1789) owe much to the radical & rational thoughts of people such as theme, Locke the Pitts, Robespierre, Hugo, Voltaire, Rousseau, Jefferson, Franklin, Paine and Hamilton etc.

What Was Happening In India During The Early Centuries?

Emperor Asoka's pillar inscriptions & rock edicts in different parts of the Mauryan Empire during 3rd century B.C are considered examples of imperial political communication to the informed & literate section of the population. Ashoka used the Prakrit language in his communication on ethics & morals as evident from his inscriptions).

The learning languages were confined to high casts, the aristocracy, priests, army personnel & landowners. Another feature of communication in ancient India was the emphasis placed on oral & aural systems. Writing was done on palm leaves using a style, but the written documents were considered too not scared to be touched or used by the lower classes. The ruling class used certain methods for coding, transmitting & decoding messages secretly through the network of spies to information about neighboring enemies.

According to historians of journalism, news was collected in a well-organized manner under Akbar the Great. In 1574, Akbar established a recording office that helped later medieval historians to gather materials for chronicles.

First Printing Press in India

The first printing press arrived in India on 6th September 1556 & was installed at the college of St.Paul in Goa.

The First Indian Newspaper

First printed newspaper of India was in English edited & published by James Augustus Hicky, an employee of East India Company. It was named Bengal Gazette which came out on 29th January 1780. Soon many other weeklies & monthlies such as Indian Gazette, Calcutta Journal, Bengal Harakaru, John Bull in the East came out during the 17th & 18th century.

The Major Indian Language Newspaper

Digdarshan was the first Indian language newspaper. It started in April 1818 by the Serampur missionaries William Carcy, Joshua Marshman & William Ward. They soon started another journal in June of the same year and named it *Samachar Darpan*. The famous Raja Ram Mohan Roy also brought out periodicals in English, Bengali & Persian. Some of Roy's papers were *Sambad Kaumadi*, *Brahmical Magazine*, *Mirat-ul-Akhbar*, and *Bangadoota* & *Bengal Herald*.

Assamese

Amnodaya, a distinguished journal in the Assamese language was started in 1846 under the editorship of the Reverend Oliver.T.Cutter.

Gujarati

The newspaper with the greatest longevity in India, *Mumbai Samachar* was also the first Gujarati Newspaper. It was established in 1822 by Farduvji Marzaban as a weekly & then became a daily in 1832.

Hindi

The first Hindi daily was *Samachar Sudhavarshan* (Calcutta, 1854). Later *Samayadant Martand*, *Banaras Akhbar*, *Shimila Akbar (Akhbaar)* & *Malwa Akhbar* came out.

Calcutta was the birth place not only of English, Bengali & Hindi journalism but also, the first Urdu newspaper was published by Urdu Akhbaar the second decade of the 19th century in Calcutta.

Kannada

Kannada Samachar was the earliest Kannada journal, according to many scholars. However some believe that the first Kannada journal was *Mangaloora Samachar*. Later *Subudhi Prakasha, Kannada Vaatika, Amnodaya, Mahilaasakhi & Sarvamitra* came out during the 18th century.

Malayalam

Mathrubhumi, Malayala Manorama, Kerala Kanmudi are the main newspapers of Kerala. The other daily newspapers are *Desabhimani, Mangalam, Madhyamam, Chandrika, Deepika* etc.

Marathi

Darpan was the first Marathi newspaper started on 6 January 1832. *Kesari & Sudarak* were other papers of the 18th century. *Induprakash* was an Anglo-Marathi daily established in 1862.

Oriya

The first Oriya magazine *Junaruna* was published by the Orissa Mission Press in 1849 under the editorship of Charles Lacey. Then came another publication from the same press *Prabhatchandrika*, under the editorship of William Lacey. *Utkal Sahitya, Bodhadayini, Baleshwar Sambad Balika* etc... started in the 18th century.

Punjabi

Although Maharaja Ranjit Singh encouraged the development of Punjabi journalism, yet the earliest Punjabi newspaper was a missionary newspaper. The first printing press in Punjab was established in Ludhiana in 1809.

Tamil

The first periodical '*Tamil Patrika*' a monthly was brought out in 1831 by the Religious Tract Society in Madras; it lasted till 1833. The next periodical weekly was the Dina Vartamani published in Madras from 1856 by the Dravidian press & edited by the Reverend P. Percival. Later *Swadeshamitran, Deshabaktan* etc were other papers.

Telugu

Kandukuri Veeresaliongam Pantulu, known as the Father of the renaissance movement in Andhra & the founder of modern Telugu, sparked a social reform movement through his weekly Vivekavardhini. He also founded separate journals for women; Satihitabodhini.

Urdu

Sir Sayyid Ahmed Khan, a great educationist, judge & social reformer did much for the development of Urdu journalism.

The Indian Press at Independence and Today

In 1947, the major English newspaper in India were the *Times of India* (Bombay), *Statesman* (Calcutta), *Hindu* (Madras), *Hindustan Times* (New Delhi), *Pioneer* (Lucknow), *Indian Express* (Bombay & Madras) *Amrita Bazaar Patrika* (Calcutta), *National Herald* (Lucknow), *Mail* (Madras) & *Hitavada* (Nagpur). Of these, the *Times of India, Statesman & Pioneer* were under British ownership till 1964, when it came under a group of Indian business.

During the long struggle for India's Independence, the major English newspaper that served the national cause were the *Hindu* (1878), *Amrita Bazaar Patrika* (1868), *Bombay Chronicle* (1913), *Free Press Journal* (1930, it became Indian Express) & Hindustan Times (1924). Among the Indian language newspapers, the prominent ones were *Aaj* (1920), *Ananda Bazaar Patrika* (1922), *Sakal* (1931), *Swadeshamitran* (1882), *Mumbai Smachar* (1822), *Malayala Manorama* (1890) & *Mathrubhumi* (1930).

Generally speaking, journalism is flourishing in India today. The Indian language newspapers have overtaken the English newspapers in number & circulation. The highest circulation till the 1990's was enjoyed by the English newspapers despite the fact that less than 5 percent of the population of India claim English as their mother tongue. English is still the medium of instruction in colleges & many prominent schools. It is also the language of administration, although state governments have introduced legislation in favor of local government.

Hindi newspapers have the largest total circulation in India. Hindi is the main language of 10 Indian states- Bihar, Chattisgarh, Delhi, Haryana, Himachal Pradesh, Jharkhand, Madhya Pradesh, Rajasthan, Uttranchal and Uttar Pradesh.

Certain trends in communication & journalism throughout the modern world prompted several sociologists & media experts to discuss the desirability of re-examining the trends in the light of basic issues. In other words, 'back to the basics' say the experts. This is where Gandhi becomes relevant. High technology is good, but if it does not enable us to solve basic problems confronting the succeed in catering to the greed of a few to the exclusion of the need of the many-as it has done through the recent decades and in all countries that experienced colonial subjugation in the past.

Chapter 13
Media Communication: International and Traditional

Media Communication: International Aspect

International communication (also referred to as global communication or transnational communication) is the communication practice that occurs across international borders. The need for international communication was due to the increasing effects and influences of globalization. As a field of study, international communication is a branch of communication, concerning with the scope of "government-to-government", "business to business", and "people to people" interactions at a global level. Currently, international communication is being taught at colleges across the United States. Due to the increasingly globalized market, employees who possess the ability to effectively communicate across cultures are in high demand. International communication "encompasses political, economic, social, cultural and military concerns

Historical Concepts

Empire and Communications

Efficient communication networks played crucial roles in establishing ancient imperial authority and international trade. The extent of empire could be used as an 'indication of the efficiency of communication'. Ancient empires such as Rome, Persia and China, all utilized writing for collecting information and dispersing, creating enormous postal and dispatch systems. As early as in fifteenth century, 'news' was popular trans-nationally in Europe. 'The wheat traders of Venice, the silver traders of Antwerp, the merchants of Nuremberg and their trading partners shared economic newsletters and created common values and beliefs in the rights of capital.

Media Ecology

According to the Media Ecology Association, the term "media ecology" can be defined as "the study of media environments, the idea that technology and techniques, modes of information and codes of communication play a leading role in human affairs." Media ecology theory centers on the principles that technology not only profoundly influences society; it also controls virtually all walks of life. It is a study of how media and communication processes affect human perception and understanding. The term was first formally introduced by Neil Postman in 1968, while the concept of the theory was proposed by Marshall McLuhan in 1964. To strengthen this theory, McLuhan and Quentin Fiore claim that it is the media of the epoch that defines the essence of the society by presenting four epochs, inclusive of Tribal Era, Literate Era, Print Era and Electronic Era, which corresponds to the dominant mode of communication of the time respectively. McLuhan argues that media act as extensions of the human senses in each era, and communication technology is the primary cause of social change.

To understand how media affect large structural changes in human outlook, McLuhan classified media as either hot or cool. Hot media refers to a high-definition communication that demands little involvement from audience whereas cool media describes media that demands active involvement from audience. McLuhan with his son Eric McLuhan expanded the theory in 1988 by developing a way to look further into the effects of technology on society. They offer the tetrad as an organized concept that allows people to know the laws of media, the past, present and future effects of media.

Media ecology is a contested term within media studies for it has different meanings in European and North American contexts. The North American definition refers to an interdisciplinary field of media theory and media design involving the study of media environments. The European version of media ecology is a materialist investigation of media systems as complex dynamic systems

Global Communication and International Relations

Changing Paradigms and Policies

The purpose of this essay is twofold: (1) to provide an overview of the impact of global communication on international relations in the theoretical discourse, military, diplomatic, economic, scientific, educational, and cultural arenas, and (2) to draw out the implications in each arena for further policy research and development.

Global communication at the turn of the 21st century has brought about many effects. On the one hand, it is blurring technological, economic, political, and cultural boundaries. Print, photography, film, telephone and telegraphy, broadcasting, satellites, and computer technologies, which developed fairly independently, are rapidly merging into a digital stream of zeros and ones in the global telecommunications networks. Economically, separate industries that had developed around each of these technologies are combining to service the new multimedia environment through a series of corporate mergers and alliances. Politically, global communication is undermining the traditional boundaries and sovereignties of nations. Direct Broadcast Satellite (DBS) is violating national borders by broadcasting foreign news, entertainment, educational, and advertising programs with impunity. Similarly, the micro-media of global communication are narrow casting their messages through audio and videocassette recorders, fax machines, computer disks and networks, including the Internet and the World Wide Web. Culturally, the new patterns of global communication are creating a new global Coca Colonized pop culture of commodity fetishism supported by global advertising and the entertainment industry.

On the other hand, global communication is empowering hitherto forgotten groups and voices in the international community. Its channels have thus become the arena for contestation of new economic, political, and cultural boundaries. Global communication, particularly in its interactive forms, has created immense new moral spaces for exploring new communities of affinity rather than vicinity. It is thus, challenging the traditional top-down economic, political, and cultural systems. In Iran, it facilitated the downfall of a monarchical dictatorship in 1978-1979 through the use of cheap transistor audiocassette recorders in conjunction with international telephony to spread the messages of Ayatollah Khomeini to his followers within a few hours of their delivery from his exile in Paris (Tehranian, 1979, 1980, 1993). In the Philippines, the downfall of the Marcos regime in 1986 was televised internationally for all to witness while alternative media were undermining his regime domestically. In Saudi Arabia, a BBC-WGBH program on "The Death of a Princess," banned by the Saudi government as subversive, was smuggled into the country by means of videotapes the day after its premier on television in London. In China, despite severe media censorship, the democracy movement in Tiananmen Square spread its message around the world in 1989 via the fax machines. In the Soviet Union, computer net workers who opposed the Moscow coup of 1991 and were sympathetic to Yeltsin, transmitted his messages everywhere despite severe censorship of the press and broadcasting (Ganley & Ganley 1987, 1989; Ganley 1992). In Mexico, the Zapatista movement managed to diffuse its messages of protest against the government worldwide in 1994 through the Internet. In this fashion, it solicited international support while embarrassing the Mexican government at a critical moment when it was trying to project a democratic image for admission to the North American Free Trade Agreement (NAFTA). In Burma or Myanmar, as it is officially known, both government and opposition have employed the Internet in their political struggles. E-mail has been used to achieve rapid global mobilization for withdrawal of Western companies from Myanmar in protest against the government's repressive policies.

Even though the examples are few yet, they demonstrate that accelerating technological advances in telecommunications and their worldwide dissemination are profoundly changing the rules of international relations. On the one hand, they are facilitating transfers of science, technology, information, and ideas from the centers to the peripheries of power. On the other, they are imposing a new cultural hegemony through the "soft power" (Nye, 1990) of global news, entertainment, and advertising. Globalizing the local and localizing the global are the twin forces blurring traditional national boundaries. The conduct of foreign relations through traditional diplomatic channels has been both undermined and enhanced by information and communication resources available to non-state actors. The emergence of a global civil society in the form of over some 30,000 non-governmental organizations (NGOs) alongside nearly some 200 state actors as well as intergovernmental organizations (IGOs), transnational corporations (TNCs), and transnational media corporations (TMCs), has added to the complexity of international relations (Commission on Global Governance 1995). Telecommunications is contributing to changes in the economic infrastructures, competitiveness, trade relations, as well as internal and external politics of states. It also affects national security, including the conduct and deterrence against wars, terrorism, civil war, the emergence of new weapons systems, command and control, and intelligence collection, analysis, and dissemination. The emergence of an international politics of cultural identity organized around religious, ethnic, or racial fetishisms suggests what the future issues in international relations might be.

Global communication is thus redefining power in world politics in ways that traditional theories of international relations have not yet seriously considered. More specifically, it is bringing about significant changes in four major arenas of hard and soft power. Hard power refers to material forces such as military and economic leverage, while soft power suggests symbolic forces such as ideological, cultural, or moral appeals. Major changes seem to be taking place in both hard and soft power conceptions and calculations. First, information technologies have profoundly transformed the nature of military power because of emerging weapons systems dependent on laser and information processing. Second, satellite remote sensing and information processing have established an information power and deterrence analogous to the nuclear power and deterrence of an earlier era. Third, global television communication networks such as CNN, BBC, and Star TV have added image politics and public diplomacy to the traditional arsenals of power politics and secret diplomacy. Fourth, global communication networks working through NGOs and interactive technologies such as the Internet are creating a global civil society and pressure groups (such as Amnesty International or Greenpeace) that have served as new actors in international relations. Although no grand theoretical generalizations on the dynamics of hard and soft power are yet possible, trends indicate that the latter is assuming increasing importance.

International Relations theory has been dominated by five major schools of thought: Realism, Liberalism, Marxism, Communitarianism (also known as Institutionalism), and Postmodernism. Table 13.1 provides a synopsis of the major propositions, principles and processes, units of analysis, and methodologies of these schools. Realists have primarily focused on the geopolitical struggles for power, employed the nation-state as their chief unit of analysis, considered international politics as devoid of moral consensus and therefore prone to violence, and argued that the pursuit of national interest in the context of a balance of power strategy is the most efficient and realistic road to international peace and security.

Realism has been the dominant school of thought, in both theory and practice focusing on peace through national strength, armament, and balance of power. For Realists, order is the primary normative value and historical analysis is the soundest methodology to pursue. Liberals, by contrast, have pointed to the integrating forces of the world market as a new reality creating considerable international interdependency in the postwar period. They have argued that increasing levels of free trade, development, deepening and broadening of interdependency, and international cooperation through intergovernmental organizations are the surest path to peace. For liberals, freedom in property ownership, politics, and trade is the primary normative value. In their studies of international relations, liberals supplement historical analysis with a variety of quantitative and qualitative methods such as time series, correlation analyses, and simulation games.

Marxists and Neo-Marxists, although in decline politically, continue to present powerful theoretical arguments that have an appeal in the peripheries of the world. They view international relations primarily in terms of class conflict within and among nations and argue that since the 16th century, capitalism has increasingly incorporated the peripheries into a world system of domination and exploitation through imperialism, colonialism, and neocolonialism. The social revolutions in Russia, China, Cuba, Vietnam, and many less developed countries (LDCs) have attempted to break away from the fetters of the world capitalist system. However, they are being reincorporated by an international regime orchestrated by the transnational corporations, the World Bank, the International Monetary Fund, the General Agreement on Tariffs and Trade (GATT), and its successor, the newly formed World Trade Organization (WTO). Marxists however further argue, internal contradictions, wars, and revolutionary struggles will continue to challenge the dominant capitalist system. For Marxists, equality is the primary normative value while historical materialism and dialectics are the dominant methodologies.

The Communitarian perspective has been articulated by a diversity of political theorists and activists (Deutsch 1966, 1988; Gandhi 1984; Khomeini 1981; Etzioni, 1993; Haas 1992; Tehranian 1990; Tehranian & Tehranian 1992). Although the ideologies of its proponents differ, the centrality of civil society as expressed through community formations, in contrast nation-state and social classes, is what unifies this theoretical perspective. As expressed in its cultural, communal, and institutional formations, civil society thus serves as the underlying unit of analysis. In the traditional literature of international relations, this school of thought is closely linked to the institutionalist perspectives emphasizing the integration processes of world and regional systems. However, it also has manifested itself in a variety of anti-colonial, nationalist, tribalist, localist, ethnic, and religious movements focused on mobilizing the common historical memories of the peripheries in waging a cultural and political struggle against the centers. The Communitarians thus emphasize the centrality of political community as a condition for a durable peace at local, national, regional, and global levels. Community is thus the primary normative value to be pursued, while institution building for world economic, political, and cultural integration are the policy recommendation.

Table 13.1: International Relations: A Typology of Normative Theories

	Major Proposition	*Axial Principle & Processes*	*Unit of Analysis*	*Methodology*
Realism	IR is a struggle for power and peace through balance of power in a political environment devoid of moral consensus and prone to use of force. In such an environment, national interest and strength must be the guiding principles.	Order: Balance of power among competing states	Nation-state	Historical Method
Liberalism	IR is struggle for power, peace, and freedom through balance of power in a political environment in which increasing interdependencies have created a need for the rule of law and cooperation through inter-governmental organizations (IGOs)	Freedom: International division of labor, trade, and development	Nation-state + IGOs	Historical Method + Game Theory+ Simulation, etc.
Marxism	IR is a class struggle for equality between those who own the means of production and those who do not. Under the world capitalist system, the struggle has been waged between the bourgeoisie and the revolutionary working classes, peasantry, and intelligentsia. As the Highest stage of capitalism, imperialism has transformed the struggle into a global class war that cuts across national boundaries.	Equality: Leveling of wealth and income through international class struggle	World System + TNCs + TMCs + Revolutionary & counter-revolutionary states & movements	Historical & Dialectical Materialism
Communitarianism	IR is a struggle for power, peace, and community through democratic cooperation and institution building from local to global in a political environment of contesting power and moral claims that need to be negotiated through global communication, adjudication, or mediation of conflicts without recourse to violence	Integration of international community through institutions of cooperation Community:	Civil Society + TNCs + TMCs + NGOs + IGOs + States	Eclectic & Multi-Disciplinary

(Contd...)

Table 13.1: Contd...

	Major Proposition	*Axial Principle*	*Unit of Analysis*	*Methodology*
Postmodernism	IR is a struggle for hegemonic power through competing truth claims, that need to be understood intertexually as negotiations of knowledge and power.	Identity: Hegemony and resistance through identity formations	Culture	Interpretive

Postmodernism is the latest theoretical perspective to have impacted international relations theory. Emanating from the poststructuralist and deconstructionist schools of thought, postmodernism is deeply imbued with linguistic analyses of knowledge and power. It therefore highlights the central importance of identity as a major principle in the globalization and localization of knowledge and power struggles and truth claims. Generally committed to radical relativism, postmodernism interprets contemporary international relations as a process of negotiation of knowledge, power, and identity through military, economic, and cultural arsenals of influence. While some tendencies in postmodernism are nihilistic, others seek out those universals in global knowledge that could unify an otherwise divided world. Plurality of meanings, tolerance of differences, fluidity of identities, and re-combinations of ideas and images from totally different eras and civilizations are thus the postmodern foci of analysis.

Although each theoretical discourse has its own unique set of assumptions and conclusions reflecting competing interests in the international community, global communication has forced them into a grudging dialogue. Table 1 confines itself to a typology of the main theoretical strands. There are many theoretical hybrids that have enriched international discourse on world order. However, it is significant to note that the axial principles of the five schools of thought together constitute the five democratic goals of order, liberty, equality, community, and identity in the modern world.

Thus, the effects of global communication on the evolution of international relations theory and its underlying international system have been two-fold. On the one hand, global communication has empowered the peripheries of power to progressively engage in the international discourse on the aims and methods of the international system. In this way, Liberalism challenged the traditional state centered, protectionist, mercantilist policies of the 16th to 18th centuries with its revolutionary doctrines of laissez-faire in international trade and protection of property and liberty in domestic life. However, it also incorporated much of the geopolitical Realist view of power politics in its justification of the colonial and imperial orders while increasingly emphasizing the role of IGOs in the management of the international system. Similarly, Marxism challenged Liberalism's dominance in the 19th and 20th centuries by its mobilization of the peripheries against the colonialist and imperialist orders. However, in practice, Communist regimes often cynically followed Realist geopolitical doctrines in favor of international proletarian solidarity. Liberalism, in turn, undermined the Communist regimes by its control of the main world capital, of trade, and of news flows through appeals to democratic values. In a world system dominated by state and corporate bureaucracies, Communitarianism is the latest phase in a continuing theoretical and ideological struggle by the peripheries to put the human rights of the oppressed on the international agenda. In its preoccupation with the collective rights of community, however, Communitarianism cannot altogether ignore the Realist focus on political order, the Liberal preoccupation with individual freedom, and the Marxist concern with social equality. Postmodernism deconstructs the truth claims of all of the foregoing schools by casting doubt on their meta-narratives. But it also posits its own meta-narrative of relativism as a truth claim.

Tensions among the five theoretical schools clearly reveal the tensions among the competing aims of democracy: order, freedom, equality, community, and identity.

On the other hand, global communication has also served as a channel for theoretical integration. Political leadership in international relations has increasingly come to mean moral leadership in such great debates as colonialism, development, population, environment, nuclear weapons, human rights, women and minority status, etc. Global communication has thus historically broadened and deepened the parameters of discourse from Realism to Liberalism, Marxism, Communitarianism, and now Postmodernism. Each school of thought has had to respond to the concerns of new layers of the international community as they have emerged from conditions of oppression and silence. International relations theory has thus progressively incorporated the new democratic claims for equality, self-determination, and cultural identity.

For example, the slogan of "New World Order" has gone through several mutations in this century. For the Axis powers in WW II, it meant a new world system making room for the imperial ambitions of Germany, Italy, and Japan. For the Allies, it meant a reorganization of the world around the United Nations principles of collective security policed by the five permanent members of the Security Council. To the Group of 77 at the United Nations calling for a New World Economic Order in a 1974 General Assembly resolution, the new order meant a revamped international economic system to redress the terms of trade in favor of the LDCs. When a small group of oil exporting countries managed to quadruple the price of crude oil in 1973 through OPEC's collective action, it appeared for a fleeting time that the raw material exporting nations might be able to redress the deteriorating terms of trade between the developed and developing countries. To UNESCO, which picked up the discourse in the 70s and 80s under the banner of a New World Information and Communication Order (NWICO), it meant balance as well as freedom in world news and information flows. The Brandt (1980, 1985) and MacBride (1980) Commission reports set out those policy agendas (Traber 1986; Galtung & Vincent 1992; Frederick 1993). Following the largely fruitless North-South negotiations of the 1980s, the discourse of the new order was resurrected and co-opted by President Bush. To mobilize international support for a war effort against Saddam Hussein, Bush employed the slogan at the wake of the Persian Gulf War in 1990-1991 with maximum effect. It now meant a new international regime of "law and order" under the aegis of the United Nations supported by the unanimity of the five permanent members of the Security Council and, whenever that fails, under alliances such as NATO or ultimately superpower action.

Views of the international system and its most urgent reform needs are thus as fractious as the world itself. The complexities of the world demand international relations theories that can focus on growing gaps and interdependencies, conflicts and cooperation, violence and peace building. They also call for policies recognizing that global communication plays a central role in problem definition and negotiation for solutions. But meaningful international communication calls for technical competence and equality of access to the means of communication, a requirement that is sorely lacking in today's world. For example, so long as the whole continent of Africa has fewer telephone lines than the city of Tokyo, global communication will continue to be largely a one way flow. Industrial countries as a whole have over 18 times more telephone lines per 100 people than all the developing countries (UNDP 1996: 193). Since telephones are the linchpin of the emerging global communication system, this situation exacerbates the existing communication gaps in the world. Theory building in international relations clearly requires greater multicultural dialogue in order to build bridges among the competing cultural constructions of world conflicts.

The Military Arena

The multiple effects of global communication are perhaps most visible in the military arena. Military technologies have become increasingly information and communication intensive. Historically, most communication technologies have immensely benefited in their research and development phase from military investments, but their introduction has often led to rapidly diffused civilian applications. They also have been quickly adopted by all adversaries. Table 2 provides a schematic view of the most important communication technologies and world orders.

Table: 13.2: Communication technologies and World orders

Dates	*Command, Communication and control systems*	*World Orders*
550 BC	Postal system equipped by realy stations with fresh horseman	Persian Empire
500BB	Postal system manned wazzu by postal manner	Chinese Empire
350 BC	Voice and fire signaling	Greek Empire
BC-476 AD	Road system from centre to peripheries, some 90,000 miles	Roman Empire
632-1259 AD	Surface mail by horses and airmail by wazzu pigeon carriers	Islamic Empire
1500-1970	Print	European Empire
1844-1914	Telegraph	British Empire
1900-1945	Radio Broadcasting	Competing Empire
1945-1989	TV, Satellites, computers and the internet	Bipolar system
1989-present	Strategic defense initiative (Star wars), cyborgs	Globalist system

Without reliable command, communication, and control, power centers cannot effectively manage their peripheries. However, every communication system also empowers the peripheries. Print, for example, facilitated the political and cultural hegemony of the West from 15th century onwards, but its spread also gave rise to increasingly potent resistance via nationalist movements throughout the world.

In the military arena, the "double sword" feature of communication technologies has led to the paradox of "more is less": More security has meant less security. A few examples illustrate the point. Nuclear weapons have been assumed to be a powerful deterrent force. However, their proliferation has created a greater probability of accidental or intentional nuclear war. Remote sensing by satellites has created a global surveillance system at the disposal of the superpowers. But commercialization of such information is now leading to its availability to those adversaries who can afford the price. Moreover, direct broadcast satellite communication through such global television networks as CNN and BBC is bringing the news of adversaries' strengths and weaknesses to each party far more quickly than was ever possible before. In warfare, technology is having two contradictory consequences. The conduct of war and resistance against domination are both becoming increasingly robotized and globalized. This is so because the technology is at once global and local as well as both powerful and vulnerable. Terrorism, as the weapon of the weak, has thus been on the ascendancy locally and globally--on the West Bank as well as at the New York World Trade Center, in the Armenia-Azerbaijan region as well as at Turkish and Armenian embassies around the world, at the Oklahoma City Federal Building as well as in Washington, DC.

The idea that stockpiling weapons of mass destruction can gain commensurately higher levels of security for those who possess them is thus proving to be problematical. As military technologies have augmented their hit/kill ratios and communication technologies have improved their powers of surveillance, conditions of permanent insecurity seem to have become more prevalent at the centers as well as at the peripheries of power. The policy implications of this phenomenon for the pursuits of power and peace are far reaching but, for reasons of space, cannot be addressed here.

The Diplomatic Arena

In addition to traditional intergovernmental diplomacy, global communication seems to have generated three new types of diplomacy, which may be labeled public, people, and virtual diplomacy. The global reach of broadcasting by such networks as CNN, the BBC, Star TV, the Voice of America, Radio Moscow, and Radio Beijing, seems to have led to a shift of emphasis from power politics to image politics. Public diplomacy has thus assumed an increasing importance in the conduct of foreign policy. Realists such as former Ambassador George Kennan (1993) and former U.S. Secretary of State James Schlesinger (1992) have, in fact, decried this tendency as tantamount to emotionalism in the policy process. James Schlesinger has argued that the U.S. policies in Kurdistan and Somalia were, in particular, driven by the impact of television images of those human tragedies. John F. Kennedy once summed it up: a videotape is more potent than ten thousand words.

Public diplomacy, however, complements rather than supplants traditional diplomacy. Steven Livingston and Todd Eachus (1995) have challenged the facile presumption of a "CNN effect" on such U.S. humanitarian interventions as those in Somalia or Kurdistan. However, the debate over the role of the media in international relations cannot be settled by a few case studies. Judging by media's role in such post Cold War crises as the Persian Gulf War, Somalia, Bosnia, and Chechnya, there seems to be a symbiosis between governments and the media in the coverage of international affairs. Governments can enhance, restrict, or manipulate the media's access to information and coverage, while the media can play a multiple role in the formation of foreign policies. In their coverage of international affairs, the media particularly commercial television tends to dichotomize, dramatize, and demonize. In this process, the media follows a pattern of story telling that has been well established in the American Westerns with enormous success at the box office, i.e. pitting the cowboys against the Indians in a dramatic struggle between the forces of good and evil.

Given government license to cover a given story, the media may legitimate prevailing policies, or accelerate, impede, or prioritize them. This is often known as the agenda setting function of the media; the media focuses us more on what to think about than tells us what to think. In the case of the Vietnam War, the first television war in history, the media initially legitimized and accelerated U.S. government policies. However, as the body bags came home and the atrocities of the war were televised into American homes, the media gradually turned against government policies, to a certain degree impeded them, and finally contributed to a change of priorities from war to peacemaking. In the case of the Gulf War the world witnessed the first government managed television war in history; about 80 percent of the American public receiving its news from television supported the war effort. Television coverage of the plight of the Iraqi Kurds and Shiites in the aftermath of the war may have accelerated the U.S. government's decision to provide relief and air cover, but it was not decisive in the adoption or execution of that policy. The media may be thus be viewed as neither powerful nor powerless but power-linked.

Public diplomacy is seen as an auxiliary instrument to traditional diplomacy. The use of television as a channel for sending messages to the opposite side by the leaders of the U.S. and Iraq during the Gulf War, the employment of CNN as a source of information and intelligence gathering by foreign and defense policy leaders, and the testing of "trial balloon" proposals via the mass media are examples of such uses of public diplomacy at times of crisis. None of these examples can conclusively suggest that, in their making of foreign policy, states have become hostages to the media. However, the examples suggest that governments are increasingly aware of the potential benefits and risks of media.

In contrast to public diplomacy, which is essentially top down, people diplomacy is a bottom-up process. Improving global transportation and telecommunications have increasingly made it possible for ordinary citizens to engage in a game that has been historically reserved for foreign policy "experts." In efforts to mediate and resolve international conflicts, such prominent citizens as Jimmy Carter, Jesse Jackson, and Ramsey Clark have provided examples of the possibilities and constraints of people diplomacy. Numerous other individuals and groups are also engaged in such efforts. The best known of such groups is Amnesty International, an organization devoted to the freedom and humane treatment of political prisoners around the world. Such interventions in the foreign policy process are often resented by the foreign policy establishments as intrusive. However, people diplomacy can serve as a corrective to the governments' narrow or nationalist objectives (Mandelbaum 1966; Hoffmann 1966).

Virtual diplomacy is of more recent vintage. Global audio, video, and computer teleconferencing has allowed numerous official and unofficial contacts on a routine basis. The institution of a hot-line between the White House and the Kremlin in the aftermath of the Missile Crisis of 1962, closed circuit video-teleconferencing by the U.S. Information Agency through its Worldnet, and other similar facilities demonstrate that diplomacy has new tools at its disposal. But the explosion of the Internet into a worldwide, interactive communication network has also provided numerous opportunities for expert groups to act as intermediaries, advocates, or advisors in international conflicts. At a Harvard University Conference on Information, National Policies, and International Infrastructure (January 1996), Henry Perritt reported on the project Virtual Magistrate jointly sponsored by the Cyberspace Law Institute and the American Arbitration Association. The project provides expert mediation services to parties at conflict through the Internet and the World Wide Web to parties in conflict. Another example of virtual diplomacy is the Internet Listserv Gulf2000 directed by Gary Sick, a retired member of the U.S. National Security Council. The list includes over 170 leading experts on the Persian Gulf. It provides both a forum for the discussion of current issues and a channel through which opinions are formed. Many other expert groups such as the Toda Institute for Global Peace and Policy Research also employ the Internet for international conflict resolution projects that aim at identifying the parties at a conflict, engaging them in a dialogue, and searching for common grounds. The possibilities for virtual diplomacy through the Internet as well as audio or video conferencing are thus immense and will no doubt be exploited further in years to come.

The Economic Arena

Expanding global transportation and telecommunications networks in recent decades have clearly enabled the TNCs (Trans-National Corporations) to decentralize their production and distribution networks while seeking higher profits in regions of the world with lower wages, rents, taxes, and government regulation. World trade and financing have also been profoundly affected by the transborder data flows that facilitate airline and hotel reservations, cash and capital transfers, and international trade in capital markets. In developing economies, the new information technologies have made technological leapfrogging possible in such world trade centers as Singapore and Hong Kong that are now among the world's highest per capita incomes and penetration of telecommunications facilities. Other Asian tigers such as South Korea, China, Taiwan, Thailand, Malaysia, and Indonesia have similarly found in telecommunications an engine of rapid technological leapfrogging and economic growth.

The economic consequences of the current worldwide information revolution are, however, less well known and more controversial. Is the information revolution leading to global leveling of wealth and income or to a new class system of information haves and have nots within and among nations? In creating and destroying jobs, is the information revolution leading to "the end of work" or to a system of structural employment prompted by the disappearance of middle management and downsizing. Is the nature of employment and career changing fundamentally from a one life one career patterns to "one life many careers and jobs".

Clearly, the transfer of industries and jobs from high to low cost areas within and among nations has led to new policy dilemmas that have been hotly debated among experts and politicians. As witnessed in U.S. presidential politics, the debate has particularly focused on internationalism versus protectionism in trade. However, issues of efficiency versus equity, national security versus. Economic freedom and development versus social and environmental health are also at stake in the debate.

The Science and Educational Arenas

The experiences of latecomers to the industrial revolution, such as Japan and China, have abundantly illustrated that the acquisition of modern science and technology is the key to catching up. In this process, the role of information technologies, from print to the Internet, cannot be overemphasized. Since the rate of obsolescence in scientific and technological knowledge is also increasing, information technologies are assuming an additional function aside from transfers of knowledge. What are the relationships between traditional educational institutions and new systems, Can scientific internationalism and technological protectionism coexist? Does leapfrogging from low tech (e. g. typewriters) to high tech (e. g. global computer networking and DBS) undermine cultural sovereignty and identity? Which is more important in the processes of economic development, financial or human capital? If the latter is more important, as the evidence tends to show (UNDP 1992 1996), what is the place of science and technology policy in an overall development strategy. What are the implications of all of this for a global science and technology policy? These questions have clearly no easy answers. But they present the beginnings of any serious international discussions on information, science, technology, and educational policies.

The Cultural Arena

The impact of global communication on international cultural life is perhaps the most visible of its effects. Traveling along the Silk Road in 1992, it was persistently followed by the CNN, BBC, and Star TV. CocaCola, Michael Jackson, and Madonna were ubiquitous wherever went from the Great Wall of China to Urumchi (capital of Sinkiang, China's Western province) to Almaty (capital of Kazakstan), Dushanbe (capital of Tajikistan), Tashkent (capital of Uzbekistan), Ashkabad (capital of Turkmenistan), Baku (capital of Azarbaijan), and Tehran (capital of Iran). In Tehran, in June 1994, courtesy of CNN and DBS, found O. J. Simpson on the run on the Los Angeles freeways. And despite Islamic edicts, MTV musical videos with their postmodern messages of sensuality, pluralism, and skepticism were reaching into the sanctity of Islamic living rooms. This was viewed by the Iranian government authorities as a cultural invasion no less menacing than the U.S. armed fleet off the coasts of the Persian Gulf.

Technological effects are always socially mediated and constructed. Each new technology has to find its own cultural space in the life of a society before it can have any meaningful impact on social relations. In the case of the media, where technologies range from the simplest to the most complex, and from the readily accessible to those accessible only by small elite, the effects are even more complex and ambiguous. A distinction between macromedia, meso-media, and micro-media might illustrate the point. The macromedia of communication (satellites, mainframe computers, the Internet, and its offshoot, the World Wide Web) seem to be acting as agents of globalization. Through global satellite and computer networks, trans-border data flows, scientific and professional electronic mailing, and commercial advertising, the macromedia are supporting the globalization of national markets, societies, and cultures. The meso-media of communication (print, cinema, and broadcasting) are primarily under the control of national governments or pressure groups and therefore function mostly as agents of national integration and social mobilization. The micro-media of communication (telephone, copying machines, audio and videocassette recorders, musical tapes, and personal computers) have primarily empowered the centrifugal forces of dissent at the peripheries of power. All three types of media are, however, closely interlinked via social networks of governments, markets, and civil societies. Without contextualizing their social and political functions in historically and cultural specific situations, media effects would therefore remain largely mystifying and incomprehensible.

Competing Paradigms and Policies

We live in a complex world, and global communication is not making it any less so. But if we view modernization as the overall theme of international relations in the last 500 years of world history and possibly the next 500, the paths to modernity may be considered to have fluctuated within four political paradigms, i.e. capitalism, communism, totalitarianism, and communitarianism. World politics has been characterized by a struggle among the proponents of these four paths. The Blues, or the pioneers of the industrial revolution (England, France, and the United States), took the liberal democratic, capitalist road with the industrial bourgeoisie leading the way, preoccupied with the rights of private property and individual freedom, following a high accumulation strategy of development and free trade policies designed to open up the markets of the rest of the world. The Reds, the communists, were led by the revolutionary working class and intelligentsia aiming at the same goal of industrial revolution through national planning with a focus on social equality, national self-sufficiency, and high mobilization strategies of development and self-sufficiency. Last but not least, the Greens have been led by the intelligentsia to argue for socially, culturally, and environmentally responsible strategies of development prizing "community" and for high integration strategies of development. The Communitarians range in perspective from Gandhian revolutionaries in the LDCs (India, South Africa, Sri Lanka) to the social Democratic and Green Parties in the West.

Capitalism in the United States (during the McCarthy era), Germany, Italy, and Japan; Communism in the Soviet Union and China; and Communitarianism in Israel (in the West Bank), India (under Hindu militancy), and Iran (under the Islamic regime), have all succumbed to that temptation. In particular, the totalitarian regimes of the latecomers to the industrial revolution (Germany, Japan, Italy, Spain, Argentina) employed their national or party colors (brown for Nazism, black for Fascism, yellow for Japanese militarism) to mobilize their societies around a new, highly repressive "order" that glorified national myths of superiority.

The global communication has already placed the democratic norms of order, freedom, equality, and community on national agendas. The central task of the media in democratic societies may be considered to be twofold: (1) to allow for the diversity of voices in society to be heard and (2) to channel that diversity into a process of democratic integration of public opinion and will formation. Without free and vigorous debate among competing views, no nation can achieve the level of integrated unity and determination necessary for democratic societies to act on public issues.

Generally speaking, media pluralism would serve these purposes better than a media system exclusively dominated by state, commercial, public, or community media. Pluralism in structures of ownership and control are therefore needed in order to obtain pluralism in perspectives and messages. However, structural pluralism is hostage to the presence of independent market institutions and voluntary associations (political parties, trade unions, religious and civic organizations). The existence of a strong civil society to counter the powers of the state and the market is therefore a precondition for media pluralism.

In formulating national communication policies, three sets of interlocking policies are at stake, namely cultural, information, and media policies. The overarching policy questions concern freedoms of conscience, speech, association, and assembly. Cultural policies include not only the question of national values, heritage, and identity but also freedoms of religion, language, and schooling. Information policies concern the production and dissemination of public information by such institutions as government agencies, public libraries, and value added networks (VANs). Media policies cover the whole gamut of mediated modes of communication, from print to cyberspace.

1. Cultural Policies

The central dilemma of how to balance cultural diversity with national unity is a perennial problem for any national cultural policy. Perhaps the most important issue in cultural policy is how a country defines itself with respect to its cultural identity, heritage, goals, and values. Although most democratic governments pay lip service to cultural diversity, national unity is often a higher priority. Even in North America and Western Europe, where cultural diversity has been accepted as a democratic value, multiculturalism has come under attack in recent years. Under communism, the Soviet Union defined itself as a bastion of the international proletariat. Composed of over 100 nationalities, however, it had to deal with the problem of nationality. Under Stalin, the Soviet empire was divided into 15 autonomous republics based on nationality. Soviet cultural policy, however, constantly vacillated between the primacy of proletarian solidarity under the banner of a Soviet culture as defined by the Soviet Communist Party and homage to the religious and ethnic diversity of its vast population. However to divide and rule, the Soviets drew the boundaries of most republics in such a way as to include significant ethnic and religious minorities. Voluntary and forced migration also significantly contributed to the multiethnic character of the population in most republics. While Soviet policies succeeded in maintaining the hegemony of the Soviet Communist Party for over 70 years, they could not destroy ethnic and religious loyalties. It is no surprise, therefore, to witness the resurgence of such loyalties to fill the vacuum that is left by the de-legitimization of the Communist ideology. As a result, in the newly independent republics, national histories, identities, goals, as well as place and family names have been revamped to fit the new circumstances. Such cultural restorations included a change from Leningrad to St. Petersburg, Leninabad to Khojand.

Competing myths and historical memories powerfully shape the cultural configurations of society. They are preserved in national monuments, libraries, national and religious rituals, textbooks, and the literature of a country. Cultural policy decides what myths and historical memories to preserve, which to discard, and what to repress. In monarchical Iran, for instance, the myths and memories of the pre-Islamic Iranian monarchy were glorified, while in Islamic Iran, they are being repressed at the same time that the Shia Islamic myths and memories are revived and embellished. The religious policy of a state thus has profound consequences for its cultural policy. Whether a state adopts a national religion, as in England, or pursues a policy of separation of church and state, as in the United States, has important implications for the type of schooling allowed or subsidized. Similarly, language policies affect educational practice. Since its independence in 1917, Finland has required Swedish language instruction in its schools. However, Finland's entry into the European Union has raised questions about the value of Swedish in contrast to English or French as bridges to the European community. By adopting bilingualism, Canada has attempted to keep Quebec within its federation. But Quebec's refusal to require bilingualism within its borders has undermined Canadian unity. To protect and enhance European identity vis-à-vis American TV programs, the European Union is imposing limits on the proportion of foreign programs on television.

2. Information Policies

The dilemma of how to reconcile freedom of information with the dictates of national security and rights of privacy seems to be at the center of any democratic national information policy. A telling example of this dilemma is the controversy in the United States on the Clipper Chip and Violence Chip. In 1993, the National Security Agency introduced a new encryption technique to be used for security and privacy on the National Information Infrastructure (NII). This new technique, commonly known as the Clipper Chip, was designed in secret by the NSA and remains classified so that its inner workings are unknown. It also has an additional "feature" the government keeps the keys for you, so if they want to wiretap anyone, they can. This proposal met with nearly universal opposition from the public and industry. In January 1994, many of the world's top cryptographers and computer security experts wrote to President Clinton asking him to withdraw it. Computer Professionals for Social Responsibility (CPSR, Internet Memo, February 3, 1994) created a listserv for an Internet Petition to Oppose Clipper. Public concern with pornography and violence has clashed with the First Amendment rights in other arena as well. The U.S. Communication Decency Act of 1995 made the dissemination of pornography on the Internet a criminal act. It also required the installation of Violence-Chips in TV sets allowing parents to control the programs their children can watch. However, in 1996, a few U.S. courts declared any infringement of freedom of speech proscribed in the Act as unconstitutional.

Cyberia, otherwise known as Cyberspace, is thus becoming a technologically visible Panoptican Society as well as a public arena for contestation among policies. In a society such as the United States committed to freedom of conscience, speech, association, and assembly, the new technologies are raising fundamental questions on how to protect the First Amendment. Questions of national security also touch on the protection of sensitive scientific and technological information. Since a great deal of science and technology development takes places at research universities, which are generally committed to academic freedom and publication, tensions between the government and institutions of higher education are clearly inevitable. In the 60s, the New York Times published the Pentagon Papers, documents relating to the War in Vietnam that were released by Daniel Ellsburg, a former member of the Central Intelligence Agency. The newspaper was sued by the government for its breach of national security. The courts decided in favor of the Times, but the question of who defines national security and its breach of information remains.

The management of news is one such issue. During the War in Vietnam, reporters had direct access to the battlefront. From the U.S. government's point of view, the consequences were disastrous. Vietnam was the first major U.S. war to be covered by television. As the images of the bloody battlefields reached American homes, the Pentagon's claims of victory were increasingly questioned and anti-war sentiments among the public forced the government to finally withdraw from that country. Government information policies on war subsequently changed. Ever since Vietnam, U.S. reporters have been denied direct access to battlefronts. In the invasions of Granada, Panama, and Iraq, reporters received most of their information through Pentagon briefings. Under these circumstances, the news media have had only limited opportunity to independently examine the veracity of Pentagon claims.

To turn to another policy arena, the question of patent and copyright protection is primarily a commercial issue. However, it has important consequences for a country's international obligations. In recent decades, the United States has been in direct conflict with a number of countries including Hong Kong, Thailand, Taiwan, China, and Singapore for their breach of copyright laws of the United States. Textbooks, computer hardware and software, and musical recordings have been systematically pirated for profit without payment of copyright royalties. However, as long as a country has not signed the Geneva Convention on Copyright, it can continue reproducing intellectual properties without compensation to the authors and publishers. The U.S. has caused considerable pressure on some of the Asian countries to sign and abide by the Geneva Convention. Some have agreed whereas others continue to refuse to sign on the grounds that their Asian heritage has been pillaged for centuries without compensation and it is now their turn to borrow or steal. In this instance, the interface between national information and foreign policies could not be any closer. Foreign policy can no longer confine itself only to the issues of security; it must also develop positions with respect to cultural identity, media freedom and protection, and information trade.

A democratic information policy would increasingly provide electronic libraries for the public and the rights of citizen access to public information. Some Sunshine Laws in the United States provide this. However, a thornier issue is the question of the rights of access of an individual to the information held about her or him. A variety of government and business files such as tax, credit, employment, and court records contain errors or facts that may be detrimental to an individual. The central policy dilemma here revolves around the question of how far the law should extend the individual rights of access and reply before government or employer rights are compromised.

3. Media Policies

Many of the dilemmas of cultural and information policies also confront those who shape national media policies. However, in multicultural societies, the dilemma of how to allow freedom of speech without encouraging hate speech is the central question. Different media philosophies would, of course, respond differently to this question. Authoritarian media policies often follow the dictates of tradition. In multinational imperial systems such as the pre-modern Islamic and European empires, the level of tolerance for religious and ethnic differences was relatively high. As long as allegiance was paid to the central authorities, each cultural community was largely left to its own devices. In the Islamic empires, in fact, the millet system ensured a high level of internal government for the Peoples of the Book (i.e., Christians, Jews, Zoroastrians, Hindus, and Buddhists).

Libertarian media policies tend to value free speech above politically correct speech. Proponents of a ban on hate speech, however, argue that it is equivalent to crying fire in a crowded theater, thus constituting a "clear and present danger." Hate speech should not be tolerated because it seriously threatens ethnic and racial peace. As the Report of the Project on Ethnic Relations in Eastern Europe and the Former Soviet Union suggests, the problem can be tackled in several different ways: First, through constitutional checks and balances. Communitarian media policies face a different set of problems in ethnic and religious conflicts. By definition, such policies value one religion or language or ethnicity over others because they consider it of vital importance to their national unity. Other religions or cultures are either repressed or not equally valued. Iran's persecution of the Baha'is, Turkey's persecution of the Kurds, Iraq's persecution of the Shi'ites and the Kurds, and Israel's persecution of Palestinians all fall within this category. Global communication can make a contribution to human rights through international censure for such systematic violations of its provisions.

National policies are often formulated in the context of global forces and policies. But who decides global policies. There is no global sovereign government comparable to national governments. Instead, we have a complex variety of players or stakeholders on the global scene each taking part in formulating policies that inevitably enhance or constrain national governments in the pursuit of their goals. States regulate, markets allocate, lobbies try to persuade, and civil societies resist and mobilize Financial institutions accumulate, and the media de/legitimate policies. It is in the context of such forces and policies as well as their respective media outlets (state, public, commercial, community, advertising, pressure group, and trade and investment media) that global, national, and local discourses take place.

Outcomes

In international relations, global communication seems to have at once encouraged globalism and its discontents, i.e., nationalism, regionalism, localism, and fundamentalism. Because of the uneven levels and rates of economic development of nations, resistance against globalism may be considered to be a chronic problem. As a force perhaps as powerful as globalism in modern history, nationalism was initially fostered by print technology. However, the other forms of resistance against globalism are also facilitated by communication technologies. Historically, the ideological thrust of nationalism in the modern nation-states has been toward uniformity in religion, language, and ethnicity. It is no wonder, therefore, that the major wars of the 20th century have revolved around clashing national identities. The Cold War temporarily shifted the emphasis to ideological issues. The post-Cold War era is clearly marked by a return to national, ethnic, and religious rivalries and conflicts. In the meantime, globalism is facilitated by expanding global communication networks with English as their lingua franca.

At the threshold of the 21st century, the world is faced with many contradictions, our awareness of which owes much to global communication. On the one hand, as Francis Fukuyama (1989) has argued, liberal capitalism appears to have triumphed to put an end to the history of ideological contestations. On the other hand, history has just begun for those marginalized nations whose growing access to the means of global communication is bringing them to the attention of the rest of the world. Some 3000 to 5000 nationalities around the world are increasingly clamoring to be subjects rather than objects of history. We may thus expect the 21st century to be an arena for competing territorial and moral claims. The hegemonic state-corporate system will continue to be challenged by sporadic but persistent acts of resistance unless the world learns to respect and celebrate diversity by devolutions of power to sub-communities of a national entity.

As Samuel Huntington (1993) has argued, a "clash of civilizations" is characterizing our own era because new economic and communication power is enabling the ancient civilizations of Asia to challenge the truth claims of the relatively new nations of Europe and America. However, thanks to global communication, a new dialogue of civilizations is also being conducted via the international communication networks. Communication technologies are enabling the past silent voices to be heard in a global Tower of Babel characterized by old and new ethnic and racial hatreds. However, global communication networks are also fostering a new ecumenicalism leading to the negotiation of new global worldviews and ethics.

In the absence of a more egalitarian world, Marshall McLuhan's global village has proved to be a place not of harmony but of colliding moral spaces and sporadic violence. The lords of the electronic castles and the rebellious serfs, shamans, and jesters surrounding them have confronted each other in physical, political, economic, cultural, and environmental encounters. In this context, global communication channels can serve the cause of world peace and reconciliation only if they can be turned into channels of international and inter-civilization dialogue. In place of exclusive national sovereignties, the global commons of outer space, ocean resources, geostationary orbit, and electromagnetic spectrum, must come under inclusive global sovereignties. In place of zones of protracted violence, such as Palestine, Kurdistan, Kashmir, and Palestine/Israel, zones of peace and shared sovereignties must be built. To turn global communication into global dialogue, however, we need to rethink the problems of sovereignty, governance, economy, human rights, civic responsibilities, and media systems in order to accommodate the human unity in diversity. That diversity can be ignored only at our own peril.

International Communication Association

The International Communication Association (ICA) is a non-profit academic association founded in 1950 as the National Society for the Study of Communication (NSSC), whose members are interested in the study, teaching, and application of all aspects of human communication." The Association maintains an active membership of more than 4,300 individuals in 70 countries, approximately two-thirds of whom are academic scholars, professors, and graduate students. Other members are in government, the media, communication technology, business law, medicine and other professions. ICA is headquartered in Washington, D.C, United States. In 2003, it was identified by the United Nations as a non-governmental organization. ICA was founded on January 1, 1950 in Austin, Texas as the National Society for the Study of Communication (NSSC), a subsidiary organization of the Speech Association of America (now the (United States) National Communication Association), when the SSA determined to exclude basic communication studies from its primary focus on rhetoric. The following year, NSSC published the first issue of its official journal, Journal of Communication. In 1967, the NSSC formally separated from the SSA and opened its membership to scholars outside of the U.S. It changed its name two years later to the International Communication Association. The association restructured into divisions to allow members to find and interact with colleagues who shared similar research interests, among the widely diverse disciplines of communication study. In the face of increasing internationalization, ICA relocated its offices from Austin, TX to the more cosmopolitan and globally accessible Washington, D.C. in 2001. The association purchased a permanent headquarters in Washington in 2006

Governance

ICA is governed by a Board of Directors, consisting of:

- the Executive Committee – President, Presidents Elect and Elect Select, the three immediate past Presidents, the least recent of whom serves as Finance Chair, and the Executive Director.
- Five Board Members at Large, representing five international regions as defined by UNESCO: the Americas; Europe; East Asia; West & South Asia; and Africa-Oceania.
- Two graduate student representatives.
- The chairs of ICA's Divisions and Interest Groups (although Interest Group representatives do not have voting rights).

The officers (except for the Executive Director) are elected by the association's members. The balloting for President and regional Members at Large is association wide; student representatives are elected by ICA's student members, while Division and Interest Group chairs are chosen by the Division/Interest Group members. Terms of office begin and end at the conclusion of each year's annual ICA conference.

The president of ICA serves a one year term; however, the Executive Committee's makeup of incoming and past presidents means that election as ICA president brings with it a six-year commitment to serve on the Executive Committee. The current (2013-2014) president is Francois Heinderyckx (Université Libre de Bruxelles), and the current President Elect is Peter Vorderer (University of Mannheim). Immediate past presidents include Cynthia Stohl (2012-2013) and Larry Gross (2011-2012).ICA maintains a series of standing and awards committees to conduct its business in relation to membership, internationalization, publications, liaison with other agencies, etc., annual awards, and ad hoc task forces. Committee members are appointed by the president. The association is administered by the six-member staff at ICA's offices in Washington, D.C. The staff is led by the Executive Director, a position held since 2000 by Michael L. Haley.

What is ICA?

- The International Communication Association (ICA) is the largest international academic association for scholars interested in the study, teaching, and application of all aspects of human and mediated communication. See the welcome message from the President and for ICA's mission statement.
- What does ICA do?
- ICA's activities fall into three main categories:
- Annual conference of some 2000 academics in the field of communication from around the world, along with occasional regional conferences.
- Publication of a series of high quality journals and related publications, both broad-ranging and specialized.

- Supports the research activities of its members. This includes a prestigious series of annual awards, representation to professional, funding and policy agencies, enabling collaboration, dissemination and publicity, and facilitating the activities of its many divisions and interest groups.

How does ICA work?

ICA is a member organization run by a Board of Directors in accordance with its bylaws. Its formal organizational structure can be seen here.

The Board of Directors consists of the President, President-Elect, President-Elect Select, Immediate Past President, the Chairpersons of Divisions enrolling 200 active members in the previous year, two Student Board Members-at-Large, and five Regional Board Members-at-Large.

A subcommittee of the Board of Directors consisting of the President, President-Elect, President Elect-Select, Immediate Past President, Past President, Finance Chair and the Executive Director is formed as the Executive Committee. The Executive Director serves as a nonvoting member of the Executive Committee and the Board, and also manages theICA Office in Washington.

ICA maintains a series of standing and awards committees to conduct its business in relation to membership, internationalisation, publications, liaison with other agencies, etc, annual awards, and ad hoc task forces.

Membership

How does ICA facilitate communication among members?

The easiest way for members to communicate to each other is through the online member search function *'Find a Colleague'*. Once a member has logged in to the ICA Members Only section, they can access *Find a Colleague* through the link provided in the list under the title, *Directories*. Search functions include looking up by first or last name, a member's section, geographic location, their keywords, research areas, and any combination of each criteria. Each ICA member creates a profile providing information others may opt to search for. For example, if a member indicates that he or she is willing to be a conference paper reviewer, someone looking for reviewers can locate the member by performing a search for those criteria.

Conferences also provide a venue for members to visit and catch up with each other. The association-wide annual conference usually takes place at the end of May or in June when held in a location outside North America. Regional conferences also provide the opportunity to meet with other members. Regional conferences take place throughout the year in various parts of the world.

With the advent of social networks, ICA is creating other internet opportunities for members and non-members to communicate. The ICA Forums is a member's only internet area where they can post questions and comments regarding just about anything! Each ICA section (division or interest group) has a site dedicated to the members of that section. As well, there is a Research Collaboration Forum available to all ICA members. The Forums can be accessed through a drop-down list on the lower right side of the My ICA members' homepage, above their list of sections. Also, ICA is establishing its presence in common social networking sites such as LinkedIn, My Space, and Facebook to name a few. In this manner, it hopes to make itself available to the world at-large providing the space for dialogue.

What fields or specialties are included in ICA?

Members can join one or more sections of the organization, depending on their interests. At present, there are 19 Divisions (sections comprising at least 200 members) and 5 Interest Groups. Together, these cover many areas of specialty within the field of communication, broadly defined. These change on occasion, reflecting the interest of members and developments within the field.

Who are the members of ICA?

As of 2007, ICA has over 4,300 individual members. Most are academics or doctoral students though some are from professions related in one way or another to the field of communication research. Members are drawn from across the social sciences and humanities, as befits the multidisciplinary nature of the field.

Members come from 76 countries world wide, though approximately 64% are from the United States, 18% from Europe, 8% in Asia, 5% in non-US Americas, and 2% in Africa and Oceania. ICA is seeking to further internationalize its membership, and welcomes new members from around the world.

ICA also has 39 institutional members (i.e. university departments or related institutions) and a growing number (11) of association members (*i.e.* communication associations around the world

Communicating Through Traditional Media

Communication is a dynamic process. It happens all the time. Your posture, the way in which you walk, talk or carry out your work, communicates something or the other. Even your dress, accessories and ornaments communicate. For example, by looking at a person, it may be possible to say that the person belongs to a certain place or that the dress worn is from a particular state. In this lesson, you will learn about the importance of traditional media as a tool of communication. You will also learn about the use of traditional media for creating awareness on social issues and the future of traditional media.

Objectives

After studying this lesson, you will be able to do the following:

- Explain the importance of traditional media as a tool of communication;
- Describe traditional media as a part of socio cultural and ritual communication;
- Identify the use of traditional media for social issues;
- Discuss the future of traditional media.

Communicating Through Traditional Media

Millions of people living in remote areas in developing countries do not have access to mass media like television. For example to watch a television you need electricity, to read newspaper literacy is required. For radio also, radio receivers and a tower are required to receive the signals. For people living without electricity and other facilities this might create problems to access information. But this is not the case with traditional media. It requires only a certain amount of space to perform. For communication to reach a common person, it must be simple and free from language barriers. Traditional media fulfills these requirements. Traditional media are vehicles of communication which are rich in variety. They are readily available and economically viable. Men and women of different age groups enjoy them. They win the confidence of rural masses, as they are live. In fact, traditional media create a 'face to face situation' between the communicator and the receiver of the message. Have you ever seen a puppet show or a street theatre performance? These are the modern forms of traditional media which have adopted various techniques and forms from traditional forms such as drama, story telling, song and music.

Do they communicate any message to you? Any media that you watch will have one or more messages. You have studied in your earlier lessons about the use of advertisements in the mass media. Traditional media are free from such advertisements. This is one of the unique characteristics of traditional media. Mass media, being funded by the advertisers might result in sometimes becoming biased or filling the media with the content which is not relevant to the common people. Traditional media being free from advertisements communicate what is always need based and purposeful. They try to share messages about various facets of a problem or issue and convey information which will benefit the audience.

Entertainment is always a part of traditional media. Along with entertainment, educating the audience on various issues is an important characteristic of traditional media. Entertainment is one of the interesting ways to involve the audience. Messages are communicated with a blend of entertainment, education and communication. They do not create any false desires or expectations.

Traditional Media as A Part of Socio-Cultural and Ritual Communication

Have you ever noticed that every festival in a village has either some performances or games related to it or some traditional motifs and paintings? You will also find that people in a village get together and participate actively in festivals. There are sometimes separate games and activities for the old and young, men, women and children. Participating in these games will not only yield relaxation but also entertainment and develop team spirit amongst the people. Have you participated in such games? These are one of the best forms of communication. You notice that communication always flows in these activities. They are intimate and personal and effective for easy adoption. Traditional media forms are a part of the way of life of a community and provide acceptable means of bringing development issues into the community on its own terms. In India, they have proven to be successful mass motivators. You have already learnt that traditional media played a significant role in the freedom movement of India. India has a history of using traditional media for communicating development messages to the masses. Let us se how this was done. Traditional performances like ritualistic dances, religious songs and mythology based rural plays which are highly popular have been used to convey messages on population, health and hygiene to the rural masses. Communicators, therefore, have to test different categories of traditional performances to identify the ones that are flexible enough to absorb development messages to meet the contemporary needs; flexibility is the most important factor. It determines the viability of a traditional medium for rural communication. Cross cultural communication hurdles are not encountered here. Though these mediums are culture and regional specific, they can communicate the solidarity and friendship with each other through non verbal communication. Performances and fine arts need no language. Have you observed communication practices in a market place, exhibition or a social gathering? When people get together at one place, communication flows among them and there is always a possibility of discussion on various issues. What do you understood here? A social, ritual or cultural gathering will initiate a discussion and communication begins. A market, festival, ritual, village fair or a celebration are all situations where traditional communication takes place. But for every situation you require a separate mode of communication. The uniqueness of traditional media helps to cater to the specific needs of these situations. Can you identify any other form of public communication places where social, cultural and ritual communication takes place? Traditional media like street plays, songs and dances have a special role in harnessing developmental messages in India. Various governmental bodies like Song and Drama Division, educational institutions and non-governmental organizations use traditional media for creating awareness amongst people. Their primary appeal is to the emotion rather than intellect. They utilize an immense variety of forms and themes to suit the communication requirements of the masses.

Applications of Traditional Media

Communication through traditional media is always respected among various communities. Do you know why? These will create a space for respect, friendship and understanding. Do you know that traditional media can be used to create awareness and sensitise people on various social issues? For example, it may not be easy to create awareness on an issue such as HIV/ AIDS using face to face communication. Such an issue can be explained through a puppet show which may be more convenient and effective. Traditional media is very useful to deal with sensitive issues of health, where face to face communication may not be suitable. During the freedom struggle also many of these performing arts have played a vital role in spreading the spirit of freedom movement. In our country, the government has been successful in spreading messages of family planning, polio immunization etc. through traditional media. Have you heard of the term 'ballad'? A ballad is a poem usually set to music and is often a story told in the form of a song. Ballads are also referred to as 'bullets' because they are very powerful and can communicate effectively. Issues relating to the environment, energy conservation, dowry etc. made simple and sung publicly using the ballad form. Such songs are often humorous also. The rhythmic play of words in a ballad make the messages work effectively on listeners. Street theatre is a combination of traditional theatre forms and uses its techniques. Due to their flexibility in staging the short play at any place at any time, they always have an audience for their messages. The mix of songs, drama and music make the messages on social issues like child labour, individual differences, environment, health issues, women etc motivate viewers and leave an unforgettable impact. While using traditional media for spreading social messages, the original form of media should not be changed. There is always the risk of distorting these forms of traditional mediums while using them for conveying messages on social issues. One must use them carefully without hurting the style, form and presentation of these forms.

Future of Traditional Media

Sustainability

When you go to a movie hall to watch a film or to a cyber café to access the internet, you have to pay some money. That money will be used for sustaining those mediums of communication. It is in a similar manner that traditional forms of media also manage to sustain themselves. After a performance, viewers generally contribute money to keep the artists and the art alive. It is the people, rather our parents and grandparents who have contributed to these art forms. The money which comes from the audience helps them to survive, grow and contribute more innovative programmes from time to time. But do you know that many of these mediums are rapidly disappearing? This amounts to losing our culture and tradition. Presently many of the performers, due to the advent of television and other modern mass media have changed their creative professions. It is a dangerous situation to lose one's own past and cultural history. These mediums are not taught in schools or universities. But they are learnt by interested young people to carry them from one generation to another generation. There any many oral traditions of which there is no written documentation. The stories and songs are orally learnt from each other and preserved in the minds of the people. If there are no takers of these mediums, then these mediums will gradually be lost. After studying the four lessons on traditional media, you might have understood how important these mediums are for our well being. But how do we make these mediums survive? Who is going to take the responsibility? It is you, and your friends. Our society needs to contribute to save these mediums. We can encourage these mediums by viewing, writing, talking, and learning about them and as a whole knowing them. The government of India is using these mediums for spreading awareness on health, environment and other social issues among masses. For revival and survival of traditional media, the government has established bodies like the Song and Drama division, Sangeet Natak Akadamy, Lalit Kala Academy, Directorate of Audio Visual Publicity etc. at the national, state and district levels. The advents of modern mass media technologies which have started recording these mediums are posing a threat to these mediums. Though mass media has its potentiality in mass communication, traditional mediums are always simple and useful for better human relations. Change is inevitable. The changing face of communication and media is limiting the space for traditional media. The debate these days is how far one can retain these mediums in their own forms? With the changing audience, these forms are also affected by modern technologies and themes. Stories of 'raja' and 'rani' or mythological stories and themes are fast disappearing. Performers are leaving their creative arts because of lack of financial support and patronage from people. As long as this media form is respected, they can be kept alive for future generations. These are all great treasures of our culture and tradition. But the challenge that lies ahead of us is to save the few traditional forms of media from fast eroding and reviving the older forms to present them for future generations

Traditional folk media

Many Africans, especially those in remote rural areas, live a long way from the global information highway. In many rural villages there are no satellite dishes, computers or televisions. But these villages contain a rich communication environment that predates modern electronic media by many, many years. Modern media may not only fail to reach many remote rural areas, but lack cultural credibility when they are available. Receptivity to outside messages can be enhanced by the use of traditional folk media that are often an integral part of rural life and that use the visual and oral expressions that are commonly understood by villagers.

Traditional folk media are cultural resources that accumulate indigenous knowledge, experiences and expressions passed down from generation to generation. Woven into proverbs and poems, songs and dances, puppet plays and stories, rhythms and beats, they are embedded with a strong sense of cultural identity which can be a potent force for development. In many cases, these media are the traditional conduits of indigenous knowledge, experience and culture. Creative use of these cultural resources in communities where they are popular and well entrenched can be a subtle and effective way of introducing development ideas and messages.

The use of traditional folk media in development is not new. For many years they have attracted the attention of communication professionals as an alternative or a complement to modern mass media. Traditional media have been used in family planning campaigns, in health care programmes, in environmental protection projects, and in adult literacy programmes, among others.

Development communicators have used traditional media to help mothers learn to prepare more nutritional food for their children, to influence attitudes towards family size and to introduce new practices to farmers. Traditional folk media offer an effective means to integrate local agricultural knowledge with new scientific knowledge from outside sources. Some of the advantages of traditional folk media are that they do not require large capital investments and there is not a dependence on expensive communication technologies that are liable to break down. Traditional media can be used "live", and are likely to have the greatest impact when audience members can interact with the performers and artists and even participate. Alternatively, they can be coupled with other media such as radio and television.

Rural radio

Rural radio is distinctive from urban radio in that it is directed specifically to rural people and their information needs. In many cases, the information needs of rural people are virtually ignored by national radio networks. A rural radio approach is an alternative to narrow, city centered urban radio. Rural radio has a rich history. The first rural radio programmes in Africa appeared in the late 1960s, growing out of educational radio efforts in Europe, Canada and the USA. One of the most significant contributors to the evolution of rural radio was the Radio Forum movement in Canada from the 1940s to the 1960s. Listening groups gathered around a radio receiver at a given time to listen to a programme on specific agricultural topics. At the end of the programme, the group discussed what they had just heard and sent their comments and questions back to the radio producers. At its height, the Radio Forum involved 1,600 groups across Canada with a total audience of 30,000 listeners.

UNESCO adapted lessons from the Canadian experience for use in India in the 1950s. In subsequent years, the use of radio as a development tool was promoted in various regions of the world, including Africa. Collective listening groups, so-called "radio clubs", were organized in various African countries, such as the Association des Radios Clubs du Niger, founded in 1962

From these early efforts, a methodology for rural radio evolved which was based initially on mixed programming (combining agriculture, health, oral tradition, music, and folk tales) and subsequently developed into a more interactive use of the medium. Today, it is widely recognized that rural radio programmes are most effective when produced with audience participation, in local languages and taking into account cultural traditions. Community participation is a fundamental characteristic of rural radio live public shows; village debates and participation in the actual management of the radio station are just a few examples. This approach empowers rural people to participate in the dialogue and decision-making processes essential for them to control their own economic, social and cultural environment and play an active part in development activities. Rural radio can fulfill a number of versatile functions. It is:

- an important mechanism for rapid diffusion of development information in a diversity of languages and to widespread, often remote geographical areas;
- a channel for interactive communication, for dialogue and debate on the major issues of rural development;
- a platform for democratic and pluralistic expression of the opinions, needs and aspirations of rural communities;
- a tool for cultural expression and entertainment, and a means of collecting, preserving and enhancing the oral and musical heritage of rural communities;
- a medium to collect local information on social issues, which is essential for defining, planning and implementing development efforts;

- a means of raising public awareness and motivation; and
- a tool which, combined with other media, can be used for training and the transfer and exchange of knowledge and technologies.

In Africa, radio is one of the most widespread and popular tools of communication. It is, therefore, a very appropriate communication technology to address food security, poverty reduction, environmental protection and a host of other areas of concern to rural Africans. Rural radio often goes beyond agricultural issues to address a wide range of related social, educational, health and cultural issues. It is excellent for motivating farmers and for drawing their attention to new agricultural production ideas and techniques. It is inexpensive, has wide coverage and is readily available, even to very remote rural populations. Programme production is relatively simple and local stations can easily create their own content.

In recent years some important changes have taken place in radio in Africa. Once a top down medium for delivery of messages, it is now becoming more interactive, with opportunities for dialogue, exchange of views and debate. Centralized radio, based in the capital or in the major urban centres, is becoming decentralized with many regional and local stations. Government controlled radio is being joined by independent, private radio stations. Deregulation and decentralization, the ending of state monopolies and the emergence of new commercial broadcasters are all creating a more positive context for rural radio - one which encourages closer and better interactive communication with African communities and which empowers rural people to actively use this important tool for their own development.

Participatory video

The popularity of small-format video as a tool for rural and agricultural development has increased dramatically in recent decades, in large part because it is a highly persuasive and effective communication medium. Video makes it possible to:

- overcome barriers of illiteracy;
- visually demonstrate new farming ideas and techniques;
- compress time (*e.g.*, a crop cycle into a short presentation);
- constrict space differences (*e.g.*, use testimonials of rural people from other locations); and
- standardize information so that it is transmitted accurately from technical sources to farmers.

Video is an effective communication tool to increase awareness about specific development problems and stimulate local discussion of possible solutions. It has become relatively cheap and easy to use. In the last decade, the cost of video equipment has decreased from about Rs.10,000 to Rs.3,000 and the weight has dropped from 30 to 3 kg. With some basic training, rural people can prepare video presentations about community development concerns in local languages. Editing can be done in the camera and through immediate playback video production results can be quickly viewed. Presentations can even be shown in areas without electricity using battery-powered monitors.

Today, CESPA is financially independent and continues to successfully produce audiovisual training material for thousands of farmers and rural residents in Mali as well as provide services to institutions and clients across West Africa. It has also become a major centre for advice and training on the use of participatory video as a development tool for Africa.

Participatory video in Mali

The Centre for Audiovisual Production Services (CESPA) was established in Mali after successful experiences training farmers in Latin America utilizing video coupled with guidebooks and group discussions.

In the early 1980s, an FAO project in Mali has as one of its objectives the training of plant sanitation agents. A large number of trained agents were needed so it was decided to carry out mass training sessions using video. This method had been successfully used in Latin America, and a Peruvian video expert was recruited to carry out the training in Mali in 1984/85.

The successes of this training approach led to an FAO project from 1988 to 1992 to establish an audiovisual production centre. The project began by training a group of "audiovisual pedagogues" who were then hired by the Centre. The project focused primarily on multimedia training of farmers, but also included were social and cultural programmes for rural people. Participatory video was an essential communication tool. Initially, CESPA operated as a non-profit, public service organization for training farmers. However, policies which emphasized private-sector initiatives, income generation and financial sustainability, resulted in the commercial evolution of the Centre and the widening of its audience. Initially, many development planners dismissed video as too "high-tech" and as an "inappropriate technology" for working with rural communities. Experience has shown otherwise. In many development settings, small-format video, in the form of consumer grade video cameras and portable monitors, has enabled rural dwellers to share their experiences with one another across vast distances; access up-to-date knowledge prepared by agricultural technicians; and speaks directly to otherwise unapproachable decisions makers. Perhaps the most significant, but frequently misunderstood, aspect of participatory video is that for the local people involved it is the process of communication that is vastly more important than the video product that is created. Instead of being used as a one-way broadcast medium, video can be used as a communication process tool which enables rural people to address their information and knowledge needs to achieve their development objectives.

In Africa, FAO is applying participatory video to rural and agricultural development based on experiences that have been successful in Latin America using a methodology called "rural audiovisual pedagogy". This methodology uses participatory video as a communication tool for mediating between rural people's needs and the possible sources of information and expertise to respond to those needs. This is achieved through the production and use, with local communities, of video documentaries and training presentations. Used in this way, video can empower rural residents to actively participate in development by articulating their ideas and taking part in decisions; by recognizing the value of the skills they already possess while gaining new knowledge; and by planning and carrying out local development activities. Experience has shown that rural people who have benefited from participatory video training are more apt to address their own development problems, use newly acquired skills and knowledge, and act as development agents for neighboring communities.

Traditional Communication Media in the Development Process - Status, Role and Relevance in New Era

Traditional media are an effective and important part of the communication system. Theses are unique in nature, as they resemble the day to day life pattern of the rural masses. These media are a source of popular entertainment for the rural audience, in addition to providing instruction and information. Our country has a rich heritage of folk arts, folk dance, folk tales, epics, ballads and plays that can be used for development work. The traditional medias have been able to aid in national development since a long time. During the freedom struggle, when the mass media were under the British control, the folk media like tamasha, Bhavai and Nautanki were used by the freedom fighters to ridicule the oppressors.. The central and stage governments have used the Nationalist Harikatha to educate the masses on family planning, developmental activities, democratic values and national integration. The institute of rural communication have successfully used the folk arts for their campaign on "environment building for literacy" in Udaipur. The folk arts like "Dholki Baris", "Loknatya", "Jatra", "Keertana", "Puppetry", etc., have been used by a number of social workers, reformists, and political leaders, developmental agencies to educate the rural masses and to disseminate new information. These media have created a tremendous impact in influencing the attitudes of the people in the past.

However, recently the developmental authorities have shifted their attention to the fast emerging electronic media. These new electronic communication gadgets hold a peculiar charm for the people involved in development work, as they feel that in a short time and less cost they can reach to every nook and corner of the country and spread their message. A critical look at the characteristics of these modern media reveals that almost all of them require a lot of infrastructure facilities like the electricity, the computer facility, Internet connection, telephone connection etc., and ours is a country wherein till today there are around thousands of Villages that do not have accessibility to electricity or telephone connection, what to talk then of computer, Internet or e-mail services Further, the soft infrastructure for the new media i.e., "literacy" is also very low in our country.

In such a situation the benefits of information revolution is to be accrued by the urban population, rural elites, progressive farmers and villagers belonging to high socioeconomic status who have accessibility to hard and soft infrastructure. The farmers in the remote rural areas are facing a communication gap or a vacuum of information as the new media have not been able to percolate into these areas and the traditional media have been shrugged off or discarded by the developmental agencies.

If we critically analyze the characteristics of some of the most famous traditional and modern media on their awareness, knowledge, attitude and mental evaluation function in the technology adoption process then we find that most of the traditional medias have a high effect in creating awareness and influencing the attitude of the audience while most of the modern media have a high effect in creating awareness and enhancing the knowledge of the audience.

Thus both the traditional and the modern media complement each other. The modern media are more effective in changing the knowledge, while the traditional media are more effective in changing the attitude. In fact the traditional media have some added advantage over the modern media like their indigenous nature, local appeal, direct personal contact, local dialect, more credibility and more cultural compatibility which make them so unique and useful.

Therefore we need to keep our traditional media alive by continuously and cautiously using them as and when possible along with the modern media. These traditional media will not only help in developmental activities but will also help in preserving and transmitting our culture, tradition and values to the next generation.

Chapter 14

Electronic Media: TV, Radio etc. for Communication of Information

Electronic Media

Electronic media is media that uses electronics or electromechanical energy for the end-user (audience) to access the content. This is in contrast to static media (mainly print media), which are most often created electronically, but don't require electronics to be accessed by the end-user in the printed form. The primary electronic media sources familiar to the general public are better known as video recordings, audio recordings, multimedia presentations, slide presentations, CD-ROM and online content. Most new media are in the form of digital media. However, electronic media may be in either analog or digital format.

Although the term is usually associated with content recorded on a storage medium, recordings are not required for live broadcasting and online networking. Any equipment used in the electronic communication process (e.g. television, radio, telephone, desktop computer, game console, handheld device) may also be considered electronic media.

History of Development

Transmission

Wire

a. Telegraph 1795-1832 tel-e-graph

It is a system of apparatus, or process for transmitting messages or signals to a distance place, especially by means of an electric device consisting essentially of a sending instrument and a distance receiving instruments connected by a conducting wire or other communications channel.

It is a mechanical apparatus for transmitting and receiving orders between the bridge of a ship and the engine room or some other part of the engineering department. In other word, send a message to (a person) by telegraph or transmit send (a message) by telegraph.

b. Facsimile 1843-1861

It is an exact copy, as of a book, painting or manuscript – it is also called fax telecommunication. It is a method or device for transmitting documents, drawing, photographs or the like, by means of radio or telephone for exact reproduction elsewhere.

c. Telephone 1849-1877

It is a system that uses wires and radio signals to send sounds such as peoples voices over long distances. The device is connected to a telephone system and that you use to listen or speak to some one who is some where else. An instrument for reproducing sounds at a distance; specifically: one in which sound is converted into electrical impulses for transmission (as by wire or radio waves).

d. Cable 1962 (Coaxial Standard)

Coaxial cable, or **coax** (pronounced 'ko.æks), is a type of cable that has an inner conductor surrounded by a tubular insulating layer, surrounded by a tubular conducting shield. Many coaxial cables also have an insulating outer sheath or jacket. The term coaxial comes from the inner conductor and the outer shield sharing a geometric axis. Coaxial cable was invented by English engineer and mathematician Oliver Heaviside, who patented the design in 1880. Coaxial cable differs from other shielded cable used for carrying lower-frequency signals, such as audio signals, in that the dimensions of the cable are controlled to give a precise, constant conductor spacing, which is needed for it to function efficiently as a radio frequency transmission line. Coaxial cable is used as a transmission line for radio frequency signals. Its applications include feedlines connecting radio transmitters and receivers with their antennas, computer network (Internet) connections, and distributing cable television signals. One advantage of coaxial over other types of radio transmission line is that in an ideal coaxial cable the electromagnetic field carrying the signal exists only in the space between the inner and outer conductors. This allows coaxial cable runs to be installed next to metal objects such as gutters without the power losses that occur in other types of transmission lines. Coaxial cable also provides protection of the signal from external electromagnetic interference.

Coaxial cable conducts electrical signal using an inner conductor (usually a solid copper, stranded copper or copper plated steel wire) surrounded by an insulating layer and all enclosed by a shield, typically one to four layers of woven metallic braid and metallic tape. The cable is protected by an outer insulating jacket. Normally, the shield is kept at ground potential and a voltage is applied to the center conductor to carry electrical signals. The advantage of coaxial design is that electric and magnetic fields are confined to the dielectric with little leakage outside the shield. Conversely, electric and magnetic fields outside the cable are largely kept from causing interference to signals inside the cable. Larger diameter cables and cables with multiple shields have less leakage. This property makes coaxial cable a good choice for carrying weak signals that cannot tolerate interference from the environment or for higher electrical signals that must not be allowed to radiate or couple into adjacent structures or circuits. Common applications of coaxial cable include video and CATV distribution, RF and microwave transmission, and computer and instrumentation data connections. The characteristic impedance of the cable is determined by the dielectric constant of the inner insulator and the radii of the inner and outer conductors. A controlled cable characteristic impedance is important because the source and load impedance should be matched to ensure maximum power transfer and minimum standing wave ratio. Other important properties of coaxial cable include attenuation as a function of frequency, voltage handling capability, and shield quality.

e. Fiber Optics 1956-1970

An **optical fiber** (or **optical fibre**) is a flexible, transparent fiber made of high quality extruded glass (silica) or plastic, slightly thicker than a human hair. It can function as a waveguide, or "light pipe", to transmit light between the two ends of the fiber. Power over Fiber (PoF) optic cables can also work to deliver an electric current for low-power electric devices. The field of applied science and engineering concerned with the design and application of optical fibers is known as fiber optics. Optical fibers are widely used in fiber-optic communications, where they permit transmission over longer distances and at higher bandwidths (data rates) than wire cables. Fibers are used instead of metal wires because signals travel along them with less loss and are also immune to electromagnetic interference. Fibers are also used for illumination, and are wrapped in bundles so that they may be used to carry images, thus allowing viewing in confined spaces. Specially designed fibers are used for a variety of other applications, including sensors and fiber lasers.

Optical fibers typically include a transparent core surrounded by a transparent cladding material with a lower index of refraction. Light is kept in the core by total internal reflection. This causes the fiber to act as a waveguide. Fibers that support many propagation paths or transverse modes are called multi-mode fibers (MMF), while those that only support a single mode are called single-mode fibers (SMF). Multi-mode fibers generally have a wider core diameter, and are used for short-distance communication links and for applications where high power must be transmitted. Single-mode fibers are used for most communication links longer than 1,000 meters (3,300 ft). Joining lengths of optical fiber is more complex than joining electrical wire or cable. The ends of the fibers must be carefully cleaved, and then carefully spliced together with the cores perfectly aligned. A mechanical splice holds the ends of the fibers together mechanically, while fusion splicing uses heat to fuse the ends of the fibers together. Special optical fiber connectors for temporary or semi-permanent connections are also available. Fiber optic (or "optical fiber") refers to the medium and the technology associated with the transmission of information as light impulses along a glass or plastic wire or fiber. Fiber optic wire carries much more information than conventional copper wire and is far less subject to electromagnetic interference. Most telephone company long-distance lines are now fiber optic. Transmission on fiber optic wire requires repeating at distance intervals. The glass fiber requires more protection within an outer cable than copper. For these reasons and because the installation of any new wiring is labor-intensive, few communities yet have fiber optic wires or cables from the phone company's branch office to local customers (known as local loop).

Wireless

a. Radio 1897-1920

Radio is the wireless transmission of signals through free space by electromagnetic radiation of a frequency significantly below that of visible light, in the radio frequency range, from about 30 kHz to 300 GHz. These waves are called radio waves. Electromagnetic radiation travels by means of oscillating electromagnetic fields that pass through the air and the vacuum of space. Information, such as sound, is carried by systematically changing some property of the radiated waves, such as their amplitude, frequency, phase, or pulse width. When radio waves strike an electrical conductor, the oscillating fields induce an alternating current in the conductor. The information in the waves can be extracted and transformed back into its original form. The transmission and reception of electric impulses or signals by means of electric waves without a connecting wire, or the use of these waves for the wireless transmission of electric impulses into which sound is converted.

b. Satellite 1958-1972

In the context of spaceflight, a satellite is an artificial object which has been intentionally placed into orbit. Such objects are sometimes called artificial satellites to distinguish them from natural satellites such as the Moon. The world's first artificial satellite, the Sputnik 1, was launched by the Soviet Union in 1957. Since then, thousands of satellites have been launched into orbit around the Earth. Some satellites, notably space stations, have been launched in parts and assembled in orbit. Artificial satellites originate from more than 50 countries and have used the satellite launching capabilities of ten nations. A few hundred satellites are currently operational, whereas thousands of unused satellites and satellite fragments orbit the Earth as space debris. A few space probes have been placed into orbit around other bodies and become artificial satellites to the Moon, Mercury, Venus, Mars, Jupiter, Saturn, Vesta, Eros, and the Sun. Satellites are used for a large number of purposes. Common types include military and civilian Earth observation satellites, communications satellites, navigation satellites, weather satellites, and research satellites. Space stations and human spacecraft in orbit are also satellites. Satellite orbits vary greatly, depending on the purpose of the satellite, and are classified in a number of ways. Well-known (overlapping) classes include low Earth orbit, polar orbit, and geostationary orbit.

About 6,600 satellites have been launched. The latest estimates are that 3,600 remain in orbit. Of those, about 1,000 are operational; the rest have lived out their useful lives and are part of the space debris. Approximately 500 operational satellites are in low-Earth orbit, 50 are in medium-Earth orbit (at 20,000 km) and the rest are in geostationary orbit (at 36,000 km). Satellites are propelled by rockets to their orbits. Usually the launch vehicle itself is a rocket lifting off from a launch pad on land. In a minority of cases satellites are launched at sea (from a submarine or a mobile maritime platform) or aboard a plane (see air launch to orbit). Satellites are usually semi-independent computer-controlled systems. Satellite subsystems attend many tasks, such as power generation, thermal control, telemetry, attitude control and orbit control.

A communications satellite or comsat is an artificial satellite sent to space for the purpose of telecommunications. Modern communications satellites use a variety of orbits including geostationary orbits, Molniya orbits, elliptical orbits and low (polar and non-polar) Earth orbits. For fixed (point-to-point) services, communications satellites provide a microwave radio relay technology complementary to that of communication cables. They are also used for mobile applications such as communications to ships, vehicles, planes and hand-held terminals, and for TV and radio broadcasting.

c. Free Space Optics 1960s

Free-space optical communication (FSO) is an optical communication technology that uses light propagating in free space to wirelessly transmit data for telecommunications or computer networking. "Free space" means air, outer space, vacuum, or something similar. This contrasts with using solids such as optical fibre cable or an optical transmission line. The technology is useful where the physical connections are impractical due to high costs or other considerations.

Typically scenarios for use are:

- LAN to LAN connections on campuses at Fast Ethernet or Gigabit Ethernet speeds
- LAN to LAN connections in a city, a metropolitan area network
- To cross a public road or other barriers which the sender and receiver do not own
- Speedy service delivery of high-bandwidth access to optical fibre networks
- Converged Voice Data Connection
- Temporary network installation (for events or other purposes)
- Re-establish high-speed connection quickly (disaster recovery)
- As an alternative or upgrade add-on to existing wireless technologies

- Especially powerful in combination with auto aiming systems, this way you could power moving cars or you can power your laptop while you move or use auto-aiming nodes to create a network with other nodes.
- As a safety add-on for important fibre connections (redundancy)
- For communications between spacecraft, including elements of a satellite constellation
- For inter and intra chip communication.

The light beam can be very narrow, which makes FSO hard to intercept, improving security. In any case, it is comparatively easy to encrypt any data traveling across the FSO connection for additional security. FSO provides vastly improved electromagnetic interference (EMI) behavior compared to using microwaves.

Advantages

RONJA is a free implementation of FSO using high-intensity LEDs.

- Ease of deployment
- Can be used to power devices
- License-free long-range operation (in contrast with radio communication)
- High bit rates
- Low bit error rates
- Immunity to electromagnetic interference
- Full duplex operation
- Protocol transparency
- Increased security when working with narrow beam(s)
- No Fresnel zone necessary

Disadvantages

For terrestrial applications, the principal limiting factors are:

- Beam dispersion
- Atmospheric absorption
- Rain
- Fog (10..~100 dB/km attenuation)
- Snow
- Scintillation
- Interference from background light sources (including the Sun)
- Shadowing
- Pointing stability in wind
- Pollution / smog

These factors cause an attenuated receiver signal and lead to higher bit error ratio (BER). To overcome these issues, vendors found some solutions, like multi-beam or multi-path architectures, which use more than one sender and more than one receiver. Some state-of-the-art devices also have larger fade margin (extra power, reserved for rain, smog, fog). To keep an eye-safe environment, good FSO systems have a limited laser power density and support laser classes 1 or 1M. Atmospheric and fog attenuation which are exponential in nature, thus limit practical range of FSO devices to several kilometres.

Internet

a. Download 1969 (file transfer protocols)

b. Live Streaming 1996 (RTP Protocol)

Display and Output

a. Information Processing

CognitiveInformationprocessingisthechange(processing)ofinformationinany manner detectable by an observer. As such, it is a process that describes everything that happens (changes) in the universe, from the falling of a rock (a change in position) to the printing of a text file from a digital computer system. In the latter case, an information processor is changing the form of presentation of that text file. Information processing may more specifically be defined in terms used by Claude E. Shannon as the conversion of latent information into manifest information (Mc Gonigle & Mastrian, 2011). Latent and manifest information is defined through the terms of equivocation (remaining uncertainty, what value the sender has actually chosen), dissipation (uncertainty of the sender what the receiver has actually received), and transformation (saved effort of questioning - equivocation minus dissipation) (Denning and Bell, 2012).

According to thefreedictionary.com, the definition of information processing is, "the sciences concerned with gathering, manipulating, storing, retrieving, and classifying recorded information". It suggests that for information to be firmly implanted in memory, it must pass through three stages of mental processing; sensory memory, short-term memory, and long-term memory. An example of this is the working memory model. This includes the central executive, phonologic loop, episodic buffer, visuo-spatial sketchpad, verbal information, long term memory, and visual information (Sternberg & Sternberg, 2012). The central executive is like the secretary of the brain. It decides what needs attention and how to respond. The central executive then leads to three different subsections. The first is phonological storage, sub-vocal rehearsal, and the phonological loop. These sections work together to understand words, put the information into memory, and then hold the memory. The result is verbal information storage. The next subsection is the visuo-spatial sketchpad which works to store visual images. The storage capacity is brief but leads to understanding of visual stimuli. Finally, there is an episodic buffer. This section is capable of taking information and putting it into long-term memory. It is also able to take information from the phonological loop and visuo-spatial sketchpad, combining them with long-term memory to make "a unitary episodic representation (Sternberg & Sternberg, 2012). In order for these to work, the sensory register takes in via the five senses: visual, auditory, tactile, olfactory, and taste. These are all present since birth and are able to handle simultaneous processing (e.g., food-tastes it, smell it, see it). In general, learning benefits occur when there is a developed process of pattern recognition. The sensory register has a large capacity and its behavioural response is very short (1–3 seconds). Within this model, sensory store and short term memory or working memory has limited capacity. Sensory store is able to hold very limited amounts of information for very limited amounts of time. This phenomenon is very similar to having your picture taken with the flash on. For a few brief moments after the flash goes off, the flash is still there. Soon after, though, it is gone and you would have never known it was there (Sternberg & Sternberg, 2012). Short term memory holds information for slightly longer periods of time, but still has a limited capacity. According to Linden (2007), "The capacity of STM had initially been estimated at "seven plus or minus two" items (Miller 1956), which fits the observation from neuropsychological testing that the average digit span of healthy adults is about seven (Cowan and others 2005). However, it emerged that these numbers of items can only be retained if they are grouped into so-called chunks, using perceptual or conceptual associations between individual stimuli." Its duration is of 5-20 seconds before it is out of the subject's mind. This occurs often with names of people newly introduced to. Images or information based on meaning are stored here as well, but it decays without rehearsal or repetition of such information. On the other hand, long-term memory has a potentially unlimited capacity (Sternberg & Sternberg, 2012) and its duration is indefinite. Although sometimes it is difficult to access, it encompasses everything learned until this point in time. One might become forgetful or feel as if the information is on the tip of the tongue.

b. Galvanometer 1832

A galvanometer is a type of sensitive ammeter: an instrument for detecting electric current. It is an analog electromechanical actuator that produces a rotary deflection of some type of pointer in response to electric current flowing through its coil in a magnetic. Galvanometers were the first instruments used to detect and measure electric currents. Sensitive galvanometers were used to detect signals from long submarine cables, and to discover the electrical activity of the heart and brain. Some galvanometers use a solid pointer on a scale to show measurements, other very sensitive types use a miniature mirror and a beam of light to provide mechanical amplification of low level signals. Initially a laboratory instrument relying on the Earth's own magnetic field to provide restoring force for the pointer, galvanometers were developed into compact, rugged, sensitive portable instruments essential to the development of electro technology. A type of galvanometer that records measurements permanently is the chart recorder. The term has expanded to include the use of the same mechanism in recording, positioning, and servomechanism equipment

c. Telegraph Sounder 1844

A telegraph sounder is a device which produces an audible sound when connected to an operating telegraph. It is similar in form to a relay.

When current flows through an electromagnet, the resulting magnetic field attracts an armature that is held up by a weight. When the current is switched on, the arm is pulled down to the electromagnet, resulting in a "click" When the current ends, the weight pulls the armature back to its resting position, resulting in a "clack." Thus, as the Telegraph key makes and breaks a contact, the sounder echoes the up and down state of the key. It is important that a sounder makes a sound both when the circuit is broken and when it is restored. This eases the process of identifying long and short presses of the key to send characters in Morse. It is important that one does not confuse this with CW. The default condition of the telegraph circuit is with electricity flowing. This enables the operator to easily detect a fault in the wire.

d. Radio/Television Tuner

A TV tuner card is a kind of television tuner that allows television signals to be received by a computer. Most TV tuners also function as video capture cards, allowing them to record television programs onto a hard disk much like the digital video recorder (DVR) does. The interfaces for TV tuner cards are most commonly either PCI bus expansion card or the newer PCI Express (PCIe) bus for many modern cards, but PCMCIA, Express Card, or USB devices also exist. In addition, some video cards double as TV tuners, notably the ATI All-In-Wonder series. The card contains a tuner and an analog (collectively known as the analog front end) along with demodulation and interface logic. Some lower-end cards lack an onboard processor and, like a Win modem, rely on the system's CPU for demodulation.

There are currently four kinds of tuner card on the market:

Analog television cards output a raw video stream, suitable for real-time viewing but ideally requiring some sort of video compression if it is to be recorded. More advanced TV tuners encode the signal to Motion JPEG or MPEG, relieving the main CPU of this load. Some cards also have analog input (composite video or S-Video) and many also provide a tuner (radio).

A tuner is a subsystem that receives radio frequency (RF) transmissions like radio broadcasts and converts the selected carrier frequency and its associated bandwidth into a fixed frequency that is suitable for further processing usually because a lower frequency is used on the output. Broadcast FM/AM transmissions are usually fed this intermediate frequency (IF) directly into a demodulator that convert the radio signal into audio-frequency signals that can be fed into an amplifier to drive a loudspeaker. More complex transmissions like PAL/ NTSC (TV),DAB (digital radio), DVB-T/DVB-S/DVB-C (digital TV) etc. uses a wider frequency bandwidth often with several sub carriers. These are transmitted inside the receiver as an intermediate frequency (IF). The next step is usually either to process sub carriers like real radio transmissions or to sample the whole bandwidth with A/Dat a rate faster than the nyquist rate that is at least 2 times the IF frequency. The term tuner can also refer to a radio receiver or standalone audio component that is part of an audio system, to be connected to a separate amplifier.

The verb tuning in radio contexts means adjusting the radio receiver to receive the desired radio signal carrier frequency that a particular radio station uses.

e. Laser Light Show 1970s

A laser lighting display or laser light show involves the use of laser light to entertain an audience. A laser light show may consist only of projected laser beams set to music, or may accompany another form of entertainment, typically musical performances. Laser light is useful in entertainment because the coherent nature of laser light allows a narrow beam to be produced, which allows the use of optical scanning to draw patterns or images on walls, ceilings or other surfaces including theatrical smoke and fog without refocusing for the differences in distance, as is common with video projection. This inherently more focused beam is also extremely visible, and is often used as an effect. Sometimes the beams are "bounced" to different positions with mirrors to create laser sculptures.

f. Computer Monitor 1950s/1976 (for PCs)

A monitor or a display is an electronic visual display for computers. The monitor comprises the display device, circuitry and an enclosure. The display device in modern monitors is typically a thin film transistor liquid crystal display (TFT-LCD) thin panel, while older monitors used acathode ray tube (CRT) about as deep as the screen size. Originally, computer monitors were used for data processing while television receivers were used for entertainment. From the 1980s onwards, computers (and their monitors) have been used for both data processing and entertainment, while televisions have implemented some computer functionality.

Signal Processing

a. Analog Encoding 1830's (morse code)
b. Electronic Modulating 1832-1927
c. Electronic Multiplexing 1853 (TDM)
d. Digitizing 1903 (PCM Telephone)
e. Electronic Encryption 1935-1945
f. Online Routing 1969
g. Electronic Programming 1943-current

Electronic Information Storage

Recording Medium

a. Punched card and Paper Tape 1725/1846
b. Phonograph Cylinder and Disk 1857-1958
c. Film 1876-1889
d. Magnetic Storage 1898-2003
e. RAM 1941-current

f. Barcodes 1952/1973 (UPC)
g. Laser Disc 1969-1978
h. Compact Disc/DVD 1982/1993-current

Content Formats

a. Content in general 1877-current
b. Audio Recording 1877-current
c. Video Recording 1952-current
d. Digital File Formats
e. Database Content and Formats 1963-current

Interactivity

a. Control Panel
b. Input Device
c. Game Controller
d. Handheld
e. Wired Glove
f. Brain computer interface (BCI)

Uses

Electronic media are ubiquitous in most of the developed world. As of 2005, there are reports of satellite receivers being present in some of the most remote and inaccessible regions of China. Electronic media devices have found their way into all parts of modern life. The term is relevant to media ecology for studying its impact compared to printed media and broadening the scope of understanding media beyond a simplistic aspect of media such as one delivery platform (e.g. the World Wide Web) aside from many other options. The term is also relevant to professional career development regarding related skill sets.

Primary Uses of Electronic Media:

a. Journalism

Journalism is a method of inquiry and literary style that aims to provide a service to the public by the dissemination and analysis of news and other information. Journalistic integrity is based on the principles of truth, disclosure, and editorial independence. Journalistic mediums can vary diversely, from print publishing to electronic broadcasting and from newspaper to television channels, as well as to the web, and to digital technology. In modern society, the news media is the chief purveyor of information and opinion about public affairs. Journalism, however, is not always confined to the news media or to news itself, as journalistic communication may find its way into broader forms of expression, including literature and cinema. In some nations, the news media is still controlled by government intervention, and is not fully an independent body.

In a democratic society, however, access to free information plays a central role in creating a system of checks and balance, and in distributing power equally amongst governments, businesses, individuals, and other social entities. Access to verifiable information gathered by independent media sources, which adhere to journalistic standards, can also be of service to ordinary citizens, by empowering them with the tools they need in order to participate in the political process.

The role and status of journalism, along with that of the mass media, has undergone profound changes over the last two decades with the advent of digital technology and publication of news on the Internet. This has created a shift in the consumption of print media channels, as people increasingly consume news through e-readers, smart phones, and other electronic devices, challenging news organizations to fully monetize their digital wing, as well as improvise on the context in which they publish news in print. Notably, in the American media landscape, newsrooms have reduced their staff and coverage as traditional media channels, such as television, grapple with declining audiences. For instance, between 2007 and 2012, CNN edited its story packages into nearly half of their original time length.

This compactness in coverage has been linked to broad audience attrition, as a large majority of respondents in recent studies show changing preferences in news consumption. The digital era has also ushered in a new kind of journalism in which ordinary citizens play a greater role in the process of news making, with the rise of citizen journalism being possible through the Internet. Using video camera equipped smart phones, active citizens are now enabled to record footage of news events and upload them onto channels like YouTube, which is often discovered and used by mainstream news media outlets. Meanwhile, easy access to news from a variety of online sources, like blogs and other social media, has resulted in readers being able to pick from a wider choice of official and unofficial sources, instead of only from traditional media organizations.

b. News

News is the communication of selected information on current events. It is shared in various ways: among individuals and small groups (such as by word of mouth or newsletters); with wider audiences (such as by publishing, either in print or online, or broadcasting, such as on television or radio); or in ways that blend those traits (such as when social media sharing starts among individuals but goes viral).

c. Marketing

Marketing is the process of communicating the value of a product or service to customers, for the purpose of selling that product or service. Marketing can be looked at as an organizational function and a set of processes for creating, delivering and communicating value to customers, and customer that also benefits the organization. Marketing is the science of choosing target markets through market analysis and market segmentation, as well as understanding consumer behavior and providing superior customer value. From a societal point of view, marketing is the link between a society's material requirements and its economic patterns of response. Marketing satisfies these needs and wants through exchange processes and building long term relationships.

d. Advertising

Advertising or advertising in business is a form of marketing communication used to encourage, persuade, or manipulate an audience (viewers, readers or listeners; sometimes a specific group) to take or continue to take some action. Most commonly, the desired result is to drive consumer behavior with respect to a commercial offering, although political and ideological advertising is also common. This type of work belongs to a category called affective labor. In Latin, ad vertere means "to turn toward". The purpose of advertising may also be to reassure employees or shareholders that a company is viable or successful. Advertising messages are usually paid for by sponsors and viewed via various old media; including mass media such as newspaper, magazines, television advertisement, radio advertisement, outdoor advertising or direct mail; or new media such as blogs, websites or text messages. Commercial advertisers often seek to generate increased consumption of their products or services through "branding", which involves associating a product name or image with certain qualities in the minds of consumers. Commercial advertisers who spend money to advertise items other than a consumer product or service include political parties, interest groups, religious organizations and governmental agencies. Nonprofit organizations may rely on free modes of persuasion, such as a public service announcement (PSA). Modern advertising was created with the innovative techniques introduced with tobacco in the 1920s, most significantly with the campaigns of Edward, which is often considered the founder of modern, Madison Avenue advertising. Internationally, the largest ("big four") advertising conglomerates are Inter public, Omnicom, Publicis, and WPP.

e. Graphic Design

Graphic design is the art of communication, stylizing, and problem solving through the use of type, space and image. The field is considered a subset of visual communication and communication design, but sometimes the term "graphic design" is used interchangeably with these due to overlapping skills involved. Graphic designers use various methods to create and combine words, symbols, and images to create a visual representation of ideas and messages. A graphic designer may use a combination of typography, visual arts and page layout techniques to produce a final result. Graphic design often refers to both the process (designing) by which the communication is created and the products (designs) which are generated. Common uses of graphic design include identity (logos and branding), publications (magazines, newspapers and books), print advertisements, posters, billboards, website graphics and elements, signs and product packaging. For example, a product package might include a logo or other artwork, organized text and pure design elements such as images, shapes and color which unify the piece. Composition is one of the most important features of graphic design, especially when using preexisting materials or diverse elements.

Commerce

a. Industry

- Industry is the production of a good or service within an economy. Manufacturing industry became a key sector of production and labour in European and North American countries during the Industrial Revolution, upsetting previous mercantile and feudal economies. This occurred through many successive rapid advances in technology, such as the production of steel and coal.
- Following the Industrial Revolution, perhaps a third of the world's economic output is derived from manufacturing industries. Many countries and many developing/semi-developed countries (People's Republic of China, India etc.) depend significantly on manufacturing industry. Industries, the countries they reside in, and the economies of those countries are interlinked in a complex web of interdependence.

b. Corporate Communications

Corporate communication is a set of activities involved in managing and orchestrating all internal and external communications aimed at creating favourable point of view among stakeholders on which the company depends. It is the messages issued by a corporate organization, body, or institute to its audiences, such as employees, media, channel partners and the general public. Organizations aim to communicate the same message to all its stakeholders, to transmit coherence, credibility and ethic. Corporate Communications help organizations explain their mission; combine its many visions and values into a cohesive message to stakeholders. The concept of corporate communication could be seen as an integrative communication structure linking stakeholders to the organization.

c. Business Presentations

A presentation is the process of presenting a topic to an audience. It is typically a demonstration, lecture, or speech meant to inform, persuade, or build good will. The term can also be used for a formal or ritualized introduction or offering, as with the presentation of a debutante.

- The process of offering for consideration or display
- A social introduction, as of a person at court
- A demonstration, lecture, or welcoming speech
- A manner or style of speaking, instructing or putting oneself forward
- The manner of presenting, especially the organization of visual details to create an overall impression
- The formal introduction of a person, as into society or at court; debut

Entertainment

a. Television

- Television, colloquially known as TV, (from French *télévision*; from Ancient Greek *(tèle)*, meaning "far", and Latin *visio*, meaning "sight") is a telecommunication medium that is used for transmitting and receiving moving images and sound. In a broader sense, television can also refer to images that are monochrome (black-and-white) or color, or images with or without accompanying sound. *Television* may also refer specifically to a television set, television program, or television transmission.

- Commercially available since the late 1920s (but in extremely limited numbers), the television has since become commonplace in homes, businesses and institutions, particularly as a vehicle for advertising, a source of entertainment, and news. During the 1950s, television became the primary medium for molding public opinion. Then, in the mid-1960s, color television became more widely available.

b) Video Games

- A video game is an electronic game that involves human interaction with a user interface to generate visual feedback on a video device. The word *video* in *video game* traditionally referred to a raster display device, but it now implies any type of display device that can produce two or three dimensional images. The electronic systems used to play video games are known as platforms; examples of these are personal computers and video game consoles. These platforms range from large mainframe computers to small handheld devices. Specialized video games such as arcade games, while previously common, have gradually declined in use. Video games have gone on to become an art form and industry.
- The input device primarily used to manipulate video games is called a game controller, and varies across platforms. For example, a controller might consist of only a button and a joystick, while another may feature a dozen buttons and one or more joysticks. Early personal computer games often needed a keyboard for game play, or more commonly, required the user to buy a separate joystick with at least one button. Many modern computer games allow or require the player to use a keyboard and a mouse simultaneously. A few of the most common game controllers are gamepads, mouse, keyboards, and joysticks. In recent years, additional methods of input have emerged such as camera-based player observation for video game consoles and touch-sensitive screens on mobile devices.
- Video games typically use additional means of providing interactivity and information to the player. Audio is almost universal, using sound reproduction devices, such as speakers and headphones. Other feedback may come via haptic peripherals, such as vibration or force feedback, with vibration sometimes used to simulate force feedback.

c) Music

Music is an art form whose medium is sound and silence. Its common elements are pitch (which governs melody and harmony), rhythm (and its associated concepts tempo, meter, and articulation), dynamics, and the sonic qualities of timbre and texture. The word derives from Greek μ (*mousike*; "art of the Muses").

The creation, performance, significance, and even the definition of music vary according to culture and social context. Music ranges from strictly organized compositions (and their recreation in performance), through improvisational music to aleatoric forms. Music can be divided into genres and subgenres, although the dividing lines and relationships between music genres are often subtle, sometimes open to personal interpretation, and occasionally controversial. Within the arts, music may be classified as a performing art, affine, and auditory art. It may also be divided among art music and folk music. There is also a strong connection between music and mathematics. Music may be played and heard live, may be part of a dramatic work or film, or may be recorded. To many people in many cultures, music is an important part of their way of life. Ancient Greek and Indian philosophers' defined music as tones ordered horizontally as melodies and vertically as harmonies. However, 20th century composer John Cage thought that any sound can be music, saying, for example, "There is no noise, only sound." Musicologist Jean and Jacques Nattiez summarizes the relativist, post-modern viewpoint: "The border between music and noise is always culturally defined which implies that, even within a single society, this border does not always pass through the same place; in short, there is rarely a consensus.

Career in Electronic Media

A Career in Electronic Media often requires you to be creative and hard working. The prime job of electronic media persons is to communicate the transmission and the production of a certain program. The young generation of today is seen seeking jobs in Electronic Media as it is regarded as one of the glamorous industries. The technology of today is rapidly growing and thus opening numerous jobs in the field of electronic media. The field of electronic media can be divided into television and the radio. These are considered as the fastest modes of media to communicate the news or a series of events to the public.

Scope of Career in Electronic Media

Many companies are in continuous need of quality professional from the field of electronic media. The want for people of the technical section is in greater demand and has better prospects. Prior qualification in film making i.e. the film and the production of the videos is an added advantage. Candidates with good creativity levels and those who can think out of the box have bright career opportunities in this field. Thus jobs in the electronic media are available in the fields of pre- production, postproduction, production assistant, broad casters, editors, animation and computer graphic designers, set designers and many more. With so many television channels and mediums of entertainment cropping up in India, the scope of a good career in Electronic Media has increased by leaps and bounds.

Television for Communication of Information

Television

Television is a telecommunication system for broadcasting and receiving moving pictures and sound over a distance. The term has come to refer to all the aspects of television from the television set to the programming and transmission. The word is derived from mixed Latin and Greek roots, meaning "far sight": Greek "tele", far, and Latin visio-n, sight.

The originals of what would become today's television system can be traced back as far as the discovery of the photoconductivity of the element selenium by Willoughby Smith in 1873 and the invention of the scanning disk by Paul Nipkow in 1884. All practical television systems use the fundamental idea of scanning an image to produce a time series signal representation. That representation is then transmitted to a device to reverse the scanning process. The final device, the television (or T.V. set), relies on the human eye to integrate the result into a coherent image.

Electromechanical techniques were developed prior to World War II, most notably by Charles Francis Jenkins and John Logie Baird. Baird gave the world's first public demonstration if a working television system that transmitted moving images with tone graduation (grayscale) on 26 January 1926 at his laboratory in London. Baird further demonstrated the world's first colour television transmission on 3 July 1928.

Completely electronic television systems relied on the inventions of Philo Taylor Fransworth, Vladimir Zworykin and others to produce a system suitable for mass distribution of television programming. Fransworth gave the world's first public demonstration of an all-electronic television system at the Franklin Institute in Philadelphia on 25 August 1934.

Regular broadcast programming occurred in the United States, the United Kingdom, Germany, France and the Soviet Union before World War II. The first regular television broadcasts with a modern level of definition (240 or more lines) were made in England in 1936, soon upgraded to the so-called "System A" with 405 lines. Large scale network broadcasting began in the United States in 1946, and television became common in American homes by the middle 1950s. While North American over-the-air broadcasting was originally free of direct marginal cost to the consumer (i.e., cost in excess of acquisition and un keep of the hardware) and broadcasters were compensated primarily by receipt of advertising revenue, increasingly United States television consumers obtain their programming by subscription to cable television systems or direct to home satellite transmissions. In the United Kingdom, on the other hand, the owner of each television must pay a license fee annually which is used to support the British Broadcasting Corporation.

Technology

Elements of a Television System

The elements of a simple television system are:

- An image source, this may be a camera for live pickup of images or a flying spot scanner for transmission of films
- A sound source
- A transmitter, which modulates one or more television signals with both picture and sound information for transmission

- A receiver (television) which recovers the picture and sound signals from the television broadcast
- A display device, which turns electrical signals into visible light
- A sound device, which turns the electrical signals into sound waves to go along with the picture
- Practical television systems include equipment for selecting different image sources, mixing images from several sources at once, insertion of pre-recorded video signals, synchronizing signals from many sources, and direct image generation by computer for such purposes as station identification. Transmission may be over the air from land based transmitters, over metal or optical cables, or by radio from synchronous satellites. Digital systems may be inserted anywhere in the chain to provide better image transmission quality, reduction in transmission bandwidth, special effects, or security of transmission from reception by non-subscribers.

Display Technology

There are several kinds of video displays used in modern TV sets:

CRT

The most common displays are direct-view CRTs for up to 40 in (100 cm) (in 4: 3) and 46 in (115 cm) (in 16: 9) diagonally. These are still the least expensive, and are a refined technology that can still provide the best overall picture quality. As they do not have a fixed native resolution, in some cases they are also capable of displaying sources with a variety of different resolutions at the best possible image quality. The frame rate or refresh rate of a typical NTSC format CRT TV is 60 Hz, and for the PAL format, it is 50 Hz. A typical NTSC broadcast signal's visible portion has an equivalent resolution of about 640 X 480 pixels. It actually could be slightly higher than that, but the Vertical Blanking Interval, or VBI, allows other signals to be carried along with the broadcast.

Rear projection

Most are very large screen TVs (up to over 100 inch (254 cm) use projection technology. Three types of projection systems are used in projection TVs: CRT based, LCD based, and DLP (reflective micro-mirror chip) based. Projection television has been commercially available since the 1970s, but at that time could not match the image sharpness of the CRT; current models are vastly improved, and offer a cost effective large-screen display. A variation is a video projector, using similar technology, which projects onto a screen.

Flat Panel LCD or Plasma

Modern advances have brought flat panels to TV that use active matrix LCD or plasma display technology. Flat panel LCDs and plasma displays are as little as 1 inch thick and can be hung on a wall like a picture or put over a pedestal. Some models can also be used as computer monitors. Each has its pros and cons. Flat panel LCD display can have narrow viewing angles and so may not suit a home environment. Rear projection screens do not perform well in natural daylight or well lit rooms and so are best suited to dark viewing areas. A complete run down of the pros and cons of each display should be sough before purchasing a single television technology.

Terminology for Televisions

Pixel resolution is the amount of individual points known as pixels on a given screen. A typical resolution of 800 X 600 means that the television display has 800 pixels across and 600 pixels on the vertical axis. The higher the resolution on a specified display the sharper the image. Contrast ratio is a measurement of the range between the brightest and darkest points on the screen. The higher the contrast ratio, the better looking picture there is in terms of richness, deepness, and shadow detail. The brightness of a picture measures how vibrant and impacting the colors are. Measured in cd/m^2 equivalent to the amount of candles required to power the image.

Transmission Band

There are various bands of frequencies on which televisions work depending upon the country. The VHF and UHF signals in bands III to V are generally used. Lower frequencies do not have enough bandwidth available for television. Although the BBC initially used Band I VHF at 45 MHz, this frequency in no longer in use for this purpose. Band II is used for FM radio transmissions. Higher frequencies behave more like light and do not penetrate buildings or travel around obstructions well enough to be used in a conventional broadcast TV system, so they are generally only used for satellite broadcasting, which uses frequencies around 10 GHz. TV systems in most countries relay the video as an AM (amplitude modulation) signal and the sound as a FM (frequency-modulation) signal. An exception is France, where the sound is AM.

Aspect Ratios

Aspect ratio refers to the ratio of the horizontal to vertical measurements of a television's picture. Mechanically scanned television as first demonstrated by John Logie Baird in 1926 used a 7: 3 vertical ratio, oriented for the head and shoulders of a single person in close up.

Most of the early electronic TV systems from the mid 1930s onward shared the same aspect ratio of 4: 3 which was chosen to match the Academy Ratio used in cinema films at the time. This ratio was also square enough to be conveniently viewed on round cathode ray tubes (CRTs), which were all that could be produced given the manufacturing technology of the time. (Today's CRT technology allows the manufacture of much wider tubes, and the flat screen technologies which are become steadily more popular have no technical aspect ratio limitations at all). The BBC's television service used a more squarish 5: 4 ratio from 1936 to 3 April 1950, when it too switched to a 4: 3 ratio. This did not present significant problems, as most sets at the time used round tubes, which were easily adjusted to the 4: 3 ratio when the transmissions changed.

In the 1950s, movie studios moved towards widescreen aspect ratios such as Cinema Scope in an effort to distance their product from television. Although this was initially just a gimmick, widescreen is still the format of choice today and square aspect ratio movies are rare. Some people argue that widescreen is actually a disadvantage when showing objects that are tall instead of panoramic, others say that natural vision is more panoramic than tall, and therefore widescreen is easier on the eye.

The switch to digital television systems has been used as an opportunity to change the standard television picture format from the old ratio of 4: 3 (1.33: 1) to an aspect ratio of 16: 9 (approximately 1.78: 1). This enables TV to get closer to the aspect ratio of modern widescreen movies, which range from 1.66: 1 through 1.85: 1 to 2.35: 1. There are two methods for transporting widescreen content, the better of which uses what is called anamorphic widescreen format. This format is very similar to the technique used to fit a widescreen movie frame inside a 1.33: 1 35 mm film frame. The image is compressed horizontally when recorded, then expanded again when played back. The anamorphic widescreen 16: 9 format was first introduced via European PALplus television broadcasts and then latter on "widescreen" DVDs; the ATSC HDTV system uses straight widescreen format, no horizontal compression or expansion is used.

Recently "widescreen" has spread from television to computing where both desktop and laptop computers are commonly equipped with widescreen displays. There are some complaints about distortions of movie picture ratio due to some DVD playback software not taking an account of aspect ratios; but this may subside as the DVD playback software matures. Furthermore, computer and laptop widescreen displays are in the 16: 10 aspect ratio both physically in size and in pixel counts, and not in 16: 9 of consumer televisions, leading to further complexity. This was a result of widescreen computer display engineers' uninformed assumption that people viewing 16 : 9 content on their computer would prefer that and area of the screen be reserved for playback controls, subtitles or their Taskbar, as opposed to viewing content full screen.

Aspect Ratio Incompatibility

The television industry's changing of aspect ratios is not without difficulties, and can present a considerable problem.

Displaying a widescreen aspect (rectangular) image on a conventional aspect (square or 4: 3) display can be shown:

- In "letterbox" format, with black horizontal bars at the top and bottom
- With part of the Image being cropped, usually the extreme left and right of the image being cut off (or in "pan an scan", parts selected by an operator)
- With the image horizontally compressed
- A conventional aspect (square or 4: 3) image on a widescreen aspect (rectangular with longer horizon) display can be shown
- In "pillar box" format, with black vertical bars to the left and right
- With upper and lower portions of the image cut off (or in "tilt and scan", parts selected by an operator)

- With the image horizontally distorted

A common compromise is to shoot or create material at an aspect ratio of 14: 9, and to lose some image at each side for 4: 3 presentations, and some image at top and bottom for 16: 9 presentation. In recent years, the cinematographic process known as Super 35 (championed by James Cameron) has been used to film a number or major movies such as Titanic, Legally Blonde, Austin Powers, and Crouching Tiger, Hidden Drag. This process results in a camera negative which can then be used to create both wide screen theatrical prints, and standard "full screen" releases for television/VHS/DVD which avoid the need for either "letterboxing" or the severe loss of information caused by conventional "pan-and-scan" cropping.

Television and communication of information

The electronic media more so radio and also television are considered a potential vehicle for disseminating agricultural technology information. The Indian had it beginning in 1959, with the objective for assessing the value of different education Television programmers suitable for group viewing in rural and urban communities. Television was first demonstrated in India in 1965 at an industrial exhibition. At the conclusion of the exhibition a part of the apparatus was few programmers designed for community viewing such as "responsibility of citizenships which include traffic and road sense, danger to community health etc.

After the success of this experiment, it was decided to extend the scope of the community viewing experiment to school or educational television, which was launched in Oct, 23, 1961 in Delhi schools. From April 1965, television general service was increased to one hour, four days a week. But from August 15, that year it becomes a daily transmission krishi Darshan Programme for better farming precautions was introduced from January 26. 1967. The rural service primarily meant to familiarize the rural viewers with the technical and scientific know how about farming, weather forecasts. The countries second television center came up at Bombay on October 2, 1972. It was quickly followed by Shringar, Amritsar 1973 Calcutta, Madras and Luck now (all three in 1975) but from August 1, 1975 Doordarshan undertook an historic and a unique steps in its march forward.

The satellite instructional television experiment popularly known as SITE, the country's most exciting and ambitious project was started by Doordarshan. In the beginning Doordarshan formed a part of all India Radio outfit. But on April 1, 1976, Doordarshan was made an independent department of the ministry of information and Broadcasting. Unit 1982 television in India has not gone colour. From August 15, 1982 colour was introduced. Between 1980 and 1984 several new high power transmitters and low power transmitter's area and are assessable to the Annual Report of information and Broadcasting Ministry.

Doordarshan now has several channels. DD_1 is the National channel available all over the country on certain fix hours. DD_2 and DD_3 are also available. There are regional channels for putting out programmers in the 13 regional languages. Krishi Darshan is the oldest well established and the best known program for the rural area and farmers the foresight and the pioneering sprit of the visionary space scientist Dr. Vikram Sarabhia and the initiative of Dr. M.S. Swaminathan.

The basic objective of the program is to familiarize the rural audience with the latest technical and scientific know how about the farming practices, rural development programs, to acquaint the viewers with the importance of health, family planning, sanitation, etc.

History of Television

The history of television records the work of numerous engineers and inventors in several countries over many decades. The fundamental principles of television were initially explored using electromechanical methods to scan, transmit and reproduce an image. As electronic camera and display tubes were perfected, electromechanical television gave way to all-electronic systems in nearly all applications.

Electromechanical Television

The beginnings of mechanical television can be traced back to the discovery of the photoconductivity of the element selenium by Willoughby Smith in 1873, the invention of a scanning disk by Paul Gottlieb Nipkow in 1884 and John Logie Baird's demonstration of televised moving images in 1926.

As a 23 year old German university student, Paul Nipkow proposed and patented the first electromechanical television system in 1884. Although he never built a working model of the system, variations of Nipkow's spinning-disk "image rasterizer" for television became exceedingly common, and remained in use until 1939. Constantin Perskyi had coined the word television in a paper read to the International Electricity Congress at the International World Fair in Paris on August 25, 1900. Perskyi's paper reviewed the existing electromechanical technologies, mentioning the work of Nipkow and others. The photoconductivity of selenium and Nipkow's scanning disk were first joined for practical use in the electronic transmission of still pictures and photographs, and by the first decade of the 20th century halftone photographs, composed of equally spaced dots of varying size, were being transmitted by facsimile over telegraph and telephone lines as a newspaper service.

However, it was not until 1907 that developments in amplification tube technology, by Lee DeForest and Arthur Korn among others, made the design practical. The first demonstration of the instantaneous transmission of still silhouette images was by Georges Rignoux and A. Fournier in Paris in 1909, using a rotating mirror-drum as the scanner and a matrix of 64 selenium cells as the receiver.

In 1911, Boris Rosing and his student Vladimir Zworykin created a television system that used a mechanical mirror-drum scanner to transmit, in Zworykin's words, "very crude images" over wires to the "Braun tube" (cathode ray tube or "CRT") in the receiver. Moving images were not possible because, in the scanner, "the sensitivity was not enough and the selenium cell was very leggy".

On March 25, 1925, Scottish inventor John Logie Baird gave the first public demonstration of televised silhouette images in motion, at Selfridge's Department Store in London. AT & T's Bell Telephone Laboratories transmitted halftone still images of transparencies in May 1925. On June 13 of that year, Charles Francis Jenkins transmitted the silhouette image of a toy windmill in motion, over a distance of five miles from a naval radio station in Maryland to his laboratory in Washington, using a lensed disk scanner with a 48-line resolution.

However, if television is defined as the live transmission of moving images with continuous tonal variation, Baird first achieved this privately on October 2, 1925. But strictly speaking, Baird had not yet achieved moving images on October 2. His scanner worked at only five images per second, below the threshold required to give the illusion of motion, usually defined as at least 12 images per second. By January, he had improved the scan rate to 12.5 images per second. Then he gave the world's first demonstration of a working television system to members of the Royal Institution and a newspaper reporter on January 26, 1926 at his laboratory in London. Unlike later electronic systems with several hundred lines of resolution, Baird's vertically scanned image, using a scanning disk embedded with a double spiral of lenses, had only 30 lines, just enough to reproduce a recognizable human face.

In 1927, Baird transmitted a signal over 438 miles (705 KM) of telephone line between London and Glasgow. In 1928, Baird's company (Baird Television Development Company/Cinema Television) broadcast the first transatlantic television signal, between London and New York, and the first shore to ship transmission. He also demonstrated an electromechanical color, infrared (dubbed "Noctovision"), and stereoscopic television, using additional lenses, disks and filters. In parallel, Baird developed a video disk recording system dubbed "Phonovision"; a number of the Phonovision recordings, dating back to 1927, still exist. In 1929, he became involved in the first experimental electromechanical television service in Germany. In November of the same year, Baird and Bernard Natan of Pathé established France's first television company, Télévision Baird Natan. In 1931, he made the first outdoor remote broadcast, of the Epsom Derby. In 1932, he demonstrated ultra-short wave television. Baird's electromechanical system reached a peak of 240 lines of resolution on BBC television broadcasts in 1936 though the mechanical system did not scan the televised scene directly. Instead a 35 mm film was shot, rapidly developed and then scanned while the film was still wet. This intermediate film system was discontinued within three months in favor of a 405-line all-electronic system developed by Marconi-EMI.

Herbert E. Ives and Frank Gray of Bell Telephone Laboratories gave a dramatic demonstration of mechanical television on April 7, 1927. The reflected-light television system included both small and large viewing screens. The small receiver had a two-inch-wide by 2.5-inch-high screen. The large receiver had a screen 24 inches wide by 30 inches high. Both sets were capable of reproducing reasonably accurate, monochromatic moving images. Along with the pictures, the sets also received synchronized sound. The system transmitted images over two paths: first, a wire link from Washington to New York City, then a radio link from Whippany, New Jersey. Comparing the two transmission methods, viewers noted no difference in quality. Subjects of the telecast included Secretary of Commerce Herbert Hoover. A flying-spot scanner beam illuminated these subjects. The scanner that produced the beam had a 50-aperture disk. The disc revolved at a rate of 18 frames per second, capturing one frame about every 56 milliseconds. (Today's systems typically transmit 30 or 60 frames per second, or one frames every 33.3 or 16.7 milliseconds respectively.) Television historian Albert Abramson underscored the significance of the Bell Labs demonstration: "It was in fact the best demonstration of a mechanical television system ever made to this time. It would be several years before any other system could even begin to compare with it in picture quality.

Meanwhile in the Soviet Union, Léon Theremin had been developing a mirror drum-based television, starting with 16 lines resolution in 1925, then 32 lines and eventually 64 using interlacing in 1926, and as part of his thesis on May 7, 1926 he electrically transmitted and then projected near-simultaneous moving images on a five foot square screen. By 1927 he achieved an image of 100 lines, a resolution that was not surpassed until 1931 by RCA, with 120 lines.

On December 25, 1926, Kenjiro Takayanagi demonstrated a television system with a 40 line resolution that employed a Nipkow disk scanner and CRT display at Hamamatsu Industrial High School in Japan. This protype is still on display at the Takayanagi Memorial Museum in Shizuoka University, Hamamatsu Campus. His research in creating a production model was halted by the US after Japan lost World War II. Mechanical scanning systems, though obsolete for the more familiar television systems, nevertheless survive in long wave infra red cameras because there is no suitable all-electronic pickup device.

Electronic Television

In 1908, Alan Archibald Campbell-Swinton, fellow of the Royal Society (UK), published a letter in the scientific journal Nature in which he described how "distant electric vision" could be achieved by using a cathode ray tube (or "Braun" tube, after its inventor, Karl Braun) as both a transmitting and receiving device, apparently the first iteration of the electronic television method that would dominate the field until recently. He expanded on his vision in a speech given in London in 1911 and reported in The Times and the Journal of the Röntgen Society. In a letter to Nature published in October 1926, Campbell-Swinton also announced the results of some "not very successful experiments" he had conducted with G. M. Minchin and J. C. M. Stanton. They had attempted to generate an electrical signal by projecting an image onto a selenium-coated metal plate that was simultaneously scanned by a cathode ray beam. These experiments were conducted before March 1914, when Minchin died, but they were later repeated by H. Miller and J. W. Strange in 1937, observing "very faint signals" from the original selenium-coated metal plate and "strong signals" when the plate was covered with zinc sulphide or selenide. Although others had experimented with using a cathode ray tube as a receiver, the concept of using one as a transmitter was novel. By the late 1920s, when electromechanical television was still being introduced, several inventors were already working separately on versions of all-electronic transmitting tubes, including Philo Farnsworth and Vladimir Zworykin in the United States, and Kálmán Tihanyi in Hungary.

On September 7, 1927, Farnsworth's Image Dissector camera tube transmitted its first image, a simple straight line, at his laboratory at 202 Green Street in San Francisco. By September 3, 1928, Farnsworth had developed the system sufficiently to hold a demonstration for the press. In 1929, the system was further improved by elimination of a motor generator, so that his television system now had no mechanical parts. That year, Farnsworth transmitted the first live human images with his system, including a three and a half-inch image of his wife Elma ("Pem") with her eyes closed (possibly due to the bright lighting required).

The problem of low sensitivity to light resulting in low electrical output from transmitting or "camera" tubes would be solved by Tihanyi beginning in 1924. His solution was a camera tube that accumulated and stored electrical charges ("photoelectrons") within the tube throughout each scanning cycle. The device was first described in a patent application he filed in Hungary in March 1926 for a television system he dubbed "Radioskop". After further refinements included in a 1928 patent application, Tihanyi was awarded patents for the camera tube in both France and Great Britain in 1928, and applied for patents in the United States in June of the following year. Although his breakthrough would be incorporated into the design of RCA's "iconoscope" in 1931, the U.S. patent for Tihanyi's transmitting tube would not be granted until May 1939. The patent for his receiving tube had been granted the previous October. Both patents had been purchased by RCA prior to their approval.

Development continued around the world. At the Berlin Radio Show in August 1931, Manfred von Ardenne gave a public demonstration of a television system using a CRT for both transmission and reception. However, Ardenne had not developed a camera tube, using the CRT instead as a flying-spot scanner to scan slides and film. Philo Farnsworth gave the world's first public demonstration of an all-electronic television system, using a live camera, at the Franklin Institute of Philadelphia on August 25, 1934, and for ten days afterwards.

In Britain the EMI engineering team lead by Isaac Shoenberg applied in 1932 for a patent for a new device they dubbed "the Emitron", which formed the heart of the cameras they designed for the BBC. On November 2, 1936, a 405-line broadcasting service employing the Emitron began at studios in Alexandra Palace, and transmitted from a specially-built mast atop one of the Victorian building's towers. It alternated for a short time with Baird's mechanical system in adjoining studios, but was more reliable and visibly superior. This was the world's first regular high-definition television service.

The original American iconoscope was noisy, had a high ratio of interference to signal, and ultimately gave disappointing results, especially when compared to the high definition mechanical scanning systems then becoming available. The EMI team under the supervision of Isaac Shoenberg analyzed how the iconoscope (or Emitron) produces an electronic signal and concluded that its real efficiency was only about 5% of the theoretical maximum. They solved this problem by developing and patenting in 1934 two new camera tubes dubbed super-Emitron and CPS Emitron. The super-Emitron was between ten and fifteen times more sensitive than the original Emitron and iconoscope tubes and, in some cases, this ratio was considerably greater. It was used for an outside broadcasting by the BBC, for the first time, on Armistice Day 1937, when the general public could watch in a television set how the King laid a wreath at the Cenotaph. This was the first time that anyone could broadcast a live street scene from cameras installed on the roof of neighbor buildings, because neither Farnsworth nor RCA could do the same before the 1939 New York World's Fair.

American television broadcasting at the time consisted of a variety of markets in a wide range of sizes, each competing for programming and dominance with separate technology, until deals were made and standards agreed upon in 1941. RCA, for example, used only Iconoscopes in the New York area, but Farnsworth Image Dissectors in Philadelphia and San Francisco. In September 1939, RCA agreed to pay the Farnsworth Television and Radio Corporation royalties over the next ten years for access to Farnsworth's patents. With this historic agreement in place, RCA integrated much of what was best about the Farnsworth Technology into their systems. In 1941, the United States implemented 525-line television. The world's first 625-line television standard was designed in the Soviet Union in 1944, and became a national standard in 1946. The first broadcast in 625-line standard occurred in 1948 in Moscow. The concept of 625 lines per frame was subsequently implemented in the European CCIR standard.

Broadcast Television

Programming is broadcast by television stations, sometimes called "channels", as stations are licensed by their governments to broadcast only over assigned channels in the television band. At first, terrestrial broadcasting was the only way television could be widely distributed, and because bandwidth was limited, i.e., there were only a small number of channels available, government regulation was the norm.

In the U.S., the Federal Communications Commission (FCC) allowed stations to broadcast advertisements beginning 1941, but required public service programming commitments as a requirement for a license. By contrast, the United Kingdom chose a different route, imposing a television license fee on owners of television reception equipment to fund the British Broadcasting Corporation (BBC), which had public service as part of its Royal Charter. Practically every country in the world now has at least one broadcast television station. Television has grown up all over the world, enabling nearly every country to share aspects of its culture and society with others.

The first regularly scheduled television service in the United States began on July 2, 1928. The Federal Radio Commission authorized C.F. Jenkins to broadcast from experimental station W3XK in Wheaton Maryland, a suburb of Washington, D.C. For at least the first eighteen months, 48-line silhouette images from motion picture film were broadcast, although beginning in the summer of 1929 he occasionally broadcast in halftones.

The first experiments in television broadcasting began in France in the 1930s, but the French were slow to employ the new technology. In November 1929, Bernard Natan established France's first television company, Télévision-Baird-

Natan. On April 14, 1931, there took place the first transmission with a thirty-line standard by René Barthélemy. On December 6, 1931, Henri de France created the Compagnie Générale de Télévision (CGT). In December 1932, Bathélemy carried out an experimental program in black and white (definition: 60 lines) one hour per week, «Paris Télévision», which gradually became daily from early 1933. The first official channel of French television appeared on February 13, 1935, the date of the official inauguration of television in France, which was broadcast in 60 lines from 8:15 to 8:30 pm.

Electromechanical broadcasts began in Germany in 1929, but were without sound until 1934. Network electronic service started on March 22, 1935, on 180 lines using telecine transmission of film, intermediate film system, or cameras using the Nipkow Disk. Transmissions using cameras based on the iconoscope began on January 15, 1936. The Berlin Summer Olympic Games were televised, using both all electronic iconoscope-based cameras and intermediate film cameras, to Berlin and Hamburg in August 1936. Twenty-eight public television rooms were opened for anybody who did not own a television set. The Germans had a 441-line system on the air in February 1937, and during World War II brought it to France, where they broadcast from the Eiffel Tower.

The first British television broadcast was made by Baird Television's electromechanical system over the BBC radio transmitter in September 1929. Baird provided a limited amount of programming five days a week by 1930. During this time, Southampton earned the distinction of broadcasting the first ever live television interview, which featured Peggy O'Neil, an actress and singer from Buffalo, New York. On August 22, 1932, BBC launched its own regular service using Baird's 30-line electromechanical system, continuing until September 11, 1935. On November 2, 1936 the BBC began broadcasting a dual system service, alternating between Marconi-EMI's 405 line standard and Baird's improved 240 line standard, from Alexandra Palace in London, making the BBC Television Service (now BBC One) the world's first regular high-definition television service. The government, on advice from a special advisory committee, decided that Marconi-EMI's electronic system gave the superior picture, and the Baird system was dropped in February 1937. TV broadcasts in London were on the air an average of four hours daily from 1936 to 1939. There were 12,000 to 15,000 receivers. Some sets in restaurants or bars might have 100 viewers for sport events.The outbreak of the Second World War caused the BBC service to be suspended on September 1, 1939, resuming from Alexandra Palace on June 7, 1946.

The first transatlantic television signal was sent in 1928 from London to New York by the Baird Television Development Company/Cinema Television, although this signal was not broadcast to the public. The first live satellite signal to Britain from the United States was broadcast via the Telstar satellite on July 23, 1962.

The Soviet Union began offering 30-line electromechanical test broadcasts in Moscow on October 31, 1931, and a commercially manufactured television set in 1932. The first experimental transmissions of electronic television took place in Moscow on March 9, 1937, using equipment manufactured and installed by

RCA. Regular broadcasting began on December 31, 1938. It was quickly realized that 343 lines of resolution offered by this format would have become insufficient in the long run, thus a specification for 441 line format was developed in 1940, superseded by a 625 line standard in 1944. This format was ultimately accepted as a national standard. The experimental transmissions in 625-line format started in Moscow from November 4, 1948. Regular broadcasting began on June 16, 1949. Details for this standard were formalized in 1955 specification called GOST 7845-55, basic parameters for black-and-white television broadcast. In particular, frame size was set to 625 lines, frame rate to 25 frames/s interlaced and video bandwidth to 6 MHz. These basic parameters were accepted by most countries having 50 Hz mains frequency and became the foundation of television systems presently known as PAL and SECAM.

The Canadian Broadcasting Corporation (CBC) adopted the American NTSC 525-line B/W 60 field per second system as its broadcast standard. It began television broadcasting in Canada in September 1952. The first broadcast was on September 6, 1952 from its Montreal, Quebec station CBFT. The premiere broadcast was bilingual, spoken in English and French. Two days later, on September 8, 1952, the Toronto, Ontario station CBLT went on the air.

Television Sets

In television's electromechanical era, commercially made television sets were sold from 1928 to 1934 in the United Kingdom, United States, and the Soviet Union. The earliest commercially made sets sold by Baird in the UK in 1928 were radios with the addition of a television device consisting of a neon tube behind a mechanically spinning disk (the Nipkow disk) with a spiral of apertures that produced a red postage-stamp size image, enlarged to twice that size by a magnifying glass. The Baird "Televisor" was also available without the radio. The Televisor sold in 1930-1933 is considered the first mass-produced set, selling about a thousand units.

The first commercially made electronic television sets with cathode ray tubes were manufactured by Telefunken in Germany in 1934, followed by other makers in France (1936), Britain (1936), and America (1938). The cheapest of the pre-World War II factory-made American sets, a 1938 image-only model with a 3-inch (8 cm) screen, cost US$125, the equivalent of US$1,863 in 2007. The most expensive model with a 12-inch (30 cm) screen was $445 ($6,633). An estimated 19,000 electronic television sets were manufactured in Britain, and about 1,600 in Germany, before World War II. About 7,000-8,000 electronic sets were made in the US before the War Production Board halted manufacture in April 1942, production resuming in August 1945.

Television usage in the United States skyrocketed after World War II with the lifting of the manufacturing freeze, war-related technological advances, the gradual expansion of the television networks westward, the drop in set prices caused by mass production, increased leisure time, and additional disposable income. In 1947, Motorola introduced the VT-71 television for $189.95, the first television set to be sold for under $200, finally making television affordable for millions of Americans. While only 0.5% of U.S. households had a television set in 1946, 55.7% had one in 1954, and 90% by 1962. In Britain, there were 15,000 television households in 1947, 1.4 million in 1952, and 15.1 million by 1968.

For many years different countries used different technical standards. France initially adopted the German 441-line standard but later upgraded to 819 lines, which gave the highest picture definition of any analogue TV system, approximately double the resolution of the British 405-line system. However this is not without a cost, in that the cameras need to produce four times the pixel rate (thus quadrupling the bandwidth), from pixels one-quarter the size, reducing the sensitivity by an equal amount. In practice the 819-line cameras never achieved anything like the resolution that could theoretically be transmitted by the 819 line system, and for color, France reverted to the 625-line CCIR system used by most European countries.

With advent of color television most Western European countries adopted PAL standard. France, Soviet Union and most Eastern European countries adopted SECAM. In North America the original NTSC 525-line standard was augmented to include color transmission with slight slowing down of frame rate.

Throughout the 1960s, television sets used exclusively vacuum tube electronics. This resulted in relatively heavy and unreliable TVs. In addition, vacuum tubes were poorly-suited to color television, as it required a large amount of tubes which caused further reliability problems. Because vacuum tubes only allowed for very simple NTSC/PAL filtering, the picture quality of early color sets was rather poor. The tint control that is still found on NTSC televisions originally was meant to correct the color burst phase's drifting when channels were changed. In addition, the large number of vacuum tubes required for color prevented the use of it in portable TVs.

By the early 1970s, solid-state electronics appeared and quickly displaced vacuum tubes in color TVs (black and white sets generally continued to be tube-based). This allowed for significantly more reliable TVs and better picture quality. 1971 was the first year when sales of color TVs in the US exceeded B&W ones. In other countries, color TV was slower to arrive and did not become common in Western Europe until the '80s.

In 1972, the FCC began requiring UHF tuners in all TVs sold in the United States. At this time, there were only 130 UHF stations in the country, mostly low-power ones that carried local programming. Up to this point, UHF support from TV manufacturers was sporadic. Most sets did not come factory-equipped with them, and often merely included an empty slot in the cabinet where an optional UHF tuner could be installed.

During the 1970s, electronic tuners began appearing in high-end TVs in place of traditional dials, and they would gradually become standard along with remote controls. Remotes had first appeared in the 1950s with Zenith's Space Command Control, but these were mechanical devices that emitted a high-pitched frequency that the TV detected. The first electronic remote controls did not appear until the 1980s.

The first LCD TVs were introduced in 1988, small, handheld devices with a B&W screen that could not display the full NTSC resolution of 480 lines. In the 1990s, 3 line digital comb filters appeared on high-end TVs. In addition, composite and S-video inputs began appearing to support devices like video games and VCRs. Analog television broadcasting in the United States ended on June 12, 2009 in favor of Digital terrestrial television or digital-only broadcasting.

Television Programme Production

As with films or other media production, production of television programme can be divided into three parts:

1. Pre-production

Pre-production begins when a script is approved. A director is chosen to plan the programme's final look. Pre-production tasks include storyboarding, construction of sets, props, and costumes, casting guest stars, budgeting, acquiring resources like lighting, special effects, stunts, etc. Once the show is planned, it must then be scheduled. Scenes are often filmed out of sequence, guest actors or even regulars might be available only at certain times. Sometimes the principal photography of different programme is to be done at the same time thereby complicating the schedule (a guest star might shoot scenes from two episodes on the same afternoon). Complex scenes are translated from storyboard to animatics to further clarify the action. Scripts are adjusted to meet altering requirements.

Some shows have a small stable of directors, but also usually rely on outside directors. Given the time constraints of broadcasting, a single show might have two or three episodes in pre-production, one or two episodes in principal photography, and a few more in various stages of post-production. The task of directing is complex enough to not allow a single director to work on more than one episode or show at a time, hence there is always a need for multiple directors.

2. Principal photography

Principal photography is the actual filming of the programme. Director, actors and crew will gather at soundstages or on location to film a scene. A scene is further divided into shots, which should be planned during pre-production. Depending on scheduling, a scene may be shot not in the chronological order of the story. Conversations may be filmed twice from different angles, often using stand-ins, so one actor might perform all their lines in one set of shots, and then the other side of the conversation will be filmed from the opposite perspective. To complete a production on time, a second unit may be filming a different scene on another set or location at the same time, using a different set of actors, an assistant director, and a second unit crew. A director of photography supervises the lighting of each shot to ensure consistency.

3. Post production

Once principal photography is complete, producers coordinate post-production tasks. Visual and digital effects are added to the film; this is often outsourced to companies specializing in these areas. Often music is performed with the conductor using the film as a time reference (other musical elements may be previously recorded). An editor cuts the various pieces of film together, adds the musical score and effects, determines scene transitions, and assembles the completed show.

4. Distribution

The show is then turned over to the network, which sends it out to its affiliates, which broadcast it in the specified timeslot. On some occasions, a series that has not attracted particularly high ratings and has been canceled can be given a reprieve if DVD sales have been particularly strong. This has happened in the cases of Family Guy in the US and Peep Show in the UK. Following are 15 basic steps required for an elaborate television production. Once you get a feel for the entire process, you can scale things down for any sized production.

Identify the Purpose of the Production

1. This is the most important step: Clearly identify the production's goals and purposes.

If there is no clear agreement on the goals and purposes of a production, it will be impossible to evaluate its success. (How will you know if you've arrived at your destination, if you didn't know where you were going in the first place?) Is the purpose to instruct, informs, or entertains or maybe to generate feelings of pride or express a social, religious, or political need? Is the real purpose to create a desire in the audience to take some action? The primary goal of most broadcasting is simply to hold the interest of an audience through the intervening commercials. Even PBS (Public Broadcasting Service), which used to be commercial free, now runs "mini commercials" for their corporate underwriters. Most productions have more than one goal and we'll elaborate on some of these later.

2. To identify and analyze the target audience

Based on things as, age, sex, socioeconomic status, and educational level, program content preferences will differ. As we've noted, we refer to audience characteristics as demographics. We can see regional demographic variations in part by differences in local programming in various areas of the country and sometimes by the films and network programming that local stations decide not to air. Sex and violence are chief among these content issues and both show a positive relationship to ratings.

In 2006, a similar controversy broke out over a vaccination that can prevent cervical cancer, which kills more than 3,500 women in the United States each year. Some social conservatives both oppose its use and the dissemination of pro-vaccination information. Although the vaccine reportedly could save thousands of lives, cervical cancer is associated with sexual activity and these people feel the vaccination could encourage premarital sex.

What is and is not broadcasted is largely determined by audience feedback and this varies greatly with demographics. But what if you're not producing programming for broadcast or general distribution?

Compared to standard broadcast television, institutional television, which includes corporate and educational video, has different needs and expectations.

But here too, predominating demographic characteristics, such as age, sex, and education, influence a production's form and content. For example, to underestimate education or experience and inadvertently "talk down to" an audience insults them. To overestimate education or experience and talk over people's heads is just as bad. Either way, you lose.

3. Check Out Similar Productions

Check out similar productions from the past. If you're going to make mistakes, at least make new ones. Ask yourself some questions: How will your proposed production differ from previous successful and unsuccessful efforts by others? Why did they work; or, maybe more importantly, why didn't they? Of course, since production styles change rapidly, you need to take into consideration differences in time, locations, and audiences. This link will take you to more information on the success and failure of TV programs.

4. Determine the Basic Value of Your Production

Next, determine the overall value of the production to a sponsor or underwriter. Obviously, underwriters and advertisers want something in return for their investment

For this, you'll need to ask yourself some questions. First, what is the probable size of the audience? In determining this, you must know if your show will be a one-shot presentation or if you can recoup production expenses over time by presenting the show to other audiences. Generally, the larger the audience the more marketable a production will be to an underwriter or advertiser. At the same time, simple numbers don't tell the full story. Let's say an advertiser has a product designed for young people athletic shoes or designer jeans. In this case, a production that draws a large percentage of this age group will be more valuable than a production that has a larger overall audience, but a lower percentage of young people. Broadcasters have canceled many TV series, not because they had a small audience, but because they had the wrong kind of audience (the wrong demographics). You'll always want to balance the potential value of a production to an advertiser or underwriter with the projected cost of producing and presenting the production. If the costs exceed the benefits, you have problems. In commercial television, the return on investment is generally in the form of increased sales and profits. But it may take other forms, such as the expected moral, political, spiritual, or public relations benefit derived from the program.

5. Develop a Treatment or Production Proposal

Next, put it down on paper. You'll recall that earlier we talked about treatments and program proposals (written summaries of what you propose to do). After the program proposal or treatment is approved, the next step is to write and submit a full script. It will be at this point that any remaining research on the content will be commissioned. The first version of a script is often followed by numerous revised versions. Throughout the rewriting process, a number of story conferences or script conferences typically take place. During these sessions audience appeal, pace, and problems with special interest groups, etc., are discussed. If it's an institutional production, you'll review the production's goals and pose questions about the most effective ways to present ideas. If the director is on board at this time, he or she should be part of these conferences.

Finally, a script version emerges that is (we can hope) more or less acceptable to everyone. Even this version, however, will probably not be final. In many instances, scene revisions continue right up to the time the scenes are shot. Typically, in a dramatic film production each new script version is issued on a different colour paper so that the cast and crew won't confuse them with earlier versions. Depending on the production, you might want to develop a storyboard. A storyboard consists of drawings of key scenes with corresponding notes on elements such as dialogue, sound effects, and music. Today, high-budget film and video productions create sophisticated storyboards with software supplied by companies such as Zebra Development.

6. Develop a Production Schedule

Next, draw up a tentative schedule. Generally, broadcast or distribution deadlines will dictate the production schedule (the written timetable listing the time allotted for each production step). Not planning things out carefully might cause you to miss a critical deadline, rendering the production useless.

7. Select Key Production Personnel

Bring on board the remaining above the line production personnel. In addition to the producer and writer, above the line personnel include the production manager, director and, in general, key creative team members. Below the line personnel, generally assigned later, include the technical staff.

8. Decide On Locations

If you're not shooting in the studio, decide on key locations.

In a major production, such as the type we are outlining in this discussion, you will hire a location scout or location manager to find and coordinate the use of the locations suggested by the script. Although it might be easier to shoot in a TV studio, it's been shown that audiences like the authenticity of "real" locations, especially in dramatic productions. Most major cities encourage TV and film production and maintain film commissions that supply photos and videotapes of interesting shooting locations in their area. They'll also provide information on usage fees and the names of people to contact. It's often necessary to make changes in the on-location settings. For instance, rooms may have to be repainted or redecorated and visible signs changed.

9. Decide On Talent, Wardrobe and Sets

Next, you'll want to make some decisions on talent, wardrobe and sets.

Depending on the type of production, auditions may take place at this point as part of the casting process (selecting people for the various roles). Once completed, you'll negotiate and sign contracts. If you're lucky enough to afford well-known actors, you'll probably have decided on them early in the preproduction process. Once you decide on the talent, you can begin wardrobe selection. These are suggested by the script, coordinated with the look of the sets and locations, and ultimately approved by the director. After a set designer is hired, he or she will review the script, possibly do research, and then discuss initial ideas with the director.

Once there's agreement, sketches of the sets can be made for final approval before actual set construction starts if there is any construction. Today, many sets exist only in computers and the actors are electronically inserted into them. If this is the case, the set sketches will be given to a computer artist. You can then schedule rehearsals, from initial table readings to the final dress rehearsal. Even though personnel may not have finished the sets at this point, the actors can start reading through the script with the director to establish pace, emphasis, and basic blocking (the positioning of sets, furniture, cameras, and actors). Once the sets are finished, the final blocking and dress rehearsals can get underway.

10. Decide on the Remaining Production Personnel

Make decisions on the remaining staff and production needs. At this point you can arrange for key technical personnel, equipment, and facilities. This includes the rental of both equipment and production facilities. Next, arrange transportation, catering (food and refreshment trucks) and on-location accommodations (for any overnight stays). Unions, which may be involved, often set minimum standards for transportation, as well as the quality of meals and accommodations. Union contracts also cover job descriptions, specific crew responsibilities and working hours, including graduated pay increases for overtime hours.

11. Obtain Permits, Insurance, and Clearances

In major U.S. cities and in many foreign countries it's not possible just to go to the location of your choice, set up your tripod, and start filming. Except for spot news and short documentary segments, you must arrange access permits, licenses, security bonds, and insurance policies. Many semipublic interior locations, such as shopping malls, require filming permits. (Yes, these things do get complicated!) Depending on the nature of the production, liability insurance and security bonds may be necessary because accidents can happen that can be directly or indirectly attributed to the production. In some locations the controlling agency will limit exterior production to certain areas and specific hours. In a street scene where traffic will be affected you'll need to arrange for special police.

We also include in this category a wide variety of clearances ranging from permission to use prerecorded music to reserving satellite time to transmit the production back to a studio. If you can't obtain clearance, you need time to explore alternatives. Are you beginning to see why list of credits in films and TV programs is so long?

12. Select Video Inserts, Still Photos, and Graphics

Arrange to shoot or acquire video and film inserts, still photos, and graphics.

To reduce production costs check out existing stock footage in film and video libraries around the country. This is generally background footage, such as general exterior scenes of an area that will be edited into the production. One example of a stock footage source is Film & Video Stock Shots in North Hollywood, California. If suitable footage is not available or does not meet the needs of the production, you may need to hire a second unit to produce needed segments.

Second unit work is production done away from the main location by a separate production crew and generally does not involve the principal talent. If part of a dramatic production calls for shots of a specific building in Cleveland, for example, a second unit can shoot the necessary exteriors in Cleveland while the primary unit works on interior shots in Southern California where the actors live. When the shots are edited together it will appear that the interior shots belong to the building in Cleveland. You will want to begin to make decisions on music at this point, including working out copyright clearances and royalties for music and visual inserts. We'll discuss these in more detail later.

13. Begin Rehearsals and Shooting

Start rehearsing and shooting. Depending on the type of production, rehearsals may take place either minutes or days before the actual shooting. Productions shot live-on-tape (without stopping, except for major problems -- whether recorded on videotape or another medium) will need to be completely rehearsed before recording starts. This includes early walk-through rehearsals, camera rehearsals, and one or more dress rehearsals. Productions that are shot single-camera, film-style (to be covered later) are rehearsed and recorded one scene at a time.

14. Begin Editing Phase

After shooting is completed, the producer, director, and video recording editor review the footage and start to make editing decisions. This has typically been done in two phases: online and offline. Briefly, in offline editing copies of the original taped footage that contains time-code number references are used to develop a kind of blueprint for final editing. In online editing the original footage is used in editing. During the final editing phase, sound sweetening (enhancing), color balancing, and visual effects are added. Because editing is so important to the creative process, we're going to devote several later chapters to the subject. If all these terms and procedures sound a bit intimidating right now, don't worry; we'll explain them in more detail later.

15. Do Postproduction Follow-Up

Although most of the production crew will conclude their work by the time production wraps (finishes), some follow-up work generally needs to be completed. Included is totaling up financial statements, paying the final bills, and determining the production's success (or failure). Ratings indicate success levels in broadcast television. In institutional television success may be determined by testing, program evaluations, and viewer feedback. Speaking of ratings those numbers often spell the life and death of TV programs. You can check your understanding of these things here. In the next module, we'll turn our attention to that important "blueprint" for the entire TV production, the script.

Radio for Communication of Information

Radio

Radio is the wireless transmission of signals, by modulation of electromagnetic waves with frequencies below those of light.

Radio waves are a form of electromagnetic radiation, created whenever a charged object (in normal radio transmission, an electron) accelerates with frequency that lies in the radio frequency (RF) portion of the electromagnetic spectrum. In radio, this acceleration is caused by an alternating current in an antenna. Radio frequencies occupy the range from a few tens of hertz to three hundred gigahertz although commercially important uses of radio use only a small part of this spectrum.

Electromagnetic radio spectrum

Other types of electromagnetic radiation, with frequencies above the RF range, are microwave, infrared, visible light, ultraviolet, X-rays and gamma rays. Since the energy of an electron from an atom, radio waves are classified as non jonizing, radiation.

Electromagnetic radiation travels (propagates) by means of oscillating electromagnetic fields that pass through the air and the vacuum of space. It does not require a medium of transport. When radio waves pass and electrical conductor, the oscillating electric or magnetic field (depending on the shape of the conductor) induces and alternating current and voltage in the conductor. This can be transformed into audio or other signals that carry information. The word 'radio' is used to describe this phenomenon, and television, radio and cell phone transmissions are all classed as radio frequency emissions.

History and Invention

Originally, radio technology was called 'wireless telegraphy', which was shortened to 'wireless'. The prefix radio-in the sense of wireless transmission was first recorded in the word radio conductor coined by the French physicist Edouard Branly in 1897 and based on the verb of radiate. 'Radio' as a noun is said to have been coined by advertising expert Waldo Warren (White 1944). The word appears in a 1907 article by Lee de Forest, was adopted by the United States Navy in 1912 and became common by the time of the first commercial broadcasts in the United States in the 1920s. (The noun 'broadcasting' itself came from an agricultural term, meaning 'scattering seeds'). The American term was then adopted by other languages in Europe and Asia, although Britain retained the term 'wireless' until the mid 20th century. In Chinese, the term 'wireless' is the basis for the term 'radio wave' although the term for the device that listens to radio waves is literally 'device for receiving sounds'.

The identity of the original inventor of radio, at the time called wireless telegraphy, is contentious. The controversy over who invented the radio can be discussed as follows.

- Nikola Tesla developed means to reliably produce radio frequencies, publicly demonstrated the principles of radio, and transmitted long-distance signals. He holds the US patent for the invention of the radio, as defined as "wireless transmission of data."
- Alexander Stepanovich Popov in 1894 built this first radio receiver, which contained a coherer. Further refined as a lightning detector, he presented it to the Russian Physical and Chemical Society on May 7, 1895.
- Guglielmo Marconi was an early radio experimenter and founded the first commercial organization devoted to the development and use of radio.
- Reginald Fessenden and Lee de Forest invented amplitude-modulated (AM) radio, so that more than one station can send signals (as opposed to spark-gap radio, where one transmitter covers the entire bandwidth of the spectrum).
- Edwin H. Armstrong invented frequency-modulated (FM) radio, so that and audio signal can avoid "static", that is, interference from electrical equipment and atmospherics.

Early radios ran the entire power of the transmitter through a carbon microphone. While some early radios used some type of amplification through electric current or battery, until the mid 1920s the most common type of receiver was the crustal set. In the 1920s, amplifying vacuum tube radio receivers and transmitters came into use.

History of Radio

The pre-history and early history of radio is the history of technology that produced instruments that use radio waves. Later in the timeline of radio, the history of dominated by programming and contents, which is closer to general history.

Various scientists proposed that electricity and magnetism while both capable of causing attraction and repulsion of objects, were linked. In 1820 Hans Christian Orsted performed a widely known experiment on man-made electric current and magnetism. He demonstrated that a wire carrying a current could deflect a magnetized compass needle. Osted's experiments discovered the relationship between electricity and magnetism in a very simple experiment. Orsted's work influenced Andre-Marie Ampere to produce a theory of electromagnetism.

In the history of radio and development of "wireless telegraphy", several people have claimed to have "invented the radio". The most commonly accepted claims are:

- Jagadish Chandra Bose
- Alexander Stepanvoich Popov
- Nikola Tesla, who developed means to reliably produce radio frequencies, publicly demonstrated the principles of radio, and transmitted long distance signals.

- Guglielmo Marconi, who equipped ships with life saving wireless communications and established the first transatlantic radio service.

In the late 19th century it was clear to various scientists and experimenters that wireless communication was possible. Various theoretical and experimental innovations lead to the development of radio and the communication system we know today. Some early work was done by local effects and experiments of electromagnetic induction. Many understood that there was nothing similar to the "ethereal telegraphy" and telegraphy by induction; the phenomena being wholly distinct. Wireless telegraphy was beginning to take hold and the practice of transmitting messages without wires was being developed. Many people worked on developing the devices and improvements.

Faraday

In 1831, Michael Faraday began a series of experiments in which he discovered electromagnetic induction. The relation was mathematically modeled by Faraday's law, which subsequently became one of the four Maxwell equations. Faraday proposed that electromagnetic forces extended into the empty space around the conductor, but did not complete his work involving that proposal. In April 1872 William Henry Ward received U.S. Patent 126356 for radio development.

Maxwell

Between 1861 and 1865, James Clerk Maxwell made experiments with electromagnetic waves. In 1873, as a result of experiments, Maxwell first described the theoretical basis of the propagation of electromagnetic waves in his paper to the Royal Society "A Dynamical Theory of the Electromagnetic Field".

Mahlon Loomis

Mahlon Loomis of West Virginia has the oldest and most documented claim of inventing radio. Loomis received U.S. Patent for a "wireless telegraph" in July 1872. This patent utilizes atmospheric electricity to eliminate the overhead wire used by the existing telegraph systems. It did not contain diagrams or specific methods. It is substantially similar to William Henry Ward's patent.

Towards end of 1875, while experimenting with the telegraph, Thomas Edison noted a phenomenon that the termed "etheric force", announcing it the press on November 28. He abandoned this research when Elihu Thomson, among others, ridiculed the idea. Between 1893 and 1894, Roberto Landell de Moura, a Brazilian priest and scientist, conducted experiments. He did not publicize his achievement until 1900.

In 1878, David E. Hughes was the first to transmit and received radio waves when he noticed that his induction balance caused noise in the receiver of his homemade telephone. He demonstrated his discovery to the Royal Society in 1880 but was told it was merely induction. In 1884, Temistocle Calsecchi-Onesti at Fermo in Italy invented a tube filled with iron filings, called a "coherer".

Between 1884 and 1886, Edouard Branly of France produced and improvement versions of the coherer. In 1885, Edison took out U.S. Patent 465971 on a system of radio communication between ships (which later he sold to Marconi).

Between 1886 and 1888, Heinrich Rudolf Hertz validated Maxwell's theory through experiment. He demonstrated that radio radiation had all the properties of waves (now called Hertzian waves), and discovered that the electromagnetic equations could be reformulated into a partial differential equation called the wave equation. However he saw no practical use for his discovery. Claims have been made that Murray, Kentucky farmer Nathan Stubblefield invented radio between 1885 and 1892, before either Tesla or Marconi, but his devices seem to have worked by induction transmission rather than radio transmission.

Wireless Radio Beginnings

In the history of radio and development of "wireless telegraphy", there are multiple claims to the invention of radio. The identity of the original inventor of radio, at the time called wireless telegraphy, is contentious. The key invention for the beginning of "wireless transmission of data using the entire frequency spectrum", known as the spark-gap transmitter, has been attributed to various individuals. Marconi equipped ships with life saving wireless communicators and established the first transatlantic radio service. Tesla developed means to reliably produce radio frequencies, publicly demonstrated the principles of radio, and transmitted long distant.

In 1893, at St. Louis, Missouri, Tesla gave a public demonstration of "wireless" radio communication. Addressing the Franklin Institute in Philadelphia and the National Electric Light Association, he described in detail the principles of radio communication. The apparatus that he used contained all the elements that were incorporated into radio systems before the development of the "oscillation valve", the early vacuum tube. Tesla was the first to apply the mechanism of electrical conduction to wireless practices. Also, he initially used sensitive electromagnetic receiver that were unlike the less responsive coherers later used by Marconi and other early experimenters. Afterwards, the principle of radio communication (sending signals through space to receivers) was publicized widely. Various scientists, inventors, and experimenters began to investigate wireless methods.

Oliver Lodge transmitted radio signals on August 14, 1894 (one year before Marconi and one year after Tesla) at a meeting of the British Association for the Advancement of Science at Oxford University.

On 19 August 1894 he demonstrated the reception of Morse code signaling via radio waves using a "coherer". Lodge improved Edouard Branly's coherer radio wave detector by adding a "trembler" which dislodged clumped filings, thus restoring the device's sensitivity.

In August 1898 he got U.S. Patent 609154, " Electric Telegraphy"' that made wireless signals using Ruhmkorff coils or Tesla coils for the transmitter and a Branly coherer for the detector. This was key to the "syntonic" tuning concept. Lodge sold the patent to Marconi in 1912.

In November of 1894, the Indian physicist, Jagadish Chandra Bose, demonstrated publicity the use of radio waves in Calcutta, but he was not interested in patenting his work. Bose ignited gunpowder and rang a bell at a distance using electromagnetic waves, proving that communication signals can be sent without using wires.

In 1894, the Russian physicist Alexander Popov built a coherer. On May 7, 1895, Popov performed a public demonstration of transmission and reception of radio waves used for communication at the Russian Physical and Chemical Society, using his coherer: this day has since been celebrated in Russia as "Radio Day". But he did not apply for a patent for this invention. Popov's early experiments were transmissions of only 600 yards.

Popov was the first to develop a practical communication system based on the coherer, and is usually considered by the Russians to have been the inventor of radio. Popov's public demonstration of radio waves, between different campus buildings, to the Saint Petersburg Physical Society (around March, 1896) was before the public demonstration of the Marconi system (around September, 1896). In 1898 his signal was received 6 miles away. In 1899 his signal was received at 30 miles away.

In 1900, Popov stated at the Congress of Russian Electrical Engineers that "the emission and reception of signals by Marconi by means of electric oscillations was nothing new, as in America Nikola Tesla did the same experiments in 1893."

Later Popov experimented with ship to shore communication. Popov died in 1905 and his claim was not pressed by the Russian government until 40 years later.

In the beginning of 1895, Tesla detected signals from his New York lab's transmissions at West Point (a distance of 50 miles). In 1895, Marconi received a telegraph message without wires, but he did not send his voice over the airwaves.

In March 1895, Popov transmitted radio waves between campus buildings in Saint Petersburg, but did not apply for a patent.

Ernest Rutherford

The New Zealander Ernest Rutherford, the first Baron Rutherford of Nelson was instrumental in the development of radio. In 1895 he was awarded an Exhibition of 1851 Science Research Scholarship to Cambridge. He arrived in England with a reputation as an innovator and inventor, and distinguished himself in several fields, initially by working out the electrical properties of solids and then using wireless waves as a method of signaling. Rutherford was encouraged in his work by Sir Robert Ball, who had been scientific adviser to the body maintaining lighthouses on the Irish coast; he wished to solve the difficult problem of a ship's inability to detect a lighthouse in fog. Sensing fame and fortune, Rutherford increased the sensitivity of his apparatus until he could detect electromagnetic waves over a distance of several hundred meters. The commercial development, though, of wireless technology was left for others, as Rutherford continued purely scientific research. Thomson quickly realized that Rutherford was a researcher of exceptional ability and invited him to join in a study of the electrical conduction of gases.

Marconi

Guglielmo Marconi was an electrical engineer and Nobel laureate known for the development of a practical wireless telegraphy system.

In 1896, Gugliemo Marconi was awarded a patent for radio with British Patent 12039, Improvements in Transmitting Electrical Impulses and Signals and in Apparatus There-for. This was the initial patent for the radio, through it used various other experiments (primarily Tesla) and resembled the instrument demonstrated by others (including Popov). During this time spark-gap wireless telegraph was widely researched. In 1896, Bose went to London on a lecture tour and met Marconi, who was conducting wireless experiments for the British post office. In 1897, Marconi established the radio station on the Isle of Wight, England. In the U.S. during 1897, Tesla applied for two key radio patents. Those two patents were issued in early 1900.

In 1898, Marconi opened a radio factory in Hall Street, Chelmsford, England, employing around 50 people. In 1899, Bose announced his invention of the "iron-mercury-iron coherer with telephone detector" in a paper presented at Royal Society, London. In 1900, Reginald Fessenden made a weak transmission of voice over the airwaves. Around 1900, Tesla opened the Wardenclyffe Tower facility and advertised services. In 1903, Wardenclyffe Tower neared completion. Various theories exist on how Tesla intended to achieve the goals of this wireless system (reportedly, a 200 kW system). Tesla claimed that Wardenclyffe, as part of transmitters, would have allowed secure multi channel transceiving of information, universal navigation, time synchronization, and a global location system. In 1904, The U.S. Patent Office reversed its decision, awarding Marconi a patent for the invention of radio, possibly influenced by Marconi's financial backers in the States, who included Thomas Edison and Andrew Carnegie. This also allowed the U.S. government (among others) to avoid having to pay the royalties that were being claimed by Tesla for use of his patents.

Early Radio Telegraphy and Telephony

British Marconi

Using various patents, the company called "British Marconi" was established and began communication between coast radio stations and ships at sea. This company along with its subsidiary American Marconi, had a stranglehold on ship to shore communication. It operated much the way American Telephone and Telegraph operated until 1983, owning all of its own equipment and refusing to communicate with non-Marconi equipped ships. Many inventions improved the quality of radio, and amateurs experimented with uses of radio, thus the first seeds of broadcasting were planted. Around the turn of the century, the Slaby-Arco wireless system was developed by Adolphus Slaby and Georg von Arco (later incorporated into Telefunken).

Fessenden

The invention of amplitude modulated (AM) radio, so that more than one station can send signals (as opposed to spark gap radio, where one transmitter covers the entire bandwidth of the spectrum) is attributed to Reginald Fessenden and Lee de Forest. On Christmas Eve of 1906, Reginald Fessenden used an Alexanderson alternator and rotary spark gap transmitter to make the first radio audio broadcast, from Brant Rock, Massachusetts. Ships at sea heard a broadcast that included Fessenden playing O Holy Night on the violin and reading a passage from the Bible.

In 1909, Marconi and Karl Ferdinand Braun were awarded the Nobel Prize in Physics for "contributions to the development of wireless telegraphy".

Charles David Herrold

In April 1909 Charles David Herrold, an electronics instructor in San Jose, California constructed a broadcasting station. It used spark gap technology, but modulated the carrier frequency with the human voice, and later music. The station "San Jose Calling" (there were no call letters), continued to eventually become today's KCBS in San Francisco. Herrold, the son of a Santa clara Valley farmer, coined the terms "narrowcasting" and "broadcasting", respectively to identify transmissions destined for a single receiver such as that on board a ship, and those transmissions destined for a general audience. (The term "broadcasting" had been used in farming to define the tossing of seed in all directions.) Charles Herrold did not claim to be the first to transmit the human voice, but he claimed to be the first to conduct "broadcasting". To help the radio signal to spread in all directions, he designed some omni directional antennas, which he mounted on the rooftops of various buildings in San Jose. Herrold also claims to be the first broadcaster to accept advertising (he exchanged publicity for a local record store for records to play on his station), though this dubious honour usually is foisted on WEAF (1922).

In 1912, the RMS Titanic sank. After this, wireless telegraphy using spark-gap transmitters quickly became universal on large ships. In 1913, the International Convention for the Safety of Life at Sea was convened and produced a treaty requiring shipboard radio stations to be manned 24 hours a day. A typical high-power spark gap was a rotating commutator with six to twelve contacts per wheel, nine inches to a foot wide, driven by about 2000 volts DC. As the gaps made and broke contact, the radio wave was audible as a tone in a crystal set. The telegraph key often directly made and broke the 2000 volt supply. One side of the spark gap was directly connected to the antenna. Receivers with thermionic valves became commonplace before spark-gap transmitters were replaced by continuous wave transmitters.

List of Radio Pioneers

Many scientists and inventors contributed to the invention of wireless telegraphy and radio. Individuals that helped to further the science include, among others: -

- Georg von Arco : European pioneer.
- Edouard Branly : invention of the Branly coherer around 1890.
- Temistocle Calzecchi-Onesti : constructed a tuning "thube".

- Amos Dolbear : Earth transmission, U.S. Patent 350299.
- Thomas Alva Edison : "Etheric Force" experiments 1875 ; U.S. Patent 465971, 1891.
- Michael Faraday: discovered electromagnetic induction.
- Reginald Fessenden: advanced "continuous" wave transmission.
- Hans Christian Osted: discovered that a magnetic field surrounds a wire carrying current.
- Joseph Henry: transmitted radiant energy from a capacitor through a coil and detected it 100 feet away, Dec. 1840.
- Charies Herrold: advanced radio broadcasting.
- David E. Hughes: early experiments with transmission and reception.
- James Clerk Maxwell: developed a set of equations expressing the basic laws of electricity and magnetism.
- Jozef Murgas: extensive work in the late 190s.
- William Henry Preece: early experiments in electromagnetism and wireless telephony.
- Augusto Righi: continued Hertz's experiments.
- Harry Shoemaker: 1901 to 1905; 40 patents.
- Adolphus Slaby: European pioneer.
- John Stone Stone: 1901 to 1904; 70 patents.
- Nathan Stubblefield: wireless telephony demonstrations around 1902 ; U.S. Patent 887357, 1908.

Audio Broadcasting (1915 to 1950s)

Crystal Sets

In the 1920s, the government publication, "Construction and Operation of a Simple Homemade Radio Receiving Outfit", showed how almost any family having a family member handy with simple tools could a build a crystal radio. This is a modern reproduction of an early receiver.

The most common type of receiver before vacuum tubes was the crystal set, although some early radios used some type of amplification through electric current or battery. Inventions of the triode amplifier, generator, and detector enabled audio radio. The invention of amplitude-modulated (AM radio), so that more than one station can send signals (as opposed to spark-gap radio, where one transmitter covers the entire bandwidth of spectra) was pioneered by Fessenden and Lee de Forest.

The First Values

During the mid 1920s, amplifying vacuum tubes (US)/thermionic valves (UK) revolutionized radio receivers and transmitters.

John Ambrose Fleming developed and earlier tube known as an "oscillation valve" (it was a diode). Lee De Forest placed a screen, the "grid" electrode, between the filament and plate electrode.

In 1920, Regular wireless broadcasts for entertainment began in Argentina, pioneered by the group around Enrique Telemaco Susini, and spark-gap telegraphy stopped. On 31 August 1920 the first known radio news program was broadcast by station 8MK, the unlicensed predecessor of WWJ (AM) in Detroit, Michigan.

In 1922 regular wireless broadcasts for entertainment began in the UK from the Marconi Research Centre at Writtle near Chelmsford, England. Early radios ran the entire power of the transmitter through a carbon microphone:

In the 1920s, the Westinghouse Company bought Lee De Forest's and Edwin Armstrong's patent. During the mid 1920s, Amplifying vacuum tubes (US)/thermionic valves (UK) revolutionized radio receivers and transmitters. Westinghouse engineers developed a more modern vacuum tube.

Licensed Public Radio Stations

The question of the 'first' publicly-targeted licensed radio station in the U.S. has more than one answer and depends on semantics. Settlement of this 'first' question might rely largely upon what constitutes 'regular' programming.

- It is commonly attributed to KDKA in Pittsburgh, Pennsylvania, which in October of 1920 received its license and went on the air as the first US licensed commercial broadcasting station. (Their engineer Frank Conrad had been broadcasting from his own station since 1976). Technically, KDKA was the first of several already-extant stations to receive a 'limited commercial' license.
- On February 17 1919, station 9XM at the University of Wisconsin in Madison had already broadcast the first human speech to the public at large. That station is still on the air today as WHA.
- 9XM sent music over the air two years earlier, was originally licensed in 1914 and sent the first transmission in 1916.
- On August 20 of 1920, at least two months before KDKA, E.W. Scripps's WBL (now WW) in Detroit started broadcasting. It has carried a regular schedule of programming to the present.
- There is the history noted above of Charles David Herrold's radio services (eventually KCBS) going back to 1909.

Broadcasting was not yet supported by advertising. The stations owned by manufactures and department stores were established to sell radios and those owned by newspapers to the 1920s, Radio was first used to transmit pictures visible as television. During the early 1930s, single sideband (SSB) and frequency modulation (FM) were invented by amateur radio operators. By 1940, they were established commercial modes.

Westinghouse was brought into the patent allies group, General Electric, American Telephone and Telegraph, and Radio Corporation of America, and became a part owner of RCA. All radios made by GE and Westinghouse were sold under the RCA label 60% GE and 40% Westinghouse. ATT's Western Electric would build radio transmitters. The patent allies attempted to set up a monopoly, but they failed due to successful competition. Much to the dismay of the patent allies, several of the contracts for inventor's patents held clauses protecting "amateurs" and allowing them to use the patents. Whether the competing manufactures were really amateurs was ignored by these competitors.

These features arose:

- Commercial (United States) or governmental (Europe) station networks
- Federal Radio Commission
- Federal Communications Commission
- CCIR
- Birth of the soap opera
- Race towards shorter waves and FM

Broadcast Communications in India

The Indian television system is one of the most extensive systems in the world. Terrestrial broadcasting, which has been the sole preserve of the government, provides television coverage to over 90% of India's 900 million people. By the end of 1996 nearly 50 million households had television sets. Internationals satellite broadcasting, introduced in 1991, has swept across the country because of the rapid proliferation of small-scale cable systems. By the end of 1996, Indians could view dozens of foreign and local channels and the competition for audiences and advertising revenues was one of the hottest in the world. In 1995, the Indian Supreme Court held that the government's monopoly over broadcasting was unconstitutional, setting the stage for India to develop into one of the world's largest and most competitive television environments.

Broadcasting began in India with the formation of a private radio service in Madras in 1924. In the same year, the British colonial government granted a license to a private company, the Indian Broadcasting Company, to open Radio stations in Bombay and Calcutta. The company went bankrupt in 1930 but the colonial government took over the two transmitters and the Department of Labour and Industries started operating them as the Indian State Broadcasting Corporation. In 1936, the Corporation was renamed All India Radio (AIR) and placed under the department of Communications. When India became independent in 1947, AIR was made a separate Department under the Ministry of Information and Broadcasting.

The early history of radio broadcasting in independent India is important because it set the parameters for the subsequent role of television in the country. At Independence, the Congress government under Jawaharlal Nehru had three major goals: to achieve political integration, economic development and social modernization. Broadcasting was expected to play an important role in all three areas.

The most important challenge the government faced at independence was that of forging a nation out of the diverse political, religious, geographic and lingual entities that composed independent India. In addition to the territories ruled directly by the British, over 500 hundred "independent" princely states had joined the new nation, some quite reluctantly. The country immediately found itself at war with Pakistan over one of those states of Kashmir. The trauma of the partition of the country into India and Pakistan and the violence between Hindus and Muslims had further weekend the political stability of the country.

Broadcasting was harnessed for the task of political nation building. National integration and the development of a "national consciousness" were among the early objectives of All India Radio. Broadcasting was organized as the sole preserve of the chief architect of this process of political integration of the State. The task of broadcasting was to help in overcoming the immediate crisis of political instability that followed Independence and to foster the long-term process of political modernization and nation building that was the dominant ideology of the newly formed state. Broadcasting was also charged with the task of aiding in the process of economic development.

The Indian Constitution, adopted in 1950, mandated a strong role for the Indian State in the economic development of the country. The use of broadcasting to further the development process was a natural corollary to this state-led developmental philosophy, broadcasting, was especially excepted to contribute to the process of social modernization, which was considered an important pre-requisite of economic development. The dominant development philosophy of the time identified the problems of development as primarily internal to developing countries. These endogenous causes, to which communication solutions were thought to exist, included traditional value systems, lack of innovation, lack of entrepreneurial ability and lack of a national consciousness. In short, the problem was one of old ideas hindering the process of social change and modernization and the role of broadcasting were to provide an inlet for the flow of modern ideas.

It was in the context of this dominant thinking about the role of broadcasting in India that television was introduced in 1959. The government had been reluctant to invest in television until then because it was felt that a poor country like India could not afford the medium. Television had to prove its role in the development process it could gain a foot-hold in the country. Television broadcasts started from Delhi in September 1959 as part of All India radio's services. Programs were broadcast twice a week for an hour a day on such topics as community health, citizens duties and rights, and traffic and road sense. In 1961 the broadcasts were expanded to include a school educational television project. In time, Indian films and programs consisting of compilation of musicals from Indian films joined the program line-up as the first entertainment programs. A limited number of old U.S. and British shows were also telecast sporadically.

The first major expansion of television in India began in 1972, when a second television station was opened in Bombay. This was followed by stations in Srinagar and Amritsar (1973), and Calcutta, Madras and Lucknow in 1975. Relay stations were also set up in a number of cities to extend the coverage of the regional stations.

In 1975, the government carried out the first test of the possibilities of satellite based television through the SITE program. SITE (Satellite Instructional Television Experiment) was designed to test whether satellite based television services could play a role in socio-economic development. Using a U.S. ATS-6 satellite and up-link centers at Ahmedabad and Delhi, television programs were beamed down for about 4 hours a day to about 2,400 villages in 6 states. The programs dealt mainly with in- and out-of-school education, agricultural issues, planning and national integration. The program was fairly successful in demonstrating the effectiveness of satellite based television in India and the lessons learnt from SITE were used by the government in designing and utilizing its own domestic satellite service INSAT, launched in 1982.

Like radio, was considered a facilitator of the development process and its introduction was justified by the role it was asked to play in social and economic development. Television was institutionalized as an arm of the government, since the government was the chief architect of political, economic and social development in the country.

By 1976, the government found it running a television network of eight television stations covering a population of 45 million spread over 75,000 square kilometers. Faced with the difficulty of administering such an extensive television system television as part of All India Radio, the government constitutes Doordarshan, the national television network, as a separate department under the Ministry of Information and Broadcasting. Doordarshan was set up as an attached office under the Ministry of Information and Broadcasting a halfway house between a public corporation and a government department. In practice, however, Doordarshan operated much like a government department, at least as far as critical issues of policy planning and financial decision making were concerned. Doordarshan was headed by a Director General appointed by the Information and Broadcasting Ministry. The Ministry itself and sometimes the office of the Director General as well, was and continues to be, staffed by members of India's civil services.

In 1982 television began to attain national coverage and develop as the government's pre-eminent media organization. Two events triggered the rapid growth of television that year. ISAT-1A, the first of the country's domestic communications satellites became operational and made possible the networking of all of Doordarshan's regional stations. For the first time Doordarshan originated as a nation-wide feed dubbed the "National Programme" which was feds from Delhi to the other stations. In November 1982, the country hosted the Asian Games and the government introduced color broadcasts for the coverage of the games. To increase television's reach, the government launched a crash program to set up low and high power transmitters that would pick-up the satellite distributed signals and re transmit them to surrounding areas. In 1983 television signals were available to just 28% of the population, this had doubled by the end of 1985 and by 1990 over 90% of the population had access to television signals.

The year 1976 witnessed a significant event in the history of Indian television, the advent of advertising on Doordarshan. Until that time television had been funded through a combination of television license and allocations from the annual budget (license were later abolished as advertising revenues began to increase substantially). Advertising began in avery small way with under 1% of Doordarshan's budget coming from advertising revenues in 1976-77. But the possibility of reaching a nationwide audience made television look increasingly attractive to advertisers after the introduction of the "National Programme" in 1982. In turn, Doordarshan began to shift the balance of its programming from educational and informational programs to entertainment programs. The commercialization of Doordarshan saw the development of soap operas, situation comedies, dramas, musical programs, quiz shows and the like. By 1990 Doordarshan's revenues from advertising were about $300 million, accounting for about 70% of its annual expenditure.

By 1991, Doordarshan's earlier mandate to aid in the process of social and economic development had clearly been diluted. Entertainment and commercial programs had begun to take center stage in the organization's programming strategies and advertising had come to be Doordarshan's main source of funding. However, television in India was still a modest enterprise with most parts of the country getting just one channel except for the major cities which received two channels. But 1991 saw the beginnings of international satellite broadcasting in India and the government launched a major economic liberalization program. Both these events combined to change the country's television environment dramatically.

International satellite television was introduced in India by CNN through its coverage of the Gulf War in 1991. Three months later Hong Kong based Star TV (now owned by Rupert Murdoch's News Crop.) started broadcasting five channels into India using the ASIASAT-1 satellite. By early 1992, nearly half a million Indian households were receiving Star TV telecast. A year later the figure was close to 2 million and by the end of 1994, an estimated 12 million households (a little less than one-fourth of all television households) were receiving satellite channels. This increase in viewer ship was made possible by the 60,000 or so small scale cable system operators who have mushroomed across the country. These systems have redistributed the satellite channels to their customers at rates as low as $5 a month. Taking advantage of the growth of the satellite television audience, a number of Indian satellite based television services were launched between 1991 and 1994, prominent among them Zee TV, the first Hindi satellite channel. By the end of 1994 there were 12 satellite based channels available in India, all of them using a handful of different satellites. This number was expected to double by the end of 1996, with a number of Indian programmers and international media companies like Turner Broadcasting, Time Warner, ESPN, CANAL 5 and Persons PLC, seriously considering the introduction of new satellite television services for India.

The proliferation of channels has put great pressure on the Indian television programming industry. Already the largest producer of motion pictures, Indian is poised to become a sizable producer of television programs as well. With Indian audience clearly preferring locally produced program over foreign programs, the new television services are spending heavily on the development of indigenous programs. The number of hours of television programming produced in India has increased 500% from 1991 to 1996 and is expected to grow at an ever faster rate until the year 2000.

Despite the rapid growth of television programming continues to be dominated by the Indian film industry. Hindi films are the staple of most national channels and regional channels rely heavily on a mix of Hindi and regional language films to attract audiences. Almost all Indian films are musicals and this allows for the development of inexpensive derivative programs. One of Doordarshan's most popular programs, Chitrahaar, is a compilation of old film songs and all the private channels, including ZeeTV and must video channels like MTV Asia and Channel V, show some variation of Chitrahaar. A number of game shows are also based on movie themes. Other genres like soap operas, talk shows and situation comedies are also gaining in popularity, but the production of these programs has been unable to keep up with demand, hence the continuing reliance on film based programming.

International satellite programming has opened up competition in news and public affairs programming with BBC and CNN International challenging Doordarshan's long standing monopoly. Most of the other foreign broadcasters, for example, ESPN and the Discovery Channel, are focusing on special interest programming. Only StarTV's STAR Plus channel offers broad-based English language entertainment programs. Most of its programs are syndicated U.S. shows, for example soap operas like The Bold and the Beautiful and Santa Barbara and talk shows like Donahue and Oprah. However, SRAR Plus has a very small share of the audience in India and even this is threatened by the launch of new channels.

A peculiar development in television programming in India has been the use of hybrid English-Hindi program formats, popularly called "Hinglish" formats, which offer programs in Hindi and English on the same channel and even have programs, including news shows that use both languages within a single telecast. This takes advantage of the audience for television (especially the audience for satellite television) which is largely composed of middle class Indians who have some knowledge of English along with Hindi and colloquially speak a language that is primarily Hindi intermixed with words, phrases and whole sentences in English.

Commercial competition has transformed Doordarshan as well and it is scrambling to cope with the changed competitive environment. Satellite broadcasting has threatened Doordarshan's audiences and self-preservation has spawned a new ideology in the network which is in the process of reinventing itself, co-opting private programmers to recapture viewers and advertising rupees lost to Zee TV and Star TV. In 1994, the government ordered Doordarshan to raise its own revenues for future expansion. This new commercial mandate has gradually begun to change Doordarshan's perception of who are its primary constituents from politicians to advertisers.

The government's monopoly over television over the years has resulted in Doordarshan being tightly controlled by successive governments. In principle, Doordarshan is answerable only to Parliament. Parliament lays down the guidelines that Doordarshan is expected to adhere to adhere to in its programming and Doordarshan's budget is debated and approved by Parliament. But the guidelines established by Parliament to ensure Doordarshan's political neutrality are largely ignored in the face of the majority that ruling parties have held in Parliament. Doordarshan has been subject more to the will of the government than the oversight of Parliament. Successive governments and ruling political parties have used Doordarshan to further their political agendas, weakening its credibility as a neutral participant in the political process. There have been periodic attempts to reconstitute Doordarshan into BBC-like public corporation, but governments have been reluctant to relinquish their hold on such a powerful medium.

The government drew its right to operate the country's broadcasting services as a monopoly from the Indian Telegraph Act of 1885 which empowers the government with the exclusive right to "establish, maintain and work" wireless services. In addition, the Constitution lists broadcasting as the sole domain of Parliament, effectively shutting out the states from making any laws with regard to television. Within the ambit of these provisions it was assumed that media autonomy or liberalization in any form was the prerogative of the government to grant.

However, the government's monopoly was challenged in the Indian Supreme Court in 1995. The Court held that the government monopoly over broadcasting was unconstitutional and while the government has the right to regulate broadcasting in the public interest, the Constitution forbids monopoly control over any medium by either individuals or the government. The Court directed the government to establish an independent public authority for "controlling and regulating" the use of airwaves. The Court's decision holds out the promise of significant structural changes in Indian broadcasting and the possibility that terrestrial television may finally free itself from government control.

It is evident that over time the State's control over television will continue to diminish. As its revenue structure begins to change and Doordarshan begins to respond to increasing commercial pressures, the character of its programming will begin to increasingly reflect the demands and pressures of the market place. In the meantime, caught between the government and the market, Doordarshan continues to struggle to maintain its mandate of public service programming. But the Supreme Court's recent decision ordering the government to establish an independent broadcasting authority to regulate television in the public interest holds the promise of allowing Indian television to escape both the stifling political control of the state and the commercial pressures of the market. There are a number of other constituencies like state governments, educational institutions, non-governmental organizations and social service agencies who can participate in a liberalized broadcast system. The Supreme Court has provided an opportunity to develop a broad based television system. How the country responds to this opportunity in the next few years will determine the future of broadcasting in India in the next century.

Information about All India Radio

Overview

All India Radio (AIR for short), officially known as Akashwani is the radio broadcaster of India and a division of Prasar Bharati (Broadcasting Corporation of India), an autonomous corporation of the Ministry of Information and Broadcasting, Government of India. All India Radio is one of the largest radio networks in the world. Its headquarters is at Akashwani Bhavan, on the Parliament Street next to the Indian parliament. Broadcasting house is an old building next to Akashwani Bhavan. The New Service Division of All India Radio under the Director General Now functions from this building. Built during the British rule, it is a very popular location and easily recognized building in New Delhi.

History

Broadcasting started in India in 1927 with two privately owned transmitters at Mumbai and Calcutta, which were taken over by the Government in 1930. These were operating under the name "Indian Broadcasting Service" until 1936 when it was given the present name "All India Radio (AIR)". It also came to be known as "Akashwani" from 1957. The Five Year Plans have given new impetus to the growth of broadcasting resulting in a phenomenal expansion from 6 stations at the dawn of Independence to around 200 stations at the close of the millennium.

Coverage

All India Radio covers 99.37% of India's populace, the largest democracy in the world with over one billion population. All India Radio maintains approximately 200 broadcasting centers around the country and transmits in 24 different languages. In spite of recent penetration by other media such as Cable Television, All India Radio remains the most common means of gaining access to information and entertainment, as the radio receivers are relatively cheap and affordable.

Services

All India Radio has many different services each catering to different regions/ languages across India. One of the most famous services of the All India Radio is the Vividh Bharati Seva (roughly translating to "Multi Indian service"). This service is the most commercial of all and is popular in all cities of India. This service offers a wide range of programmes including news, film music, comedy shows, etc. The Yuv-vani (The Voice of Youth) service of All India Radio provides an enriching and novel radio-experience by encouraging youth participation and experimenting with varied script ideas. With shows like "Mehfil" (The Party), "In the groove" and "The Roving Microphone" which have been around for more than three decades, Yuv-vani still holds a firm ground of its own. All India Radio, after launching the news-on-phone service on 25th February 1998 from New Delhi, is now running the service from Chennai, Mumbai, Hyderabad and Patna also. The service is accessible through STD, ISD and local telephone calls. The service is going to be started from 10 more cities like Ahmedabad, Banglore, Guwahati, Imphal, Jaipur, Kolkata, Lucknow, Raipur, Simla and Thiruvanthapuram shortly.

The main projects that describe the growth of educational radio are:

School Broadcast Project

This project was commissioned in 1937 and the target group was School students. This programme started from Delhi, Calcutta, Madras and Bombay. In the beginning the school programme were not strictly governed by the curriculum. With the passages of time and acquisitions of more experience, the AIR tried to make its radio broadcasts more curriculum oriented, but in absence of common syllabi and timetables in schools, even within the same state, it could not succeed in its aim.

Adult education and community development project (Radio Forum)

Commenced in 1956, the Villagers of 144 villages in the vicinity of Poona (in Maharastra state), were the main beneficiaries of this project. This was agriculture based project, which was originally designed and tried out in Canada. With the help of UNESCO, it was tried in 144 villages of Poona and was named as ' Radio forums Project' (defined as a listening cum-discussion-cum-action group). The members of the forum could listen thirty-minute radio programme on some agricultural or community development programme, then discuss and decide regarding its adoption in their own village. This project was a great success. Many action programmes were planned and put into practice.

Farm and Home Broadcast Project

This project was commenced in 1966 and again targeted at farmers and villagers. These broadcasts were designed to provide information and advice on agricultural and allied topics. The aim was to educate the farmers and provide them assistance in adopting innovative practices in their fields as per the local relevance. The experts also conducted occasional farm radio schools, which proved to be very effective.

University broadcast project

This project for University students was initiated in 1965, with an aim to expand higher education as widely as possible among the different strata of society. The Programme consisted of two types- 'General' and 'enrichment'. The general programmes included topics of public interest and enrichment programmes supported correspondence education offered by universities in their respective jurisdictions. School of Correspondence studies, University of Delhi and the Central Institute of English and Foreign Languages, Hyderabad is well known for preparation and broadcast of their programmes through All India Radio.

Language Learning Programme

The project, popularly known as 'Radio Pilot Project' was started in 1979-80 jointly by All India Radio and Department of Education, Government of Rajasthan, with an aim to teach Hindi to School going children as first language in 500 primary schools of Jaipur and Ajmer districts on experimental basis.

IGNOU: All India Radio Broadcast

In collaboration with IGNOU, All India Radio stations of Mumbai, Hyderabad and Shillong started radio broadcasts of IGNOU Programmes from January 1992. Main target group of this project were students of Open / Conventional Universities.

IGNOU-AIR Interactive Radio Counselling (IRC)

Started in 1998 for students of Open / Conventional Universities, this project is also very successful. In order to bridge the gap between Institutions and learners by instantly responding to their queries and also to provide Academic Counselling in subject area, IGNOU in collaboration with All India Radio Bhopal started this project in May 1998 as an experimental programme for one year (Sharma, 2002a). With the success of the experiment, it was extended to 8 other All India Radio stations (Lucknow, Patna, Jaipur, Shimla, Rohtak, Jalandhar, Delhi and Jammu). Presently Interactive Radio counseling is being provided on every Sunday for one hour (4:00 PM - 5:00 PM) from 186 radio stations of All India Radio. Toll free telephone facility is available from 80 cities (effective from February 2001) enabling the learners to interact with experts and seek clarification, without paying for their telephone calls.

Gyan-Vani: (Educational FM Radio Channel of India)

This project is recently launched (in year 2001) and again the target group is students of Open / Conventional Universities. Gyan Vani (Gyan = Knowledge, Vani = aerial broadcasting) is Educational FM Radio Channel of India, a unique decentralized concept of extending mass media for education and empowerment, suited to the educational needs of the local community (Sharma, 2002b). It is operating presently through Allahabad, Bangalore and, Coimbatore FM stations of India on test transmission mode. The network is slotted to expand to a total of 40 stations by June-2002. Gyan Vani, in addition to giving the hardcore education also deals with awareness programmes including the ones for Panchayati Raj Functionaries, Women Empowerment, Consumer Rights, Human Rights, the Rights of the Child, Health Education, Science Education, Continuing Education, Extension Education, Vocational Education, Teacher Education, Non formal Education, Adult Education, Education for the handicapped, Education for the down trodden, education for the tribals and so on. Gyan Vani is available through commercial FM radio set.

Radio-Vision (Multimedia through Digital Radio)

Pioneered by the BBC, the technique of radio vision allows the subject matter to be presented through two channels, the audio and the visual. The visuals are presented in the form of still filmstrips, charts, slides, models, etc, while the explanation is given through recorded narration. Educational institutions use this as a substitute for educational television. Radio-vision has its own advantages:

a. It is economical

b. It can cater to different categories of learners

c. It is easy to produce such programmes at the institutional level or at the learning centers

d. It provides visual support to the concept that is taught.

Radio Programme Production

If you listen to radio at least once in a while, you may remember a few programmes. You remember them because you liked them. You liked them because they were interesting. You do not remember a majority of programmes because what is heard is fast forgotten. We have already discussed about this feature of radio in the lesson on 'characteristics of radio'. To make a programme interesting, it has to be produced well. Radio production is a very vast subject. Due to constant change in technology, the techniques of radio production keep changing. But the basics remain the same. You may have the best of ideas, the best of scripts and the best of voices; but if the programme is not produced properly it will not be interesting to listen to.

OBJECTIVE:

- describe the qualities required for a radio producer;
- explain the main elements of radio production;
- distinguish between three different types of microphones;
- identify music and sound effects for radio production;
- classify the different stages of radio program production;
- state the importance of a script in a radio programme.

Qualities of a Radio Producer

Anybody can be a cook, but only the food prepared by some turnout to be very tasty. Similarly all people who produce radio programmes may not come out with good and interesting programmes.

Ability to see and hear what is happening: This would mean a keen desire to know more about things. We may call this inquisitiveness. Unless a person is inquisitive, he or she cannot be a good radio producer. This is because you need ideas to produce interesting programmes. How do you get ideas? Let us think. Firstly, we need to observe things to get ideas. Remember the story of Isaac Newton who observed an apple falling? This observation led him to wonder why the apple fell down from the tree and why it did not go up.

(a) Drawing from experiences: We have good and bad experiences in life. People who have written stories, novels, poems and plays draw their ideas mostly from their experiences. The great poet Kalidasa saw the dying bird and its pair which was sad and took to writing poems. The English poet Wordsworth observed flowers (daffodils) dancing in the air and wrote one of the best pieces of poetry. So from our own experiences or going through the experiences of others, we get ideas.

You have just read about the experiences of others. How do you come to know about them? Well, you need to talk to others to find out and understand their experiences.

(b) Ability to conceptualise ideas: You may get any number of ideas, but you need to turn them into concepts and then in to radio scripts.

(c) Creativity: What is creativity? Are you creative? This is a quality which is found in almost everyone. But you need to be creative in putting together an idea in a manner which can attract people .Creativity would mean, doing something new or different. The same idea can be made into a radio programme differently by different people. If everyone does it in the same way, they will all be the same and not interesting.

(d) A good user of words: You need the ability to use the right word at the right time in the right manner. Radio programmes depend on a good script and that depends on your ability to write well. For this you need a very good knowledge or stock of words or a good vocabulary.

Elements of Radio Production

The ingredients of a radio format are:

a. Human voice or spoken word.

b. Music

c. Sound effect

The various elements of radio production flow from the above ingredients of a radio format. So let us look at the main elements of radio production.

(i) **Studio:** The radio studio is a room where radio programmes are recorded. For producing a radio programme, you need a 'sound proof' studio where human voice can be recorded or broadcast in the best manner.

(ii) **Microphones:** For our voice to be recorded in a studio, we use a microphone. They amplify or in other words, increase the volume of your voice. When you speak before a microphone, you don't have to shout. You speak normally and it will be made louder if you use a loudspeaker to listen to. When we think of radio, the microphone is the most important element using which you present your programme.

There are basically three types of microphones and they are known by their directivity.

(a) **Uni-directional microphone**: As the name suggests, this microphone picks up sound from one direction. As you speak in front of it, your voice is picked up. If you speak from the other side of the microphone, your voice will not be picked up properly. In a radio studio, the announcers, presenters and newsreaders use this type of a microphone. A microphone is very sensitive and you need to use it carefully. You should be at the right distance from it when you speak. Otherwise, your voice will not sound good. Even if you turn a paper or breathe heavily, the microphone will pick up that sound and your programme or your voice will be affected.

(b) **Bi-directional microphone:** Here again as the name (bi) suggests, the voice or sound is picked up from two directions. If you are recording an interview in a radio studio, you may use this type of a microphone.

(c) **Omni-directional microphone:** You may be familiar with the word omni.

We say god is omnipresent, which means 'present everywhere'. In the case of an omni-directional microphone, it picks up sound from all directions. This type of microphone is used when a number of voices are used in a single Programme like a radio discussion or a radio drama.

There are many other types of microphones which come in different sizes and lengths. If you watch television programmes, you may find a small microphone clipped on the collar. This is called a lapel microphone which is actually a uni-directional microphone. These microphones are not normally used in radio. Then there are long microphones called gun microphones used in sports production. These microphones are often omni directional ones. There are also cordless microphones. You might have seen them being used in stage shows. They do not have any cables or wires attached to them. They have a small transmitter in them which can send the sounds to an amplifier.

(iii) **Sound effects:** Sound effects in a radio programme give meaning and sense of location. It adds realism to a programme and helps a listener to use imagination. Think of a crowded market or temple. If you are creating that scene in a radio programme, you do not have to go to a crowded market or temple to record. Well, you can record those sounds and use them. But in most cases, you use sound effects which are already recorded. Sound effects can be used in two ways:

a. Spot effects or effects that are created as we speak and

b. Recorded sound effects.

If you are recording a radio programme in which someone knocks at the door, you can make a knocking sound either on a door or a wooden partition. Or you want to show that someone is pouring water from a bottle into a glass; here again you can use the actual sounds produced on the spot. But if you want a lion roaring or a dog barking, you probably cannot bring a lion or a dog to the studios! Here we use recorded sounds which are kept on tapes or discs. Almost all sounds are available on CDs which you can try and use. There are also certain types of computer software available for this. You can also create sound effects. You can use two coconut shells to produce the sound effects of the sounds of horses' hooves. Take a piece of cellophine paper or aluminium wrapper and crush them in front of a microphone. Record the sound and hear. It will sound as if fire is raging. You can think and create many such sound effects. However, there is a word of caution. If you record an actual door opening, you many not get the real feeling of a door opening when you record it. What matters is what it sounds like and not what it is.

(iv) **Music:** Music is the soul of radio. It is used in different ways on radio as already discussed in the earlier lesson. Film songs and classical music programmes are independent programmes on radio. Music is also used as signature tunes or theme music of various radio programmes.

Let us see what music does to any programme.

a. Music adds colour and life to any spoken word programme.

b. Music can break monotony.

c. Music is used to give the desired effect of happy or unhappy situations, fear or joy.

d. Music can suggest scenes and locations. For example, you have to create a bright early morning situation. This can be done by playing a pleasing note on the flute along with the sound of chirping birds.

(v) **Artificial echo:** If you enter an empty building or fort and shout, your voice will come back to you. This is called on echo. An echo is used in radio programmes. This is a technical input.

(vi) **Filter or distort:** If you listen to someone speaking to you on phone, the Voice would not sound normal. This sort of effect called distort is produced using technology. Some times distort is used along with echo. Think of someone speaking from a mine 100 feet below the earth. To make it realistic, distort and echoes are used.

(vii) **Human voice:** The main stay in any radio programme is the human voice. Think of the voice of an announcer or newsreader on radio. You often find them very pleasant and nice to listen to. That is because of the quality of their voice and the proper use of it. There are two aspects of the use of human voice in radio production. Firstly, there has to be a well written script to be spoken and then someone has to speak or read it before a microphone in a studio.

Radio Programme Production

Now let us learn about the different stages of radio programme production. Whatever we do, there has to be a clear plan. In the case of radio production also, there is a well accepted process of production which is carried out in three stages.

Let us list them out.

a) Pre-production

b) Production and

c) Post-production

A) Pre-production

As the title suggests, this is the first stage before the actual production.

a. **An idea is born:** This phase includes how a programme is born as an idea and its conceptualization. The topic or subject matter is decided.

b. **Plan of action:** After the topic is decided, a plan of action is worked out. Here the format of the progamme is determined i.e. whether the programme is a talk, discussion, interview, drama, documentary etc. Once that is clear, the right person for writing the script and the performers are decided. The plan of action would also determine the equipment for outside recordings that are required. The time and venue of the recording are also worked out.

c. **The script** is examined to make it suitable for broadcast. It is examined according to the principles of writing for radio or in other words 'for the ear'.

d. **Paper work:** If people who are not working in the radio station are involved for writing or providing voice for the programmes, they have to be invited with an agreement to accept the job or assignment. This type of agreement is referred to as a contract. Similarly, permission is often required to interview certain people if the programme is based on such interviews.

e. **Rehearsing:** The voices of speakers is also part of this stage.

f. **B) Production:** This is the actual process of recording and editing a radio programme. Proper studios, microphones and computers are required to record and edit the programme.

g. **C) Post production:** Writing to inform people involved in the production is a major activity during this phase. The progamme has to be given publicity both on radio and in other media. This is done to ensure that people know about the programmes and also listen to them. The announcements for the presentation of the programme are also written and provided for the actual broadcast of the programme.

Writing for Radio

The script is examined according to the principles of 'writing for radio'. Let us discuss about the importance of the script in a radio programme. What happens when you listen to your favourite announcer speak on radio? You may say that you would like to continue to listen to the announcer. But why? Well, because you like what they say and the way they say that. Don't you?

What they say is in most cases drawn from a script. The way in which they say it is about the whole idea of using their voice. We shall now discuss about the language we use in radio. When we refer to the language used in radio, we are not referring to any particular language like English or Hindi but the style of any language used in radio. Here in radio, you find that it is different from the language used in newspapers and magazines or what is referred to as print media. In the print media, everything is written down or printed and they exist in the pages of the newspaper or magazine. You can keep them for as long as you want. You can take your own time to read them. If you do not understand them in one reading, you can read them again and understand. If you do not know the meaning of any word, you can refer to a dictionary to understand the meaning.

Now think of radio. You hear programmes as they are broadcast. You hear them only once. Of course, you may say that you can record them and listen to them. Record it and listen to it later. Well, that generally does not happen. In our day to day conversations, we speak, someone else listens, or someone speaks and we listen. It is a momentary experience. You may or may not remember what you said or heard. Similarly, what you hear on radio is heard only once and then forgotten in a majority of cases. Do you remember all that you had spoken yesterday? Do you remember all that you heard from others yesterday? Of course, the answer is a big 'no'. In our day to day conversation, we do not use the words and expressions found in a newspaper editorial or a feature article. Have you wondered why this happens?

It is because those words and expressions are neither friendly nor easy to understand. Now compare that with your daily conversation. You use simple words, sentences and familiar expressions. You also take into account the person to whom you are speaking. Your gestures and movements add to what you speak. Now let us go back to your favourite announcer. You don't see the person. But yet you feel that the person is speaking to you. You may not say anything but you may feel that there is some familiarity with that person. Interestingly, in most cases, what you hear on radio is written down or scripted. It is spoken and is heard only once. From what you have read so far, we can come to certain conclusions.

- First of all, when it comes to radio, you listen to a person who is not seen.
- The person speaks from a script written down earlier.
- You feel that the person is talking to you.
- The person sounds friendly and you have no problem in understanding what is being said.
- You can call the language or the words used in a radio script as the spoken word as against the written word or the printed word used in the print media.
- Unlike the printed word which is written for the eye, the spoken word is written for the ear.

The main characteristics of spoken word

A. Though it is written, it is spoken

B. It is written for the ear not the eye.

C. It is heard only once. The listener normally does not get a second chance to listen.

   a. It is conversational and should therefore sound like one and have the following qualities:

   b. It should be simply worded without any difficult or unfamiliar words.

   c. The sentences should be short and simple and not complex.

   d. There should be only one idea in a sentence and not many ideas.

   e. Though there are thousands of listeners, what is written should be meant for just one listener? At the listening end, there are generally only one or two persons and not a crowd. So it should be speaking to one person.

   f. The words chosen should denote the exact meaning and not be vague or abstract.

   g. The words should make pictures in the minds of the listeners.

   h. Abbreviations or short forms should be avoided. If an abbreviation is used, then its full form should be given.

   i. If the script consists of big numbers it can be rounded off to the nearest whole number.

   j. While referring to more than one person, avoid using 'he' or 'she'. It can confuse the listeners.

Chapter 15

Cyber Extension, ICT and GIS: A New Area of Communication

Emerging Issues in ICT

In the 2008 US presidential election, Barack Obama mobilized and coordinated an unprecedented number of people and resources through innovative uses of mobile phones and online social networking tools. In early 2008, Georgia began disseminating information through Google blogs, to counter a Russian denial of service (DoS) attacks against Georgian websites. In 2001, more than a million Philippine citizens protested the then President Estrada, mobilized by waves of text messages eventually toppling the government. More recently, Egyptians have begun to organize through Facebook to protest against the government. Indian state governments (Tamil Nadu , Kerala , etc.) have mandated open standards and Brazil's public sector is now required by law to use non-proprietary, open source software. In Philippines, indigenous communities are using a participatory geographic information systems methodology to claim their ancestral domain and manage their own human and natural resources in a more sustainable way. In Mexico, participatory budgeting (a process through which citizens can influence or contribute to decisions on the use and distribution of public resources) has been facilitated by the growth of the Internet. Some of the world's greatest universities, such as MIT, have now made their curriculum openly available on the Internet. Similarly, a consortium of 18 South African and international organizations work together to create freely accessible educational resources and course design guidance for teachers in Sub-Saharan Africa. What do all of these anecdotes have in common? First, they are all predicated on the emergence of new information and communication technologies (ICTs) that have greatly enhanced possibilities for communication and collaborative action. Second, all of these activities have taken advantages of a relatively open structure of technology and content to push towards effect social change.

The world is changing and new information and communication technologies (ICTs) are powerful contributors to this change. Many aspects of society are shifting from an industrial model with vertical hierarchical structures to a networked society with increasingly horizontal organizational structures. This change has been underpinned by the Internet and interconnected devices for computation and communication that have massively improved communication and collaboration possibilities. As the cost of new technologies decreases these trends should continue facilitating the diffusion of powerful (and smart) networked devices that can access the expanding network infrastructure.

- **Education**: If a central area of future ICT4D activities is based on open activities involving sharing, mobilization, collaboration and innovation, then the extent and success of that venture will be dependent upon the skills and networks of individuals in developing countries to engage in these activities. Furthermore, as new innovations and adaptations arise the importance of critical thinking skills and flexibility become even more essential. These skills will be needed to harness the value of new social capital networks that have both economic and intellectual value. This raises important questions concerning what type of education are best and how should we deliver that education?
- **Socio-economic divide**: Given the crucial role of social factors, it is clear, as would be anticipated, that the social, political, and economic factors that determine different individuals' relative resource sets and capacities for action will be influential in determining their ability to take advantage of these networking possibilities (for example, one's social capital). While the increased ubiquity of mobiles might make universal access a possibility, these other constraints will still play a major role. What social-economic factors are most critical for realizing Open ICT4D activities to generate the maximum social benefit?
- **Institutional change**: A movement from closed to more open and participatory institutions (such as with e-government and open business models) generally require fundamental changes to organizational processes and structures. For example, universities originally sought to develop proprietary models of courseware distribution over the Internet, until MITs Open Course Ware (OCW) finally broke the mould. This was a radical shift requiring universities to see the value in open-educational resources and to institutionalize the process of online courseware provision and all that entails. How can institutions best negotiate these institutional changes? What forms will the new institutional structures take?

- **Sustainability/new business models**: As many "open" ICT4D activities involve content production outside of the typical funding models of the market and state, issues of funding/incentives/and sustainability become increasingly important. Thus, central questions for future open activities are: What are the sustainable business models for content production and provision? What types of content should/could be provided by the government or market? What are the pitfalls and benefits of market or state based hosting of information? What sacrifices are made if hybrid models are used, for example, supporting free content through private sector advertisements?
- **Intellectual property rights**: The fundamental and changing role of property rights in the digital and network era cannot be overlooked. Evidence is emerging that property rights, as currently conceived, do not work as theorised (especially with respect to digital content), and even might be constraining innovation. What property rights regimens are best suited for particular Open ICT4D activities?
- **Complexity**: Openness is not a binary concept, it is scalar; content is neither open nor closed, rather it ranges from highly open to highly closed depending upon who makes it, who accesses or uses it, and how it is owned. The degree of openness determines the amount of flexibility you have with a particular piece of content (broadly defined). For example, software APIs are partially open whereas open source is more open. This is potentially a significant understanding for the following two reasons: (1) Understanding the level of openness potentially helps us to anticipate what types of outcomes are possible (the less open the more constrained, the more open the more possibilities); (2) This suggests that there are different levels of complexities related to different open activities. This has implications for research, for example, that the greater the complexity the harder it might be to generate big effects making research focusing on outcomes more difficult.
- **Filtering and accreditation**: The movement towards more horizontal structures of organization brings a different set of issues to the fore, such as how and who will validate, filter and organize data? For some activities, intermediaries in the hierarchical structures with special expertise/skills may still be necessary (e.g., for certain government services, or for peer micro-lending) while some may be possible without intermediation by replacing expertise with collective intelligence such as peer rating systems. Both of these situations require further research: for what technologies and in what circumstances do different approaches to peer-based validation of content work (or not) and how do we avoid the downsides (such as the spread of mis- or dis-information)?

ICT: Needs for Farming Community

Agriculture is an important sector with the majority of the rural population in developing countries depending on it. The sector faces major challenges of enhancing production in a situation of dwindling natural resources necessary for production. The growing demand for agricultural products, however, also offers opportunities for producers to sustain and improve their livelihoods. Information and communication technologies (ICT) play an important role in addressing these challenges and uplifting the livelihoods of the rural poor. The agricultural sector plays a crucial role in the India's development. Food grain production quadrupled during the post-independence era; this growth is projected to continue. The impact of climate change on agriculture could result in problems with food security and may threaten the livelihood activities upon which much of the population depends. Climate change can positively and negatively), as well as the types of crops that can be grown in certain areas, by impacting agricultural inputs such as water for irrigation, amounts of solar radiation that affect plant growth, as well as the prevalence of pests. While the magnitude of impact varies greatly by region, climate change is expected to impact on agricultural productivity and shifting crop patterns. The policy implications are wide-reaching, as changes in agriculture could affect food security, trade policy, livelihood activities and water conservation issues impacting large portions of the population.

ICTs and Dissemination of Agricultural Technologies

The need for improved agricultural extension especially in developing world has never been greater, hence rural extension continue to be in a phase of transition. In developing world, farming vulnerability, changes in natural resources quality, lack of coping and adaptation, level of decision making to climate change are well documented. Although, the globalization of commodities trade offer a mix of opportunities as well as challenges. The role of extension and support systems in this background is undergoing profound changes while no unified alternative framework has emerged. ICT is an umbrella term is often used in its plural sense (ICTs) to mean a range of technologies instead of a single technology. From the perspective of agricultural knowledge and information systems (AKIs), ICTs can be seen as useful in improving linkages between the research and the extension sub systems. The experience of rural tele-centres in the developing world shows that ICT can help in enabling rural development workers to gather, store, retrieve, adapt, localize and disseminate a broad range of information needed by rural families (Davison *et al* 2005). The ICTs in extension can lead to the emergence of knowledge workers that will result in the realization of a bottom-up, demand-driven paradigm for technology generation, assessment, refinement and transfer.

The linkages between agriculture and climate are pronounced and often complex. Crops and livestock are sensitive to climate change in both positive and negative ways. Agricultural systems are most sensitive to extreme climatic events such as droughts, floods and hailstorms, and to seasonal variability and changing rainfall patterns. Against this backdrop, farmer adaptations are influenced by many factors, including agricultural policy, prices, technology research and development, and agricultural extension services. The poor often bear a disproportionate burden of direct damage from catastrophes and climate change as concluded by most studies in developing countries (IPCC, 2001). The role of inadequate institutional support is frequently cited in the literature as a hindrance to adaptation. Many observers of rural development in recent times have commented on the frequent manifestations of unsatisfactory extension performance. Feder *et al* (2001) have suggested interrelated characteristics of extension systems in the developing world that jointly result in deficient performance, namely low staff morale, reduced efficiency and financial stress etc.

One more such key factor is the number of clients and vast spectrum of information/ services needed to be covered by extension systems. Policy makers in the developing world have reacted to this with the deployment of more extension personnel which has continued the emphasis on a more centralized, hierarchical and top-down management systems. The requirement for combining a bottom up approach with the conventional extension process is yet to be fulfilled and the limitations on the extension process to influence issues such as credit availability, input supplies, market linkages and logistics facilitation continue without change. In effect, there has been no visible impact due to such changes within the extension system in many parts of the developing world.(Sulaiman and Hall. 2006) have described a range of extension initiatives from the public and private sectors that explain the way extension agenda is expanding as embodied in the concept of "extension plus" and have pleaded for new experiments in extension. Pluralistic institutional arrangements are emerging and are finding wider acceptance and this is mainly because developing countries have realized the need for extension to engage in a wider range of issues beyond merely disseminating production-oriented technologies.

Extension pluralism is at the core of farmer adaptation strategies and ICT's can offer new advantages in enabling reliable and rapid access to expert information support which is much needed in the realization of adaptation strategies on a large scale.

ICT within the Agricultural Sector

Climate change is one of the most complex challenges that humankind has to face in the coming decades. As the change process seems to be irreversible, it became urgent to develop sound adaptation processes to the current and future shifts in the climate system. In particular, it is likely that the biggest impacts of changes will be on agricultural and food systems over the next few decades. At the crossroads between climate change and agriculture there are several tools available, because of the high number of crops and because of the complexity of replicating the same conditions across different regions. Every tool allows analyzing different processes of the agricultural sector, from local crop modelling under climate change conditions to the management of economic impacts of climate change on the agriculture sector (demand and supply, production, etc.), and so on.

As many tools exist, it's interesting to focus on their common aspects rather than their specific peculiarities. Some of the tools allow simulating the growth of specific crops, verifying their variations under different climate change scenarios. Usually these tools are site-specific, but they can be applied at national and/or regional level through a link to an appropriate Geographic Information System (GIS).

The first step of the applications happen with the definition of boundary conditions (which include data on crop calendar, soil status, etc.) and input climate parameters and data (such as: temperature, precipitations, wind speed, global radiation, soil moisture, air humidity, water flows...); some of the tools include also data related to crop management conditions. The second step is the development of the growth simulation in a specific state of potential crop production (e.g. with a certain fixed amount of water resources and nitrogen production) for different management options and for a chosen climate change scenario, through the link to an appropriate GCM or an ad hoc expert system. The general output of this kind of software is the assessment of crop production under given scenarios, facilitating decision making at farm level up to a whole crop system.

Opportunities and Challenges for the Agricultural Sector

Agriculture in developing countries is facing a major challenge of enhancing production in a situation of dwindling natural resources necessary for production. The growing demand for agricultural products, however, also offers opportunities for producers to sustain and improve their livelihoods. The agricultural sector is confronted with the major challenge of increasing production to feed a growing and increasingly prosperous population in a situation of decreasing availability of natural resources. Factors of particular concern are; water shortages, declining soil fertility, effects of climate change and rapid decrease of fertile agricultural lands due to urbanization. However, the growing demand - including for higher quality products - also offers opportunities for improving the livelihoods of rural communities. Realizing these opportunities requires compliance with more stringent quality standards and regulations for the production and handling of agricultural produce. New approaches and technical innovations are required to cope with these challenges and to enhance the livelihoods of the rural population. ICT play an important role in addressing these challenges and uplifting the livelihoods of the rural poor. This includes the use of computers, internet, geographical information systems, mobile phones, as well as traditional media such as radio or TV. Although it is a relatively new phenomenon, evidence of the contribution of ICT to agricultural development and poverty alleviation is becoming increasingly available.

a) Enhancing agricultural production

Increasing the efficiency, productivity and sustainability of small scale farms is an area where ICT can make a significant contribution. Farming involves risks and uncertainties, with farmers facing many threats from poor soils, drought, erosion and pests. Key improvements stem from information about pest and disease control, especially early warning systems, new varieties, new ways to optimize production and regulations for quality control.

b) Improving market access

Awareness of up to date market information on prices for commodities, inputs and consumer trends can improve farmers' livelihoods substantially and have a dramatic impact on their negotiating position. Such information is instrumental in making decisions about future crops and commodities and about the best time and place to sell and buy goods. Simple websites to match offer and demand of agricultural produce are a start of more complex agricultural trade systems. These sites tend to evolve from local selling/buying websites and price-information systems, to systems offering marketing and trading functions. Typically, price information is collected at the main regional markets and stored in a central database. The information is published on a website, accessible to farmers via information centres. To reach a wider audience, information is broadcast via rural radio, TV or mobile phone, thereby creating a 'level playing field' between producers and traders in a region. In Sri Lanka, the Govi Gnana project displays prices on light boards at major markets. The sustainability of these systems requires attention, with an important role for the private sector and organized producer groups. Web-based trading platforms offering one-stop shop facilities are emerging, especially for main commodities.

In India the private sector led Agriwatch (www.agriwatch.com) and e-Choupal programme (www.itcportal.com/ruraldevp_philosophy/ echoupal.htm) support several million farmers with price information, tender and transaction facilities. In recent years, short message and text services have taken up and effectively deliver prices and trading information via mobile phone to farmers, for instance in Senegal, Benin, and Zambia.

The set-up of price and market information systems has been piloted by IICD in Bolivia, Uganda, Tanzania and Ghana. Partner organizations are supported in adding ICT to core processes. In Ghana, International Institute for Communication and Development (IICD) supports the Social Enterprise Foundation of West Africa (SEND) in linking rural soybean producers to mills, through the use of satellite, databases and mobile phones, thereby ensuring a fair income for producers and a steady supply of raw materials for the mills.

c) Capacity-building and empowerment

Communities and farmer organizations can be helped through the use of ICTs to strengthen their own capacities and better represent their constituencies when negotiating input and output prices, land claims, resource rights and infrastructure projects. ICT enables rural communities to interact with other stakeholders, thus reducing social isolation. It widens the perspective of local communities in terms of national or global developments, opens up new business opportunities and allows easier contact with friends and relatives.

A role is also played by ICT in making processes more efficient and transparent. It helps in making laws and land titles more accessible. Global Positioning Systems (GPS) linked to Geographical Information Systems (GIS), digital cameras and internet, help rural communities to document and communicate their situation. Rural communities benefit from better access to credit and rural banking facilities. Recent mobile banking initiatives offer further scope to reduce costs and stimulate local trade. The Indian AMUL programme automates milk collection and payments for its 500,000 members, thereby enhancing transparency of the milk volume and quality collected and ensuring fair payments to farmers.

Prerequisite for effective ICT in Extension

For information and knowledge to be effectively used by rural communities several conditions should be met.

i) ICT in agriculture sector plans

Increasingly governments realize the necessity to link ICT and agriculture and incorporate ICT in agricultural sector policies and programmes. For instance: IICD has been supporting policy processes in Bolivia, Ghana and Jamaica in which national ICT for Development (ICTD) networks play a key role.

ii) Need for relevant agricultural information

It is generally accepted that information to sustain and increase agricultural production is spread over different agencies, notably farmers, universities, research institutes, extension services, commercial enterprises, and non-governmental organizations (NGOs). However, this knowledge is often poorly documented or hard to access. IICD promotes documentation of local agricultural practices in indigenous communities in Bolivia. Other organizations attempt to capture local knowledge, such as the Honey Bee programme in India. If the information does not exist, intermediary organizations can help to generate it, make it accessible and influence research agendas. The Tele-Support project active in two Indian states collects farmers' questions. Answers from local research institutes and universities are then repackaged on video and in local language, stored online (www.telesupport.org) and fed back into the local community.

iii) Timely information available in appropriate formats

Information on technologies is predominantly only available in hardcopy form or in stand-alone databases. Data are often incomplete or not compatible with other sources. Local knowledge on good practices and lessons learned about innovations is generally not captured. Economies of scale can be realized through the use of shared platforms using common standards. Information should be presented in an appropriate format in order to be effectively used by rural communities. Messages through videos in local languages have proved to be effective.

At this point in time, combining old and new media is most successful, such as videos of good practices, rural theatre, TV and radio broadcasts, which all provide input for local innovation. For example, a Ugandan IICD partner project effectively uses a drama group to convey messages to rural communities. IICD also supports the Zambian Ministry of Agriculture and Co-operatives in making research information available by combining textual and visual media in English and local languages. As information is increasingly stored online, it is accessible time and again.

iv) Institutional mechanisms and human capacity to link rural communities

There is a huge gap between information residing in agricultural knowledge centres and rural communities. At local level, multi-stakeholder mechanisms are important to make relevant information accessible to end users. Intermediary organizations have to connect rural communities to available knowledge. Users will increasingly want tailor-made, quality answers to their questions. In the Agricultural Clinics in India and Chile's online advisory service, customers get answers within one to two days. Mobile Q&A services are being piloted in India. At national level, mechanisms need to be in place to ensure learning and information sharing. In nine countries, IICD supports national ICT4D networks that play an important role in knowledge sharing, bringing various stakeholders together, and engaging in policy dialogue.

v) Rural access and exchange mechanisms: connectivity and tele-centres

The type of ICT used by local communities is subject to rapid change. However, broadband internet access is seen as central for societal innovation because storing of large datasets and live communication requires good connectivity. Until recently, connectivity in rural areas was limited to slow dial-up lines. Satellite connections now make broadband access possible in remote areas. Use of mobile phones has seen an enormous increase in recent years, especially in rural areas in Africa. Nevertheless, big differences still exist in broadband access between developed and developing countries, with Africa having only three per cent of global broadband users. New wireless technologies such as MESH and WiMAX, and new generation mobile phone networks, will provide high speed internet services at sharply reduced costs, thereby dramatically increasing internet coverage in rural areas. Various access tools are converging, becoming cheaper and more flexible.

New mobile phones and laptops provide omnipresent access with ample functionality for communication, transactions and transfer of data. The National Alliance for Mission 2007 and the Common Service Centre Scheme to establish tele-centres country-wide in India is clear examples of the government's dedication to enhance rural access. In Ecuador and Bolivia IICD has supported NGOs to set-up tele-centres in remote areas based on very small aperture terminals (VSATs) with shared cost of access. This is replicated and up scaled at national level. Ensuring sustainability is a major challenge.

Cost-sharing arrangements between local stakeholders, such as health centres, farmers' organizations, schools and local governmental bodies, are taking place. Also, payments for local services can generate revenues to sustain tele-centres. Price information projects in India report that farmers are willing to pay for price information from the gains made through access to it. Rural information centres also provide a learning environment for farmer groups on the use of ICT but also on jointly solving problems in their livelihoods.

What is Human Resources Information System (HRIS)?

A **HRIS, or Human Resource Information System**, is a software solution for small to mid-sized businesses to help automate and manage their HR, payroll, management and accounting activities. A HRIS generally should provide the capability to **more effectively plan, control and manage HR costs**; achieve improved efficiency and quality in HR decision making; and improve employee and managerial productivity and effectiveness. **A HRIS offers HR, payroll, benefits, training, recruiting and compliance solutions. Most are flexibly designed with**

integrated databases, a comprehensive array of features, and powerful reporting functions and analysis capabilities that you need to manage your workforce. This can give back hours of the HR administrator's day previously spent attending to routine employee requests. A **HRIS also facilitates communication processes** and saves paper by providing an easily-accessible, centralized location for company policies, announcements, and links to external URL's. Employee activities such as time-off requests and W-4 form changes can be automated, resulting in faster approvals and less paperwork. An affordable Human Resource Information System (HRIS), for example Sage HRMS's e-capabilities, **allows companies to manage their workforce** through two powerful main components: HR & Payroll. In addition to these essential software solutions, HRIS offers other options to help companies understand and fully utilize their workforce's collective skills, talents, and experiences.

Some of the most popular modules are:

- **Organization charts** (Create professional looking, dynamic organization charts)
- **Employee self service** (Employees can update personal information and view benefits elections, absence transactions, time-off balances and payroll information)
- **Benefits Administration** (Save paper and postage, take weeks off the benefits open enrollment period, reduce administration time, and improve data accuracy)
- **Track training for employees**

Selecting the right HRIS is important. Your company will need to make sure that they can customize the system to meet its specific and unique needs and that it is a system that will grow with your company.

Human Resources Management System (HRMS) or Human Resources Information System (HRIS)

A Human Resources Management System (HRMS) or Human Resources Information System (HRIS) refers to the systems and processes at the intersection between human resource management (HRM) and information technology. It merges HRM as a discipline and in particular its basic HR activities and processes with the information technology field, whereas the programming of data processing systems evolved into standardized routines and packages of enterprise resource planning (ERP) software. On the whole, these ERP systems have their origin from software that integrates information from different applications into one universal database. The linkage of its financial and human resource modules through one database is the most important distinction to the individually and proprietary developed predecessors, which makes this software application both rigid and flexible.

The function of human resources (HR) departments is administrative in nature and is common to all organizations. Organizations may have formalized selection, evaluation, and payroll processes. Management of "human capital" progressed to an imperative and complex process. The HR function consists of tracking existing employee data which traditionally includes personal histories, skills, capabilities, accomplishments and salary. To reduce the manual workload of these administrative activities, organizations began to electronically automate many of these processes by introducing specialized human resource management systems. HR executives rely on internal or external IT professionals to develop and maintain an integrated HRMS. Before client–server architectures evolved in the late 1980s, many HR automation processes were relegated to mainframe computers that could handle large amounts of data transactions. In consequence of the high capital investment necessary to buy or program proprietary software, these internally developed HRMS were limited to organizations that possessed a large amount of capital. The advent of client–server, application service provider, and software as a service (SaaS) or human resource management systems enabled higher administrative control of such systems. Currently human resource management systems encompass:

1. Payroll
2. Time and attendance
3. Performance appraisal
4. Benefits administration
5. HR management information system
6. Recruiting/Learning management
7. Performance record
8. Employee self-service
9. Scheduling
10. Absence management
11. Analytics

The **payroll module** automates the pay process by gathering data on employee time and attendance, calculating various deductions and taxes, and generating periodic pay cheques and employee tax reports. Data is generally fed from the human resources and time keeping modules to calculate automatic deposit and manual cheque writing capabilities. This module can encompass all employee-related transactions as well as integrate with existing financial management systems.

The **time and attendance module** gathers standardized time and work related efforts. The most advanced modules provide broad flexibility in data collection methods, labor distribution capabilities and data analysis features. Cost analysis and efficiency metrics are the primary functions.

The **benefits administration module** provides a system for organizations to administer and track employee participation in benefits programs. These typically encompass insurance, compensation, profit sharing and retirement.

The **HR management module** is a component covering many other HR aspects from application to retirement. The system records basic demographic and address data, selection, training and development, capabilities and skills management, compensation planning records and other related activities. Leading edge systems provide the ability to "read" applications and enter relevant data to applicable database fields, notify employers and provide position management and position control. Human resource management function involves the recruitment, placement, evaluation, compensation and development of the employees of an organization. Initially, businesses used computer based information systems to:

- produce pay checks and payroll reports;
- maintain personnel records;
- pursue talent management.

Online **recruiting** has become one of the primary methods employed by HR departments to garner potential candidates for available positions within an organization. Talent management systems typically encompass:

- analyzing personnel usage within an organization;
- identifying potential applicants;
- recruiting through company-facing listings;
- recruiting through online recruiting sites or publications that market to both recruiters and applicants.

The significant cost incurred in maintaining an organized recruitment effort, cross-posting within and across general or industry-specific job boards and maintaining a competitive exposure of availabilities has given rise to the development of a dedicated applicant tracking system, or 'ATS', module.

The **training module** provides a system for organizations to administer and track employee training and development efforts. The system, normally called a "learning management system" (LMS) if a stand alone product, allows HR to track education, qualifications and skills of the employees, as well as outlining what training courses, books, CDs, web based learning or materials are available to develop which skills. Courses can then be offered in date specific sessions, with delegates and training resources being mapped and managed within the same system. Sophisticated LMS allow managers to approve training, budgets and calendars alongside performance management and appraisal metrics.

The **employee self-service module** allows employees to query HR related data and perform some HR transactions over the system. Employees might question their attendance record from the system without asking the information from HR personnel. The module also lets supervisors approve O.T. requests from their subordinates through the system without overloading the task on HR department.

Many organizations have gone beyond the traditional functions and developed human resource management information systems, which support recruitment, selection, hiring, job placement, performance appraisals, employee benefit analysis, health, safety and security, while others integrate an outsourced applicant tracking system that encompasses a subset of the above.

The **Analytics** module enables organizations to extend the value of an HRMS implementation by extracting HR related data for use with other business intelligence platforms. For example, organizations combine HR metrics with other business data to identify trends and anomalies in headcount in order to better predict the impact of employee turnover on future output.

Information and communications technology (ICT) is often used as an extended synonym for information technology (IT), but is a more specific term that stresses the role of unified communications and the integration of telecommunications (telephone lines and wireless signals), computers as well as necessary enterprise software, middleware, storage, and audio-visual systems, which enable users to access, store, transmit, and manipulate information. The phrase ICT had been used by academic researchers since the 1980s, but it became popular after it was used in a report to the UK government by Dennis Stevenson in 1997 and in the revised National Curriculum for England, Wales and Northern Ireland in 2000. As of September 2013, the term "ICT" in the UK National Curriculum has been replaced by the broader term "computing".

The term ICT is now also used to refer to the convergence of audio-visual and telephone networks with computer networks through a single cabling or link system. There are large economic incentives (huge cost savings due to elimination of the telephone network) to merge the audio-visual, building management and telephone network with the computer network system using a single unified system of cabling, signal distribution and management.

The term Info-communications is sometimes used interchangeably with ICT. In fact Info-communications is the expansion of telecommunications with information processing and content handling functions on a common digital technology base.

The application of information and communications technology (ICT) in agriculture is increasingly important. E-Agriculture is an emerging field focusing on the enhancement of agricultural and rural development through improved information and communication processes. More specifically, e-Agriculture involves the conceptualization, design, development, evaluation and application of innovative ways to use information and communication technologies (IT) in the rural domain, with a primary focus on agriculture. E-Agriculture is a relatively new term and we fully expect its scope to change and evolve as our understanding of the area grows. Agriculture is one of the action lines identified in the declaration and plan of action of the World Summit on the Information Society (WSIS). The "Tunis Agenda for the Information Society," published on 18 November 2005 and emphasizes the leading facilitating roles that UN agencies need to play in the implementation of the Geneva Plan of Action. The Food and Agriculture Organization of the United Nations (FAO) has been assigned the responsibility of organizing activities related to the action line of ICT Applications on E-Agriculture.

The main phases of the agriculture industry include crop cultivation, water management, fertilizer application, fertigation, pest management, harvesting, post-harvest handling, transport of food products, packaging, food preservation, food processing/value addition, quality management, food safety, food storage, and food marketing.

All stakeholders of agriculture industry need information and knowledge about these phases to manage them efficiently. Any system applied for getting information and knowledge for making decisions in any industry should deliver accurate, complete, concise information in time or on time. The information provided by the system must be in user-friendly form, easy to access, cost-effective and well protected from unauthorized accesses.

a. Record text, drawings, photographs, audio, video, process descriptions, and other information in digital formats,

b. Produce exact duplicates of such information at significantly lower cost,

c. Transfer information and knowledge rapidly over large distances through communications networks.

d. Develop standardized algorithms to large quantities of information relatively rapidly.

Types of ICT tools

There are various ICT tools or applications available to simplify the operation for general purpose use to mission critical applications. The general purpose applications can be used to simplify the day-to-day operations in the organization. The following range of ICT tools (applications) is available for the general purpose use.

a. Word processing applications - used for preparing text documents;

b. Presentation software - used for preparing presentations;

c. Spreadsheets - used for calculation and preparing tabular data with calculations;

d. Databases - used to store the information in the form of records and retrieving information using query mechanism;

e. Multimedia – used to organize the information more meaningful manner with the help of text, images, animation, audio and videos;

f. Web browsers and E-mail – used to searching information and sending or receiving mail using Internet.

Computer

A computer is a machine that is able to take information (input), do some work on or make changes to the information, to make new information (output). Computers have existed for much of human history. Examples of early computers are the astrolabe and the abacus. Modern computers are very different from early computers. They are now very powerful machines that are able to do billions of calculations every second. Most people have used a personal computer in their home or at work. Computers are useful for many different jobs where automatic tasks are useful. Some examples are controlling traffic lights, vehicle computers, security systems, Washing machines and Digital Televisions.

A person (called a user) can control a computer by telling it to do things. Some ways of controlling a computer are with a keyboard, mouse, buttons, and touch screen. Some very new computers can also be controlled with voice commands or hand gestures. Computers can be designed to do anything with information. Computers are used to control factories, which in the past were controlled by humans. They are also in homes, where they are used for things such as listening to music, reading the news, and writing.

Modern computers are electronic machines. A computer is only useful if it has both hardware and software. Hardware is the physical parts the computer is made of - for example keyboard, mouse, screen, tower, and the circuits inside it. Software is the computer programs (mathematical instructions). The software uses the hardware, when the user gives it instructions and changing it in to useful output.

Many modern computers do billions of calculations each second. They do mathematical arithmetic very quickly but computers do not really "think". They only follow the instructions in their software programs.

Computer programs are designed or written by computer programmers. A few programmers write programs in the computer's own language called machine code. Machine Code has only a few instructions and is based on logic and Mathematical arithmetic. Using machine code is difficult for most human programmers.

Today, most programmers write programs using a programming language like C++, Java or BASIC. These programming languages are more like the language you talk and write with every day. A program called a compiler translates the user's instructions into binary code (machine code) and turns it into useful information that a user can easily understand, and then the computer will understand the instructions and do what is needed.

History of computers

The Jacquard loom was one of the first programmable devices.

Definition

A computer is an electronic machine which helps solving problems quickly and easily. It solves problems according to instructions given to it by the computer user.

Invention

Nobody knows who built the first computer. This is because the word "computer" used to mean a person who did mathematics as their job (a human computer). Because of this, some people say that humans were the first computers. Human computers got bored doing the same calculation over and over again, and made tools (mostly mechanical calculating devices like abacuses) to help them get the answers to their problems. It is said that the inventor of the 'modern' computer was Charles Babbage.

Automation

Humans have a problem with calculations. To show this, try doing 584 x 3,220 in your head. It is hard to remember all the steps. People made tools to help them remember where they were in a calculations problem. The other problem people have is that they have to do the same problem over and over and over again. A cashier used to make change every day in her head or with a piece of paper. That took a lot of time and people made mistakes. So people made machines that did those same things over and over. This part of computer history is called the "history of automated calculation," which is a fancy phrase for "the history of machines that make it easy for me to do this same maths problem over and over without making mistakes."

Programming

Some people did not want a machine that would do the same thing over and over again. For example, a music box is a machine that plays the same music over and over again. Some people wanted to be able to tell their machine to do different things. For example, they wanted to tell the music box to play different music every time. They wanted to be able to program the music box- to order the music box to play different music. This part of computer history is called the "history of programmable machines" which is a fancy phrase for "The history of machines that I can order to do different things if I know how to speak their language."

One of the first examples of this was built by Hero of Alexandria. He built a mechanical theater which performed a play lasting 10 minutes and was operated by a complex system of ropes and drums. These ropes and drums were the language of the machine- they told what the machine did and when. Some people argue that this is the first programmable machine.

Most historians agree that the "castle clock", an astronomical clock invented by Al-Jazari in 1206, is the first known programmable analog computer. It showed the zodiac, the solar and lunar orbits, a crescent moon-shaped pointer traveling across a gateway that made some doors to open every hour, and five robotic musicians who play music when levers hit them. The length of day and night could be changed (AKA re-programmed) every day in order to account for the changing lengths of day and night throughout the year. Ada Lovelace is considered to be the first programmer.

The Computing Era

At the end of the Middle Ages, people in Europe thought math and engineering were more important. In 1623, Wilhelm Schickard made a mechanical calculator. Other Europeans made more calculators after him. They were not modern computers because they could only add, subtract, and multiply- you could not change what they did to make them do something like play Tetris. Because of this, we say they were not programmable.

In 1801, Joseph Marie Jacquard used punched paper cards to tell his textile loom what kind of pattern to weave. He could use punch cards to tell the loom what to do, and he could change the punch cards, which means he could program the loom to weave the pattern he wanted. This means the loom was programmable.

Modern computers were made when someone (Charles Babbage) had an idea. He wanted to make a machine that could do all the boring parts of math, (like the automated calculators) and could be told to do them different ways (like the programmable machines.) Charles Babbage was the first to make a design of a fully programmable mechanical computer. He called it the "The Analytical Engine". Because Babbage did not have enough money and always changed his design when he had a better idea; he never built his Analytical Engine.

As time went on, computers got more and more popular. And that stands out at the beginning. This is because people get bored easily doing the same thing over and over. Imagine spending your life writing things down on index cards, storing them, and then having to go find them again. The U.S. Census Bureau in 1890 had hundreds of people doing just that. People got very bored and very frustrated, and would say, "There HAS to be an easier way to do this." Then a clever person worked out how to make machines do a lot of the work. Herman Hollerith figured out how to make a machine that would automatically add up information that the Census bureau collected. The Computing Tabulating Recording Corporation (which later became IBM) made his machines, and everyone was happy. At least, they were happy until their machines broke down, got jammed, and had to be repaired. This is when the Computing Tabulating Recording Corporation invented tech support. Because of machines like this, new ways of talking to these machines were invented, and new types of machines were invented, and eventually the computer as we know it was born.

Analog and Digital Computers

In the first half of the 20th century, scientists started using computers, mostly because scientists had a lot of calculations to figure out and wanted to spend more of their time thinking about the secrets of the universe instead of spending hours adding numbers together. For example, if they had to launch a rocket ship, they needed to do a lot of math to make sure the rocket did not blow up.

So they put together computers. These computers used analog circuits, which made them very hard to program. Then, in the 1930s, they invented digital computers, which made them easier to program.

High-scale computers

Scientists figured out how to make and use digital computers in the 1930s and 1940s. They made a lot of digital computers, and as they did, they figured out how to ask them the right sorts of questions to get the most out of them. Here are a few of the computers they built:

EDSAC was one of the first computers that remembered what you told it even after you turned the power off. This is called (von Neumann) architecture.

Konrad Zuse's electromechanical "Z machines". The Z3 (1941) was the first working machine that used binary arithmetic. Binary arithmetic means using "Yes" and "No." to add numbers together. You could also program it. In 1998 the Z3 was proved to be Turing complete. Turing complete means that it is possible to tell this particular computer anything that it is mathematically possible to tell a computer. It is the world's first modern computer.

The non-programmable Atanasoff–Berry Computer (1941) which used vacuum tubes to store "yes" and "no" answers, and regenerative capacitor memory.

The Harvard Mark I (1944), A big computer that you could kind of program.

The U.S. Army's Ballistics Research Laboratory ENIAC (1946), which could add numbers the way people do (using the numbers 0 through 9) and is sometimes called the first general purpose electronic computer (since Konrad Zuse's Z3 of 1941 used electromagnets instead of electronics). At first, however, the only way you could reprogram ENIAC was by rewiring it.

Several developers of ENIAC saw its problems. They invented a way to for a computer to remember what they had told it, and a way to change what it remembered. This is known as "stored program architecture" or von Neumann architecture. The first of these was completed in Great Britain. The first to be demonstrated working was the Manchester Small-Scale Experimental Machine (SSEM or "Baby"), while the EDSAC, completed a year after SSEM, was the first really useful computer that used the stored program design. Shortly afterwards, the machine originally described by von Neumann's paper EDVAC was completed but was not ready for two years. Nearly all modern computers use the stored-program architecture in some form. It has become the main concept which defines a modern computer. Most of the technologies used to build computers have changed since the 1940s, but many current computers still use the von-Neumann architecture.

Microprocessors are miniaturized devices that often implement stored program CPUs. In the 1950's computers were built out of mostly vacuum tubes. Transistors replaced vacuum tubes in the 1960's because they were smaller and cheaper. They also need less power and do not break down as much as vacuum tubes. In the 1970s, technologies were based on integrated circuits. Microprocessors, such as the Intel 4004 made computers smaller and cheaper. They also made computers faster and more reliable. By the 1980s, computers became small and cheap enough to replace mechanical controls in things like washing machines. The 1980s also saw home computers and personal computer. With the evolution of the Internet, personal computers are becoming as common as the television and the telephone in the household.

In 2005 Nokia started to call some of its mobile phones (the N-series) "multimedia computers" and after the launch of the Apple i-Phone in 2007, many are now starting to add the smart-phone category among "real" computers. In 2008, if the category of smart-phones is included in the numbers of computers in the world, the biggest computer maker by units sold is no longer Hewlett-Packard, but rather Nokia.

Kinds of Computers

There are six types of computers: personal computers, mainframes, laptop computers, mini computers, super computers and embedded computers.

A "desktop computer" is a small machine that has a screen (which is not part of the computer). Most people keep them on top of a desk, which is why they are called "desktop computers." "Laptop computers" are computers small enough to fit on your lap. This makes them easy to carry around. Both laptops and desktops are called personal computers, because one person at a time uses them for things like playing music, surfing the web, or playing video games.

There are bigger computers that many people at a time can use. These are called "Mainframes," and these computers do all the things that make things like the internet work. You can think of a personal computer like this: the personal computer is like your skin: you can see it, other people can see it, and through your skin you feel wind, water, air, and the rest of the world. A mainframe is more like your internal organs: you (hopefully) never see them, and you barely even think about them, but if they suddenly went missing, you would have some very big problems.

There is another type of computer, called an embedded computer. An embedded computer is a computer that does one thing and one thing only, and usually does it very well. For example, an alarm clock is an embedded computer: it tells the time. Unlike your personal computer, you cannot use your clock to play Tetris. Because of this, we say that embedded computers cannot be programmed, because you cannot install programs like Tetris on your clock. Some mobile phones, automatic teller machines, microwave ovens, CD players and cars are examples of embedded computers.

All-in-one PC

All-in-one computers are desktop computers that have the CPU and speaker in the same case as the monitor. Apple has made several popular examples of all-in-one computers, such as the original Macintosh of the mid-1980s and the iMac of the late 1990s and 2000s.

- Common uses of home computers
- Playing computer games
- Writing
- Solving calculations
- Looking for things on the Internet
- Watching TV and movies
- Listening to music
- Communicating with other people
- Downloading
- Sending e-mail

- Sorting
- Recording
- Calculating
- Editing
- Taking Pictures (photography)
- Drawing

Working methods

Computers store data and the instructions telling them what to do with the data as numbers, because computers can do things with numbers very quickly. These data are stored as binary symbols (1s and 0s). A 1 or a 0 symbol stored by a computer is called a bit, which comes from the words binary digit. Computers can use many bits together to represent instructions and the data that these instructions use. A list of instructions is called a program and is stored on the computer's hard disk. Computers work through the program by using a central processing unit, and they use fast memory called RAM as a space to store the instructions and data while they are doing this. When the computer wants to store the results of the program for later, it uses the hard disk because things stored on a hard disk can still be remembered after the computer is turned off.

An operating system tells the computer how to understand what jobs it has to do, how to do these jobs, and how to tell people the results. It tells the electronics inside the computer, or "hardware", how to work to get the results it needs. This lets most computers have the same operating system, or list of orders to tell it how to talk to the user, while each computer can have its own computer programs or list of jobs to do what its user needs. Having different programs and operating systems makes it easy to learn how to use computers for new things. When a user needs to use a computer for something different, the user can learn how to use a new program.

The Internet

One of the most important jobs that computers do for people is helping with communication. Communication is how people share information. Computers have helped people move forward in science, medicine, business, and learning, because they let experts from anywhere in the world work with each other and share information. They also let other people communicate with each other, do their jobs almost anywhere, learn about almost anything, or share their opinions with each other. The Internet is the thing that lets people communicate between their computers.

Computers and waste

A computer is now almost always an electronic device. It usually contains materials that will become toxic waste when disposed of. When a new computer is bought in some places, laws require that the cost of its waste management must also be paid for. This is called product stewardship.

Computers can become obsolete quickly, depending on what programs the user runs. Very often, they are thrown away within two or three years, because newer programs require a more powerful computer. This makes the problem worse, so computer recycling happens a lot. Many projects try to send working computers to developing nations so they can be re-used and will not become waste as quickly, as most people do not need to run new programs. Some computer parts, such as hard drives, can break easily. When these parts end up in the landfill, they can put poisonous chemicals like lead into the ground water. Hard drives can also contain secret information like credit card numbers. If the hard drive is not erased before being thrown away, an identity thief can get the information off of the hard drive, even if the drive doesn't work, and use it to steal money from the previous owner's bank account.

Main hardware

Computers come in different forms, but most of them have a common design.

- All computers have a CPU.
- All computers have some kind of data bus which lets them get inputs or output things to the environment.
- All computers have some form of memory. These are usually chips (integrated circuits) which can hold information.
- Many computers have some kind of sensors, which lets them get input from their environment.
- Many computers have some kind of display device, which lets them show output. They may also have other peripheral devices connected.

A computer has several main parts. When comparing a computer to a human body, the CPU is like a brain. It does most of the 'thinking' and tells the rest of the computer how to work. The CPU is on the Motherboard, which is like the skeleton. It provides the basis for where the other parts go, and carries the nerves that connect them to each other and the CPU. The motherboard is connected to a power supply, which provides electricity to the entire computer. The various drives (CD drive, floppy drive, and on many newer computers, USB flash drive) act like eyes, ears, and fingers, and allow the computer to read different types of storage, in the same way that a human can read different types of books. The hard drive is like a human's memory, and keeps track of all the data stored on the computer. Most computers have a sound card or another method of making sound, which is like vocal cords, or a voice box. Connected to the sound card are speakers, which are like a mouth, and are where the sound comes out. Computers might also have a graphics card, which helps the computer to create visual effects, such as 3D environments, or more realistic colors, and more powerful graphics cards can make more realistic or more advanced images, in the same way a well trained artist can.

Computer Speaker

Computer speakers, or multimedia speakers, are speakers external to a computer that disable the lower fidelity built-in speaker. They often have a low-power internal amplifier. The standard audio connection is a 3.5 mm (approximately 1/8 inch) stereo jack plug often color-coded lime green (following the PC 99 standard) for computer sound cards. A plug and socket for a two-wire (signal and ground) coaxial cable that is widely used to connect analog audio and video components. Rows of RCA sockets are found on the backs of stereo amplifier and numerous A/V products. The prong is 1/8" thick by 5/16" long. A few use an RCA connector for input. There are also USB speakers which are powered from the 5 volts at 500 milliamps provided by the USB port, allowing about 2.5 watts of output power.

Computer speakers range widely in quality and in price. The computer speakers typically packaged with computer systems are small, plastic, and have mediocre sound quality. Some computer speakers have equalization features such as bass and treble controls. The internal amplifiers require an external power source, usually an AC adapter. More sophisticated computer speakers can have a subwoofer unit, to enhance bass output, and these units usually include the power amplifiers both for the bass speaker, and the small satellite speakers.

Some computer displays have rather basic speakers built-in. Laptops come with integrated speakers. Restricted space available in laptops means these speakers usually produce low-quality sound. For some users, a lead connecting computer sound output to an existing stereo system is practical. This normally yields much better results than small low-cost computer speakers. Computer speakers can also serve as an economy amplifier for MP3 player use for those who wish to not use headphones, although some models of computer speakers have headphone jacks of their own.

BIOS Error Codes

The PC speaker is generally the first output device to be activated during the boot process. Since it is active before the graphics card, it can be used to communicate error codes related to problems that prevent the much more complex initialization of the graphics card to take place. For example the Video BIOS usually cannot activate a graphics card unless working RAM memory is present in the system, while beeping the speaker is doable with just ROM and the CPU registers. Usually, different error codes will be signaled by specific beeping patterns, such as e.g. "one beep; pause; three beeps; pause; repeat". These patterns are motherboard specific and are usually documented in the technical manual of the motherboard.

Games

The PC speaker was often used in very innovative ways to create the impression of polyphonic music or sound effects within computer games of its era, such as the LucasArts series of adventure games from the mid-1990s, using swift arpeggios. Several games such as *Space Hulk* and *Pinball Fantasies* were noted for their elaborate sound effects; Space Hulk in particular even had full speech.

Other Programs

Several programs, including MP (Module Player, 1989), Scream Tracker, Fast Tracker, Impulse Tracker, and even device drivers for Linux and Microsoft

Windows, could play pulse-code modulation (PCM) sound through the PC speaker using special techniques explained later in this article. Modern Microsoft Windows systems have PC speaker support as a separate device with special capabilities - that is, it cannot be configured as a normal audio output device. Some software uses this special sound channel to produce sounds. For example, Skype can be used as a reserve calling signal device for situations when the primary audio output device cannot be heard (for instance when volume is set to the minimum level or the amplifier is turned off).

Headphones

Headphones are a pair of small loudspeakers, or less commonly a single speaker, held close to a user's ears and connected to a signal source such as an audio amplifier, radio, CD player or portable media player. They are also known as stereo-phones, headsets or, colloquially, cans. The in-ear versions are known as earphones or ear-buds. In the context of telecommunication, the term headset is used to describe a combination of headphone and microphone used for two-way communication, for example with a telephone.

The telephone earpiece such as the one pictured at the right was common at the beginning of the 20th century. Headphones originated from the earpiece, and were the only way to listen to audio signals before amplifiers were developed. The first truly successful set was developed by Nathaniel Baldwin, who made them by hand in his kitchen and sold them to the U.S. Navy.

Very sensitive headphones such as those manufactured by Brandes around 1919 were commonly used for early radio work. These early headphones used moving iron drivers, either single ended or balanced armature. The requirement for high sensitivity meant no damping was used, thus the sound quality was crude. They also had very poor comfort compared to modern types, usually having no padding and too often having excessive clamping force to the head. Their impedance varied; headphones used in telegraph and telephone work had an impedance of 75 ohms. Those used with early wireless radio had to be more sensitive and were made with more turns of finer wire; impedance of 1,000 to 2,000 ohms was common, which suited both crystal sets and triode receivers.

In early powered radios, the headphone was part of the vacuum tube's plate circuit and had dangerous voltages on it. It was normally connected directly to the positive high voltage battery terminal, and the other battery terminal was securely earthed. The use of bare electrical connections meant that users could be shocked if they touched the bare headphone connections while adjusting an uncomfortable headset.

Applications

Headphones may be used both with fixed equipment such as CD or DVD players, home theater, personal computers and with portable devices (e.g. digital audio player/mp3 player, mobile phone, etc.). Cordless headphones are not connected with a wire, receiving a radio or infrared signal encoded using a radio or infrared transmission link, like FM, Bluetooth or Wi-Fi. These are actually made of powered receiver systems of which the headphone is only a component, these types of cordless headphones are being used more frequently with events such as a Silent disco or Silent Gig.

In the professional audio sector headphones are used in live situations by disc jockeys with a DJ mixer and sound engineers for monitoring signal sources. In radio studios, DJs use a pair of headphones when talking to the microphone while the speakers are turned off, to eliminate acoustic feedback and monitor their own voice. In studio recordings, musicians and singers use headphones to play along to a backing track. In the military, audio signals of many varieties are monitored using headphones.

Wired headphones are attached to an audio source. The most common connection standards are 6.35mm (") and 3.5mm TRS connectors and sockets. The larger 6.35mm connector tending to be found on fixed location home or professional equipment. Sony introduced the smaller, and now widely used, 3.5mm "minijack" stereo connector in 1979, adapting the older monophonic 3.5mm connector for use with its Walkman portable stereo tape player and the 3.5mm connector remains the common connector for portable application today. Adapters are available for converting between 6.35mm and 3.5mm devices.

Types

The particular needs of the listener determine the choice of headphone. The need for portability indicates smaller, lighter headphones but can mean a compromise in fidelity. Headphones used as part of a home hi-fi do not have the same design constraints and can be larger and heavier. Generally, headphone form factors can be divided into four separate categories: *circumaural*, *supra-aural*, *ear bud*, and *in-ear*.

Circumaural

Circumaural headphones (sometimes called full size headphones) have circular or ellipsoid ear-pads that encompass the ears. Since these headphones completely surround the ear, circumaural headphones can be designed to fully seal against the head to attenuate any intrusive external noise. Because of their size, circumaural headphones can be heavy and there are some sets which weigh over 500 grams (1 lb). Good headband and ear-pad design is required to reduce discomfort resulting from weight.

Supra-aural

Supra-aural headphones have pads that sit on top of the ears, rather than around them. They were commonly bundled with personal stereos during the 1980s. This type of headphone generally tends to be smaller and more lightweight than circumaural headphones, resulting in less attenuation of outside noise.

Open or Closed Back

Circumaural and supra-aural headphones can both also be further differentiated by the type of earcups: Open-back headphones have the back of the earcups open. This leaks more sound out of the headphone and also lets more ambient sounds into the headphone, but gives a more natural or speaker-like sound and more spacious "soundscape" - the perception of distance from the source. Closed-back styles have the back of the earcups closed. Depending on the model they may block 8-32db of ambient noise, but have a smaller soundscape, giving you a perception that the sound is coming from within your head.

Ear-Fitting Headphones

Ear-buds

Ear-buds or earphones are very small headphones that are fitted directly in the outer ear, facing but not inserted in the ear canal; they have no band or other arrangement to fit over the head. Such earphones are portable and convenient, but many people consider them to be uncomfortable and prone to falling out. Various models are available, starting at very low prices. They provide hardly any acoustic isolation and leave room for ambient noise to seep in; users may turn up the volume dangerously high to compensate, at the risk of causing hearing loss. From about 1990 ear-buds have commonly been bundled with personal music devices.

In-ear headphones

In-ear headphones, like ear-buds, are small and without headband, but are inserted in the ear canal itself. They are sometimes known as canalphones. Price and quality range from relatively inexpensive to very high; the better ones are called in-ear monitors (IEMs) and are used by audio engineers and musicians as well as audiophiles. Canalphones offer portability similar to ear-buds, block out much environmental noise by obstructing the ear canals, and are far less prone to falling out. When used for casual portable use they block out sounds which can be important for safety (*e.g.*, approaching vehicles).

- Universal canalphones provide one or more stock silicone rubber, elastomer, or foam sleeves to fit various ear canals, for correct placement and best noise isolation.
- Custom canalphones are fitted to the ears of the individual user: castings of the ear canals are made, and the manufacturer uses the castings to create custom-molded silicone rubber or elastomer plugs that provide added comfort and noise isolation. Because of the individualized labor involved, custom IEMs are more expensive than universal IEMs; resale value is very low as they are unlikely to fit other people.

Headset

A headset is a headphone combined with a microphone. Headsets provide the equivalent functionality of a telephone handset with hands-free operation. The most common uses for headsets are in console or PC gaming, Call centers and other telephone-intensive jobs and also for personal use at the computer to facilitate comfortable simultaneous conversation and typing. Headsets are made with either a single-earpiece (mono) or a double-earpiece (mono to both ears or stereo). The microphone arm of headsets is either an external microphone type where the microphone is held in front of the user's mouth, or a voice tube type where the microphone is housed in the earpiece and speech reaches it by means of a hollow tube.

Telephone Headsets

Telephone headsets connect to a fixed-line telephone system. A telephone headset functions by replacing the handset of a telephone. All telephone headsets come in a standard 4P4C commonly called an RJ-9 connector. For older models of telephones, the headset microphone impedance is different from that of the original handset, requiring a telephone amplifier for the telephone headset. A telephone amplifier provides basic pin-alignment similar to a telephone headset adaptor, but it also offers sound amplification for the microphone as well as the loudspeakers. Most models of telephone amplifiers offer volume control for loudspeaker as well as microphone, mute function and headset/handset switching. Telephone amplifiers are powered by batteries or AC adaptors.

Benefits and Limitations

Headphones may be used to prevent other people from hearing the sound either for privacy or to prevent disturbance, as in listening in a public library. They can also provide a level of sound fidelity greater than loudspeakers of similar cost. Part of their ability to do so comes from the lack of any need to perform room correction treatments with headphones. High quality headphones can have an extremely flat low-frequency response down to 20 Hz within 3dB. Marketed claims such as 'frequency response 4 Hz to 20 kHz' are usually overstatements; the product's response at frequencies lower than 20 Hz is typically very small.

Headphones are also useful for video games that use 3D positional audio processing algorithms, as they allow players to better judge the position of an off-screen sound source (such as the footsteps of an opponent).

Although modern headphones have been particularly widely sold and used for listening to stereo recordings since the release of the Walkman, there is subjective debate regarding the nature of their reproduction of stereo sound. Stereo recordings represent the position of horizontal depth cues (stereo separation) via volume and phase differences of the sound in question between the two channels. When the sounds from two speakers mix, they create the phase difference the brain uses to locate direction. Through most headphones, because the right and left channels do not combine in this manner, the illusion of the phantom centre can be perceived as lost. Hard panned sounds will also only be heard only in one ear rather than from one side.

Binaural recordings use a different microphone technique to encode direction directly as phase, with very little amplitude difference below 2 kHz, often using a dummy head, and can produce a surprisingly life-like spatial impression through headphones. Commercial recordings almost always use stereo, rather than binaural, recording, because loudspeaker listening has been more popular than headphone listening.

It is possible to change the spatial effects of stereo sound on headphones to better approximate the presentation of speaker reproduction by using frequency-dependent cross-feed between the channels, or better still a Blumlein shuffler (a custom EQ employed to augment the low-frequency content of the difference information in a stereo signal). While cross-feed can reduce the unpleasantness that some listeners find with hard panned stereo in headphones, a dummy head recording with artificial pinnae, played through headphones, can give the experience of hearing the performance as though situated in the position of the dummy head. Optimal sound is achieved when the dummy head matches the listener's head, since pinnae vary greatly in size and shape.

Headsets can have ergonomic benefits over traditional telephone handsets. They allow call centre agents to maintain better posture without needing to hand-hold a handset or tilt their head sideways to cradle it.

Over time, headphone cables fail, usually by copper wires breaking at junction points on the cord (at the TRS jack, or at the point of connection to the headphone) after repeated flexing. These sites are usually fitted with some kind of strain relief, but eventually fail.

Trends in Agricultural Information Management

Virtual Community

A virtual community, e-community or online community is a group of people that primarily interact via communication media such as letters, telephone, email or Usenet rather than face to face. If the mechanism is a computer network, it is called an online community. Virtual and online communities have also become a supplemental form of communication between people who know each other primarily in real life. Many means are used in social software separately or in combination, including chat rooms and forums that use voice, video and/ or text.

Virtual Information Networks

A number of virtual information networks with the objective of linking agricultural institutions for facilitating better access to information resources have been initiated successfully. While some of these network agricultural institutions within a country, link to agricultural institutions around the globe. Some of these are profiled here.

International Initiatives

a. Agriculture Network Information Centre (AgNIC) www.agnic.org

AgNIC is an Internet based network of public and private agricultural libraries and information centres, coordinated by the National Agriculture Library (NAL), USA. The network aims to provide global access to agricultural information. AgNIC Members represent 40 land-grant universities and other national and international partners including universities and research institutions, government agencies, and a non-profit organization. Through its website the network provides access to a network of electronic sources on research and teaching in agriculture, food, renewable natural resources, forestry, and physical and social sciences. AgNIC is a distributed discipline-oriented source of agricultural information in electronic form on the Internet. The goals are to: identify major collections of agriculture-related information; provide mechanisms to facilitate access/ retrieval from these information resources; create mechanisms to encourage organizations to collaborate in creating/ using AgNIC. AgNIC partners select important information sources for inclusion in the system.

b. Agricultural Libraries Network (AGLINET)

The Agricultural Libraries Network (AGLINET- www.fao.org/ library/ info_services EN/ aglinet EN.htm) coordinated by the Food and Agriculture Organization of the United Nation (FAO), is a world-wide, voluntary network of international agricultural libraries with regional/country coverage and other comprehensive or specialized subject resource collections. All member libraries provide, upon request, access to the literature originating in the country or region or for a given specialization. AGLINET aims at combining resources for mutual and rational use through delivery of primary documents, by means of inter library loan provision of reproductions, bibliographic information; with appropriate regional and subject specialization.

c. International Information System for Agricultural Sciences and Technology (AGRIS) http://www.fao.org/agris

The Food and Agriculture Organization (FAO) of the United Nations initiated agricultural information networking through AGRIS. Established in 1974, AGRIS facilitates information exchange and brings together world literature dealing with all aspects of agriculture and related subjects. AGRIS is a cooperative system in which 240 national, international and intergovernmental centres in participating countries input references to the literature produced within their countries and draw on the information provided by the other participants.

d. Aggregate (www.agrigate.edu.au)

Coordinated by the National Library of Australia is a project of the libraries of the Universities of Melbourne, Adelaide and Queensland, and the Commonwealth Scientific and Industrial Research Organization (CSIRO) in Australia. This is a subject information gateway for resources, both online and offline, in agricultural research. The purpose is to support the identification and dissemination of quality research materials selected by an editorial review process consisting of members of the agricultural research community. The content covers agriculture and related areas of value to the Australian agricultural research community. The resources selected have been reviewed by subject specialist librarians. The majority of resources identified in the database are available online.

e. AGROWEB CEE network (http://www.iaaldcee.hu/awnt.html)

AGROWEB CEE network has been established by organizations and individuals involved in agricultural and rural development in Central and Eastern Europe. This collaborative network is ICTs: Concepts and Relevance maintained by representatives of participating countries, and facilitated by International Association of Agricultural Information Specialists (IAALD) and FAO.

The AgroWeb national portals provide Internet links, to these national institutions and to other relevant national web sites. In addition to the national portal pages, cross-cutting subject areas have been identified and a number of thematic sectors of the network are being coordinated by regional focal points.

f. ELIN - Entomology Library and Information Network

ELIN project hosted by the Natural History Museum, London, at http:// www.nhm.ac.uk/ hosted_sites/elin/ aims to establish a global Entomology Library and Information Network to provide entomologists with a co-ordinated and structured electronic information resource including a gateway to a wide range of entomological information. A mailing list was launched in 1998 and has over 100 members from libraries and information centres from around the world, which has encouraged international communication between specialists working with entomological information. The ELIN project provides up-to-date, detailed information about entomological libraries around the world.

Information Network on Post-harvest Operations (INPhO) www.fao.org/inpho/

INPhO is an FAO databank project of the Post-harvest Management Group. It is an international collaborative effort by the FAO, GTZ and CIRAD which aims to support the collection and dissemination of information on proven technologies and products in post harvest systems. Components of the network include a comprehensive collection of information on post harvest issues, communication/ interactive services; links with other databases. The website includes full text documents of training and technical publications, country profiles, crop profiles (main focus is on cereals and grains, fruits and vegetables, oil seeds, roots and tubers). Recipients of this information include people working in the agricultural production and marketing sector viz., producers, researchers, policy makers, private investors and donors (FAO, 2006).

Philippine Agricultural Libraries and Information Services Network (PhilAgriNet) http://www.philagrinet.org.ph/

PhilAgriNet aims to create and maintain a central electronic database of Philippine technical agricultural literature and make this accessible to agricultural scientists worldwide. Membership is open to agricultural institutions. The database intends to cover all technical publications on agriculture, generated by member institutions, both published and unpublished, and written in any language or dialect, regardless of format. Affiliation is open to libraries of: Government agencies and corporations engaged in agricultural research; state colleges and universities, whose curriculum includes agriculture and independent or private agricultural agencies.

Virtual Extension Research Communication Network (VERCON)

Virtual Extension and Research Communication Network (VERCON) established by FAO aims to use the potential of the Internet and apply it to strengthening and enabling linkages among the research and extension components of the national agricultural knowledge and information system. VERCON aims at improving linkages between and within agricultural research and extension institutions through human and technological components. The human component is a network of staff of research and extension institutions, faculty of agricultural education, NGO workers, in some cases agro producers committed to strengthening collaboration, communicating, sharing information and supporting improved agricultural production. In order to link the human component there is the technological component, which allows members of the network to communicate and develop, share, store and retrieve information (FAO, 2006).

The Egyptian Government has established a pilot VERCON to support research and extension interaction and the flow of information between research, extension and farmers in Egypt.

Web-based Information Service for Agricultural Research for Development

(WISARD) http://www.wisard.org/wisard/home.html

WISARD is a web-based information platform that provides searchable information on experts, organizations, outputs and projects in the fields of Agricultural Research for Development (ARD), Natural Resource Management (NRM) and Sustainable Development (SD) from the mid-nineties till date. The system can be used at organisation, network, national and international levels. WISARD allows decentralized data input and management through focal points at organization, network or national level

Present partners include among others, CGIAR, IPM-Europe, Rice-Wheat Consortium for the Indo-Gangetic Plains, FAO, donor agencies, international and national NGOs and European agencies and national agricultural research systems of India, Pakistan, Nepal and Bangladesh.

Electronic Publishing in Agriculture

Electronic Publishing (e-Publishing) is quickly becoming an important part of publishing mainstream. Recent innovations in this area have made it possible to publish on the information super highway. E-document can be accessed at the computer. It makes much easier for readers to search the information. It is very quick and easy for a reader to browse the table of the contents of previous issues, to jump directly to a particular section of a document or even particular section of an article.

Advantages of e- Publishing

E-Publishing (EP) difference lies in the new levels of value it provides through features not possible in traditional media. EP products may differ to an even greater degree than print products. Nevertheless, there are some common features to distinguish EP from print Publishing in terms of value to end-users. ICTs: Concepts and Relevance EP products create additional value for the user with regard to following three dimensions: content availability, content transparency and interactivity and content format.

Content availability means that EP products can be delivered and accessed with more independence of time and place than can be traditional print products and that their delivery is less limited with regard to quantity. Content availability includes: time of delivery - available any time; location of delivery - consumption anywhere; amount of information - end of traditional boundaries imposed by paper volume and price. Content transparency and interactivity refers to new tools and opportunities concerning information navigation. The main features are: Interactivity- contextual hyperlinks open new dimensions of information retrieval and lead to new types of information behaviour browsing, etc. the possibility to integrate content and services; and search tools across one or thousands of documents - interactive information processing.

Electronic Publishing in Agriculture in India

The Directorate of Information and Publications of Agriculture, (DIPA), New Delhi, is the official publication wing of the ICAR through which the research and other activities are revealed to the world. DIPA brings out a variety of publications in English and Hindi languages for the use of scientists, researchers, students, policy planners, extension personnel, farmers and the general public. The e-publications of DIPA include - "Handbook of Horticulture", ICAR Research Projects Information – Research Project Files (RPF) Database, ICAR Vision 2020 Document etc. Some other important institutions bringing out e-publications in Agriculture in India include, National Institute of Agricultural Extension Management, MANAGE, Hyderabad (www.manage.gov.in), Indian Agricultural Research Institute, New Delhi (www.iaripusa.org) , Indian Farmers Fertilizer Cooperative Limited (IFFCO), (www.iffco.nic.in), Krishak Bharati Co-operative Limited (KRIBHCO), (www.kribhco.net), National Bank for Agricultural and Rural Development (NABARD) (www.nabard.org), National Agricultural Co-operative Marketing Federation of India Ltd. (NAFED) (www.nafed-india.com), Indian Agricultural Statistics Research Institute (IASRI), New Delhi (www.iasri.res.in) and State Agriculture Universities of Tamilnadu, Uttarakhand, Punjab, Haryana and Andhra Pradesh. Two major open universities namely Indira Gandhi National Open University (IGNOU), Delhi and Yeshwantrao Chavan Maharashtra Open University (YCMOU), Nashik and National Institute of Agricultural Extension Management, MANAGE, Hyderabad have taken open and distance learning in Agriculture on a national scale, and all these institutions are in the process of bringing out a number of e-publications in Agriculture.

Expert System

Concepts and Importance of Expert Systems

The branch of computer science known as Artificial Intelligence (AI), covers a number of different fields of application. Expert system is one such field that has attracted significant attention in recent years. Expert systems have been developed and applied to many fields such as office automation, medicine and agriculture. Knowledge representation is an issue that arises in both cognitive science and artificial intelligence. In cognitive science, it is concerned with how people store and process information. In AI, the primary aim is to store knowledge so that programs can process it and achieve the verisimilitude of human intelligence. AI researchers have borrowed representation theories from cognitive science. Thus, there are representation techniques such as frames, rules and semantic networks, which have originated from theories of information processing. Since knowledge is used to achieve intelligent behaviour, the fundamental goal of knowledge representation is to represent knowledge in a manner as to facilitate inferencing, i.e. drawing conclusions from knowledge. Knowledge engineers are concerned with the representation chosen for the expert's knowledge declarations and with the inference engine used to process that knowledge. A user can use the knowledge acquisition component of the expert system to input the several characteristics known to be appropriate to a good inference technique. They are:

a. A good inference technique is independent of the problem domain.
b. In order to realize the benefits of explanation, knowledge transparency, and reusability of the programs in a new problem domain, the inference engine must not contain domain specific expertise.
c. Inference techniques may be specific to a particular task, such as diagnosis of hardware configuration. Other techniques may be committed only to a particular processing technique.
d. Inference techniques are always specific to the knowledge structures.
e. Successful examples of rule processing techniques are forward chaining and backward chaining.

Importance of Expert Systems

The complexity of problems faced by farmers are yield loses, soil erosion, selection of crop, increasing chemical pesticides cost, pest resistance, diminishing market prices from international competition, and economic barriers hindering adoption of farming strategies. The farmer may not become expert manager of all these aspects of farming operations. On the other hand, agricultural researchers need to address problems of farm management and discover new management strategies to promote farm success. Numerical methods have failed to provide better solution because understanding about crop systems is qualitative based on experience and cannot be mathematically represented. Expert systems are computer programs that are different from conventional computer programs as they solve problems by mimicking human reasoning processes, relying on logic, belief, rules of thumb opinion and experience. The experience and knowledge of scientists will be used to develop expert system on various issues of agriculture, which in turn will provide advisory support to the farmers. In agriculture, expert systems are capable of integrating the perspectives of individual disciplines such as plant pathology, entomology, horticulture and agricultural meteorology into a framework that best addresses the type of ad hoc decision-making required of modern farmers. Expert systems can be one of the most useful tools for accomplishing the task of providing growers with the day-today integrated decision support needed to grow their crops.

Cases of expert systems in agriculture

Rice-Crop Doctor

National Institute of Agricultural Extension Management has developed an expert system to diagnose pests and diseases for rice crop and suggest preventive or curative measures. The subject matter expert knowledge on rice pathology and entomology has been obtained from Scientists of Directorate of Rice Research, Hyderabad and Acharya N G Ranga Agricultural University. The rice crop doctor illustrates the use of expert systems broadly in the area of rice production through development of a prototype, taking into consideration a few major pests and diseases and some deficiency problems limiting rice yield. The following diseases and pests have been included in the system for identification and suggesting preventive and curative measures. The diseases included are rice blast, brown spots, sheath blight, rice tungro virus, false smut fungi, bacterial leaf blight, sheath rot and zinc deficiency disease. The pests included are stem borers, rice gall midge, brown plant hopper, rice leaf folder, green leafhopper and Gundhi bug.

a. The brief logic flow of the expert system is as follows:

b. The part of the plant where symptoms have been observed is given by the extension officer.

c. The basic symptoms are given as input:

d. Considering these symptoms the user is expected to give further information based on other visual symptoms.

e. At this step the disease and pest are identified.

d. The user is then given the option to either stop or further diagnose and other disease / pest or get preventive or curative measures on these.

CALEX

This is a blackboard based integrated expert decision support system for agricultural management, developed at University of California. CALEX can be used by growers, pest control advisors, consultants and other managers.

LEY Expert Systems

This is a RF-telemetry based computer-controlled, automated, remote, real-time weather data acquisition and reporting system in Washington State. Cooperation among Washington State University, the National Weather Service (NWS) the U.S. Bureau of Reclamation and several private grower organizations have made this system possible. Data is collected, processed and transmitted to the NWS hourly. These hourly updates of actual conditions are broadcast on the NOAA weather band during the spring frost season to assist fruit growers with frost protection. Real-time weather data are also being used in applications such as irrigation scheduling, crop protection and pest management.

Prototype decision support system

A prototype decision support system by P.H. Heinemann is being used to assist managers with the development, implementation, and management of frost protection. The program is divided into four main modules: frost protection strategies, operational management, forecasting, and instrumentation. E. Satellite downlink facility using a unique satellite downlink facility at by R.R. Getz Auburn University, real time meteorological data are being processed on a network of advanced computer workstations. This data feeds a series of computer models that generate site specific predictions of temperature, dew point, wet bulb temperature, and other parameters used to alert Alabama fruit, vegetable, and nursery growers. Model outputs are distributed to Extension specialists and others over a campus-wide fiber optic communications system and to all county Extension offices over the ACENET computer network. During critical freeze situations, information can be disseminated live via satellite to media and rural satellite dish owners using the Auburn University uplink facility. A pilot network of automated weather stations incorporating a 10-m tower for monitoring inversions has been installed to enhance the program.

Application of ICT

e – Governance

Empowering the citizen via electronic governance has been gaining momentum across liberal economies in the world where democratic principles underlie the government constitutional structures. This has been further enabled by the growing use of new information and communication technologies (ICTs), particularly designed for strengthening democratic governance around the world. E-governance is widely acknowledged as the application of new ICTs, especially the Internet and World Wide Web (www), as channels for engaging citizens in government as well as facilitating delivery of government information and services.

ICT has a vital role to play in all transactions that the government undertakes. It helps the government cut red-tapism, avoid corruption, and reach citizens directly. This realization is wide spread among all the states of the country. Almost every state has an IT policy in place with the aim of evolving itself from being an IT-aware to an IT-enabled government. State governments are fast recognizing the benefits of an IT-enabled working environment and IT-enabled service delivery. Effective use of ICTs in government or e-Governance is perceived as a facilitator. Governments, at central and state level, in general feel necessity of that being accessible online 24/7 to the average citizen. Why e-Governance? The basic objective of e-Governance is to support and simplify governance for government, citizens and businesses. The use of ICTs can connect these three segments and support processes and activities. Other objectives are to make government administration more transparent, speedy and accountable, while addressing the society's needs and expectations through efficient public services and effective interaction between the people, businesses and government.

Concept and Definition of e-Governance

Definitions of e-governance vary and efforts to incorporate all of them in one single statement seem impossible. However, a common thing is that e-governance refers to the use of Information and Communication Technology (ICT), particularly web based applications, to provide access to and deliver information / service to the public, business, other agencies and governmental entities faster, cheaper, easier and more efficiently. This in turn enhances relationships, enlarges the overall customer base and improves core business operations through re-examination of internal processes. Other definitions go beyond this main core.

Wimmer and Traunmuller (2000) consider e-Government as a guiding vision towards modern administration and democracy. For them it is concerned with the transformation that government and public administration have to undergo in the coming decades. Hence, e-government is not an option. Instead it is a must for the governments to use the benefits of the information technology in order not to fall behind. Lawson sees e-government as "transferring power to people" by operating by a one stop, non-stop way and doing more for less. e-Governance is defined by M. Backus (Backus, M. , 2001), as the "application of electronic means in (a) interaction between government and citizen and between government and businesses as well as (b) in internal government operations to simplify and improve democratic, government and business aspects of governance".

The term 'electronic means' above stands for Internet and other ICT applications; the term 'interaction stands' for delivery of government products and services, exchange of information, communication, transactions, and system integration; Government consists of levels and branches. Government levels include- Central, State, District, Block (Taluka), Mandal level government agencies/ departments/ institutions. Government branches include- Parliament, Judiciary, Police, Civil Services and all branches of administration. Government operations include all back-office processes and inter-governmental interactions within the total government body.

E-Governance and E-Government

E-Governance and e-Government are two very distinct terms. Whereas e-Governance is a broader topic that deals with the whole spectrum of the relationship and networks within government regarding the usage and application of ICTs, e-Government is a narrower discipline dealing with the development of online services to the citizen, *i.e.* government services, such as e-agriculture, e-passport, e-tax, e-transportation, e-health, etc. e-Governance is a wider concept that defines and assesses the impact that technologies are having on the practice and administration of governments and the relationships between public servants and the civil society, and dealings with the elected bodies or outside groups, such as not-for-profit organizations, NGOs or private sector corporate entities. E-Governance encompasses a series of necessary steps for government agencies to develop and administer, in order to ensure successful implementation of e-government services to the public at large.

E-Government

E-Government or electronic government is the delivery of more convenient, customer oriented and cost effective public services and sharing of information through electronic media. Specifically, e-Government harnesses information and communication technologies (such as Wide Area Networks, the internet, and mobile computing) to transform relations with citizens, businesses, and other arms of government. e-government is an administration where citizens can avail of government services such as land records, filing of tax, etc, e-Government are also defined as the use of Information and Communication Technologies to provide better government services. This definition usually refers to more efficient, more transparent and potentially two-way information flow between the government and citizens. But the fact e-Governance is that in most of the cases e-Government is not more than digitalization of the operational functions of the government. The Global e-Government Readiness Report says the role of government has to be reoriented and redefined from "that of managerial authority to that of leadership in a multi-centered government, the culture of governance needs to change from being bureaucratic to participatory from authoritative to accountable, from being monopolistic to being competitive and innovative, from being closed to being participatory, from being autocratic to democratic, and from being exclusive to inclusive". E-Governance is socially inclusive governance with the promise of good governance and effective participation and inclusion of citizens.

E-Governance

E-Governance is the development, deployment and enforcement of the policies, laws and regulations, necessary to support the functioning of a Knowledge Society as well as of e-Government. E-Governance brings about 'anytime anywhere access' to the citizen. E-Governance is about transforming the relationship between governments and their citizens through the use of information technologies, and that e-Government is about delivering government services using information technologies.

Benefits of e-Governance

E-Governance was initiated worldwide to improve speed, transparency, efficiency and reliability of the Government services. The basic premise to undertake e-Governance initiatives was to harness the advantages of ever-enlarging network and reach of Internet for improving reach, timeliness, credibility and consistency of Government services. E-governance offers the potential to bring citizens closer to their governments. Regardless of the type of political system that a country has, the public benefits from interactive features that facilitate communication between citizens and government. The scope as well as potential benefits of e-Governance are increasing with enrichment of e-services by various departments/agencies. The following are direct benefits of e-Governance:

a. Faster, easier, more convenient service.
b. Convenience, Near to Home, more services at one place.
c. Faster processing, shorter wait, shorter queues at Government offices.
d. Less number of trips to government offices.
e. Better interaction environment; No harassment.
f. Reduced transport cost.
g. Avoids wage loss for daily-wage earners.
h. Better quality service.
i. More accurate and legible documents, easy recovery from errors.
j. Lesser corruption, increased transparency.

k. Improved access to offices (nearer home, 24X7 or at least 12X6, in place of limited number of working hours or only on working days) and functionaries, no intermediaries.

There are some derived benefits of e-Governance services, as the back-end procedures and databases are also linked to services. This streamlining of data-networking helps the concerned departments to handle queries and complaints more effectively. These benefits include:

a. Improved complaint handling, complaint tracking.
b. Introduces transparency in data, decisions/actions, rules, procedures and performance of Govt. Agencies.
c. Automates processes to take away discretion.
d. Reduces arbitrariness in decision making.
e. Entry point for simplification of rules and reengineering processes.
f. Makes decisions traceable- tracks actions.
g. Builds accountability- greater access to information through web publishing-role of civil society.
h. Provides documentation to citizens for follow up.
i. Tracks procedural/ human delays.
j. Easier to fix responsibility/ accountability.
k. Modularizes Tasks Making Outsourcing Possible.
l. Introduces competition amongst delivery channels and departments.
m. Standardized documentation of comments/ objections leads to effective supervision- through comparative indicators.
n. Centralizes data for better audit and analysis Integration of data across applications-provides improved intelligence.
o. Enables unbiased sampling for audit purposes.

Further, there are social benefits of e-Governance projects and services as it induces increased skill and knowledge among the young and curious citizenry. E-Governance creates more job opportunities at grass-root level. Just like the PCO (Public Call Offices) revolution in telecommunications created over a million jobs in 90s, the Community Services Centres (CSCs) are potential creator of over two million jobs in next five years, if the 1,00,000 CSCs are made operational by Ministry of Information Technology, as planned.

Geographical Information System (GIS)

An emerging technology paradigm, "Geographical Information System (GIS)" has emerged as powerful tool which has potential to organize complex spatial environment with tabular relationships. The emphasis is on developing digital spatial database, using the data sets derived from precise navigation and imaging satellites, aircrafts, digitization of maps and transactional databases. The power and potential of GIS is limited only by ones imagination. However to exploit the technology to benefit the planning process there is need to initiate a process for integration of data, based on Interoperable open standards, specifications & formats. The enormous demand for the storage, analysis and display of complex and voluminous data has led, in recent years, to the use of Geographic Information Systems for effective data handling and also for analyzing and geographically transferring the information around the world. NIC offers its users, GISNIC, a software designed to provide a complete state-of-the-art desktop GIs solution for retrieval, projection, transformation and analysis of both spatial and non-spatial data, so that the User is able to manipulate and manage coordinate (locational) and attribute (thematic) data and produce thematic maps as well as tabular reports.

Through its GISNIC software, NIC offers to undertake the tasks to generate and plot various types of thematic maps (Flow Lines maps, Graduated Symbol maps, Choropleth maps, Pie charts and Bar charts). The various tasks related to geographic analysis such as creation of buffer zones, performing polygon overlay and manipulating tabular data, are also carried out using GISNIC. Also, the policy framework provided by IT Task Force for India to emerge as Global Info tech Superpower in this millennium emphasizes the availability of Spatial data to GIS user community and industry and provide a direction for the liberalization to this vital sector. With the evolution of this framework, it is now possible to have Spatial Decision-Support Information Systems Network, with distributed Network-Centric Web-Enabled GIS Application Services using NICNET facilities.

Concept

A Geographical Information System (GIS) is a system for capturing, storing, analyzing and managing data and associated attributes, which are spatially referenced to the Earth. The geographical information system is also called as a geographic information system or geospatial information system. It is an information system capable of integrating, storing, editing, analyzing, sharing, and displaying geographically referenced information. In a more generic sense, Geographical Information System is a software tool that allows users to create interactive queries, analyze the spatial information, edit data, maps, and present the results of all these operations. Geographical Information System technology is becoming essential tool to combine various maps and remote sensing information to generate various models, which are used in real time environment. Geographical information system is the science utilizing the geographic concepts, applications and systems.

Geographical Information System can be used for scientific investigations, resource management, asset management, environmental impact assessment, urban planning, cartography, criminology, history, sales, marketing, and logistics. For example, agricultural planners might use geographical data to decide on the best locations for a location specific crop planning, by combining data on soils, topography, and rainfall to determine the size and location of biologically suitable areas. The final output could include overlays with land ownership, transport, infrastructure, labour availability, and distance to market centres.

Components of Geographical Information System (GIS)

Software

Geographical Information System (GIS) software provides the functions and tools needed to store, analyze, and display geographic information. Key software components are (a) a database management system (DBMS) (b) tools for the input and manipulation of geographic information (c) tools that support geographic query, analysis, and visualization (d) a graphical user interface (GUI) for easy access to tools. Geographical Information System software are either commercial software or software developed on Open Source domain, which are available for free. However, the commercial software is copyright protected, can be expensive and is available in terms number of licensees.

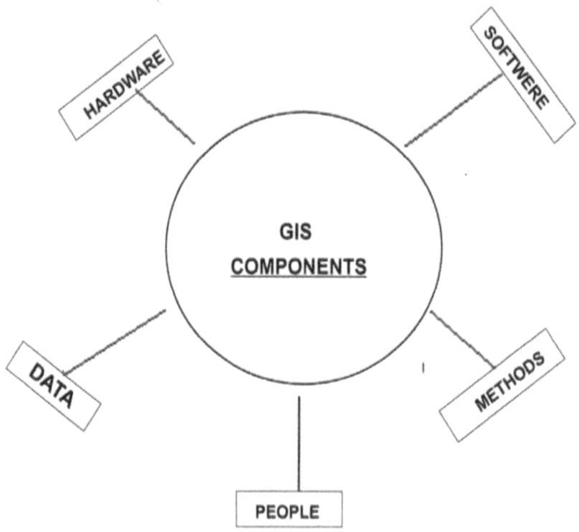

Currently available commercial Geographical Information System software includes ArcGIS, GeoMedia, Geomatica MapInfo, Gram++ etc. Out of these ArcGIS is the most popular software package. And, the open source software is AMS/MARS etc.

Hardware

Hardware is the computer on which a Geographical Information System operates. Today, Geographical Information System runs on a wide range of hardware types, from centralized computer servers to desktop computers used in stand-alone or networked configurations. Minimum configuration required to run ArcGIS 9.0 application is as follows:

Platform	:	PC-Intel
Operating System	:	Windows XP Professional Edition, Home Edition
CPU Speed	:	800 MHz minimum, 1.0 GHz recommended or higher
Processor	:	Pentium or higher

Memory/RAM	:	512 MB or higher
Display Properties	:	Greater than 256 colour depth
Swap Space	:	300 MB minimum
Disk Space	:	Typical 605 MB NTFS, Complete 695 MB FAT32 + 50 MB for installation
Browser	:	Internet Explorer 6.0 Requirement

(Some features of Arc Info Desktop 9.0 require a minimum installation of Microsoft Internet Explorer Version 6.0.)

Data

The most important component of a Geographical Information System is the data. Geographic data or spatial data and related tabular data can be collected in-house or bought from a commercial data provider. Spatial data can be in the form of a map/remotely-sensed data such as satellite imagery and aerial photography. These data forms must be properly geo-referenced (latitude/longitude). Tabular data can be in the form attribute data that is in some way related to spatial data. Most Geographical Information System software comes with inbuilt Database Management Systems (DBMS) to create and maintain a database to help organize and manage data.

People

Geographical Information System technology is of limited value without the users who manage the system and to develop plans for applying it. Geographical Information System users range from high level technical specialists to common users using these to do their everyday work. These users are largely interested in the results of the analyses and may have no interest or knowledge of the methods of analysis. The user-friendly interface of the Geographical Information System software allows the non technical users to have easy access to Geographical Information System analytical capabilities without needing to know detailed software commands. A simple User Interface can consist of menus and pull-down graphic windows so that the user can perform required analysis with a few key presses without needing to learn specific commands in detail.

Methods

A successful Geographical Information System operates according to a well-designed plan and business rules, which are the models and operating practices unique to each organization.

Functions of Geographical Information System

General-purpose Geographical Information System software performs six major tasks such as input, manipulation, management, query, analysis, and visualization.

Input

The important input data for any Geographical Information System is digitized maps, images, spatial data and tabular data. The tabular data is generally typed on a computer using relational database management system software. Before geographic data can be used in a Geographical Information System it must be converted into a suitable digital format. The DBMS system can generate various objects such as index generation on data items, to speed up the information retrieval by a query. Maps can be digitized using a vector format in which the actual map points, lines, and polygons are stored as coordinates. Data can also be input in a raster format in which data elements are stored as cells in a grid structure (the technology details are covered in following section).

The process of converting data from paper maps into computer files is called digitizing. Modern Geographical Information System technology has the capability to automate this process fully for large projects; smaller jobs may require some manual digitizing. The digitizing process is labour intensive and time-consuming, so it is better to use the data that already exist. Today many types of geographic data already exist in Geographical Information System compatible formats. These data can be obtained from data suppliers and loaded directly into a Geographical Information System.

Manipulation

Geographical Information System can store, maintain, distribute and update spatial data associated text data. The spatial data must be referenced to geographic coordinate systems (latitude/longitude). The tabular data associated with spatial data can be manipulated with help of data base management software. It is likely that data types required for a particular Geographical Information System project will need to be transformed or manipulated in some way to make them compatible with the system. For example, geographic information is available at different scales (scale of 1:100,000; 1:10,000; and 1:50,000). Before these can be overlaid and integrated, they must be transformed to the same scale. This could be a temporary transformation for display purposes or a permanent one required for analysis. And, there are many other types of data manipulation that are routinely performed in Geographical Information System. These include projection changes, data aggregation, generalization and weeding out unnecessary data.

Management

For small Geographical Information System projects it may be sufficient to store geographic information as computer files. However, when data volumes become large and the number of users of the data becomes more than a few, it is advised to use a database management system (DBMS) to help store, organize, and manage data. A DBMS is a database management software package to manage the integrated collection of database objects such as tables, indexes, query, and other procedures in a database. There are many different models of DBMS, but for Geographical Information System use, the relational model database management systems will be highly helpful. In the relational model, data are stored conceptually as a collection of tables and each table will have the data attributes related to a common entity. Common fields in different tables are used to link them together with relations. Because of its simple architecture, the relational DBMS software is being used widely. These are flexible in nature and deployed in applications both within and without Geographical Information System.

Query

The stored information either spatial data or associated tabular data can be retrieved with the help of Structured Query Language (SQL). Depending on the type of user interface, data can be queried using the SQL or a menu driven system can be used to retrieve map data. For example, you can begin to ask questions such as:

a. Where are the soils suitable for sunflower crop?

b. What is the dominant soil type for Paddy?

c. What is the groundwater available position in a village/block/district?

Both simple and sophisticated queries utilizing more than one data layer can provide timely information to officers to have overall knowledge about situation and to take a more informed decision.

Analysis

Geographical Information System systems really come into their own when they are used to analyze geographic data. The processes of geographic analysis often called spatial analysis or geo-processing uses the geographic properties of features to look for patterns and trends, and to undertake "what if" scenarios. Modern Geographical Information System have many powerful analytical tools to analyse the data. The following are some of the analysis which is generally performed on geographic data.

A. Overlay Analysis

The integration of different data layers involves a process called overlay. At its simplest, this could be a visual operation, but analytical operations require one or more data layers to be joined physically. This overlay, or spatial join, can integrate data on soils, slope, and vegetation, or land ownership. For example, data layers for soil and land use can be combined resulting in a new map which contains both soil and land use information. This will be helpful to understand the different behaviour of the situation on different parameters.

B. Proximity Analysis

Geographical Information System software can also support buffer generation that involves the creation of new polygons from points, lines, and polygon features stored in the database. For example, to know answer to questions like; How much area covered within 1 km of water canal? What is area covered under different crops? And, for watershed projects, where is the boundary or delineation of watershed, slope, water channels, different types of water harvesting structures required, etc.

Visualization

Geographical Information System can provide hardcopy maps, statistical summaries, modelling solutions and graphical display of maps for both spatial and tabular data. For many types of geographic operation the end result is best visualized as a map or graph. Maps are very efficient at storing and communicating geographic information. Geographical Information System provides new and exciting tools to extend the art of visualization of output information to the users.

Functioning of Geographical Information System:

Geographical Information System software also contains programmes for verifying the correctness of all geometric, topological and attribute data, e.g. making certain that all graphical data is suitably defined, that attribute data does not exceed expected ranges and that impossible combinations of attributes do not occur. Data may be copied, deleted, moved, joined, altered, etc. Any of these data editing functions should be capable of being performed on both the graphical and the textual data. If data is not carefully verified, and errors remain, then manipulations of the data at a later processing stage will cause error propagation and multiplication, thereby invalidating, or at least making less useful, any final GIS output. Further details on the importance of data validation and error correction can be found in Burrough (1986), Goodchild and Gopal (1989), Dunn et al (1990), Chrisman (1991), Thapa and Bossler (1992) and Rybaczuk (1993).

Structure Conversion

For many manipulations it may also be preferable to convert data from a raster to vector structure, or vice versa. This is necessary since there are still no truly integrated Geographical Information System's which are able to handle both raster and vector data with equal ease. It is important to note that in the vector to raster conversion (rasterizing) there will be an inevitable loss of accuracy, a factor which would be exacerbated both with increasing sinuosity of the lines and with increasing raster cell size. In the raster to vector conversion (vectorising) the Geographical Information System software programme performs a vectorising process which "threads" a line through groups of pixels using a special "thinning" algorithm. There will be a consequent need for topological information to be constructed and for individual features to be identified. These latter requirements can call for considerable operator intervention, but there are Geographical Information System functions which automatically compute new nodes and links and compile topology tables. Star and Estes (1990) explain in some detail the advantages of certain raster to vector data structure conversions, i.e. according to the type, and the proposed use for, the data being handled.

Geometric Conversion

When performing manipulations on mapped digital data, it is important that, if the data is to be merged in any way, then it should all conform to the same geometric reference system. Latitude and longitude co-ordinates are frequently used in small scale mapping, although the most frequently used co-ordinate system in Geographical Information System is the Universal Transverse Mercator (UTM) system. Nearly all Geographical Information System software allows for the possibility of converting the map referencing system used to a wide range of possible map projections or from one co-ordinate system to another. Such processes are sometimes called transformations or rectification. Transformations are based on the mathematical relationships that exist between the various map projections, i.e. relative to angles, areas, direction and distances. A more basic type of geometric conversion is called registration. This is simply changing one mapped view to line up with another, *i.e.* irrespective of any referencing system. Scale changes are easily accommodated via a simple multiplier function and maps can easily be rotated to particular orientations. A more complex function which most GIS's can achieve is the correction for distortions (rectification). These distortions may occur in the original source data for a number of reasons, e.g.

a. Aerial photographs or RS satellite images have varying scales due to platform tilt and the curvature of the Earth.

b. The angle of view or relief differences also causes variations in scale.

c. Photographs or maps variably shrink with age.

d. Maps on paper can easily suffer from stretching.

f. Distortions in the optic systems being used.

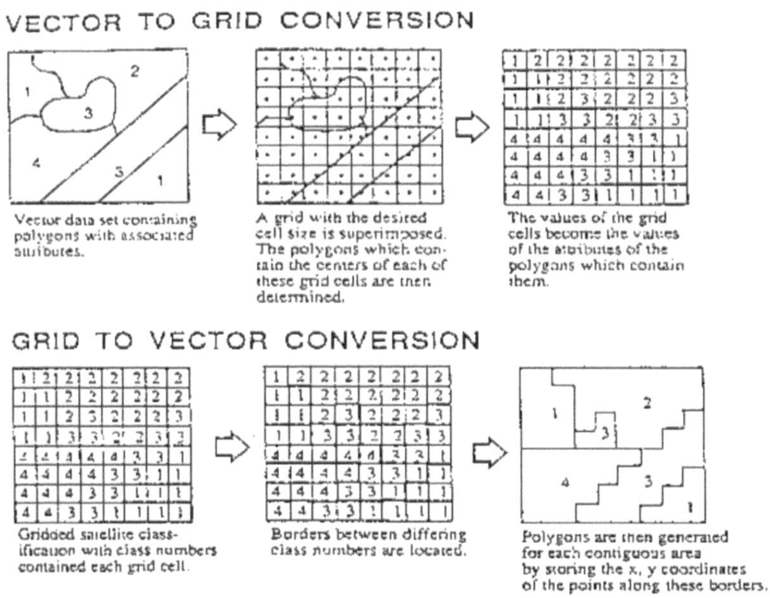

Summary of Vector to Raster and Raster to Vector Structure Conversion (from Robinson Barker, 1988)

Distortions are manipulated by "rubber sheeting" methods, i.e. by treating the distorted image as an elastic sheet which can be stretched or compressed until it exactly fits a selected base map on which there are positively identifiable ground control points.

Generalization and Classification

Under this general heading a large number of manipulations can be performed, all of which are designed to change the data in some way such that it can be more easily used for a particular purpose, *e.g.*

a. Adding to data, or deleting undesirable data.

b. Aggregating or disaggregating numerical or attribute data.

d. Classifying or reclassifying data into user defined, or GIS suggested, classifications. This usually involves deciding upon attribute value classes or changes to existing classes.

e. Using data reduction algorithms to generalize or smooth linear data, e.g. to thin out coordinates in digitized lines (in order to greatly reduce the amount of data storage). The figure illustrates how generalization can be applied to a well known map outline.

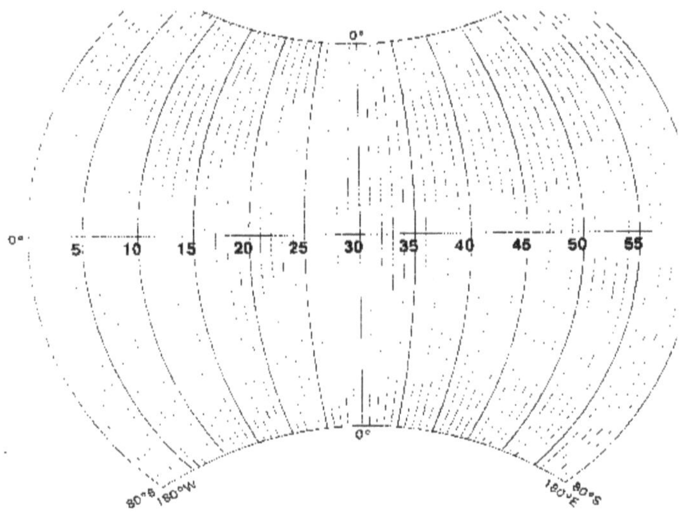

The Universal Transverse Mercator (UTM) Zones

a. Lines can be deleted or "dissolved" in order to simplify mapped surfaces.

b. New attributes can be assigned to spatial points, lines or polygons.

c. Annotations can be added to maps using labels, text, legends or cartographic symbols.

Integrations

These manipulations involve the creation of new or revised mapped surfaces by (in one way or another) joining two or more previously defined maps. Perhaps the most frequently performed manipulation under this heading is the merging by overlaying of two or more mapped layers, i.e. any number of raster or vector map layers can be progressively added or subtracted to produce a desired map. Existing data on a single theme can be merged, e.g. a water qualitative map could be progressively built up by merging maps of perhaps water temperature, pH, dissolved oxygen, salinity, etc. By merging these layers a new map would be created which consisted of numerous polygons, each of which may have a different combination of water qualitative factors. The overlaying of maps can give rise to major problems. For instance, it is likely that many new polygons will be created, some of whose boundaries should correspond between the two maps being overlaid. Invariably the original data sources will be different, or the digitizing will be inaccurate, and the new map will exhibit numerous so called "slivers", i.e. polygons arising from a poor match between boundaries or other lines Here the overlaying of map a) on to map b) created map c) with its numerous slivers. There are now "intelligent" algorithms in some GIS's which can automatically counteract these slivers. Other algorithms will also be needed to establish new topology and new attribute tables. It can be seen that overlay procedures are quite complex and they may take up a lot of computing time.

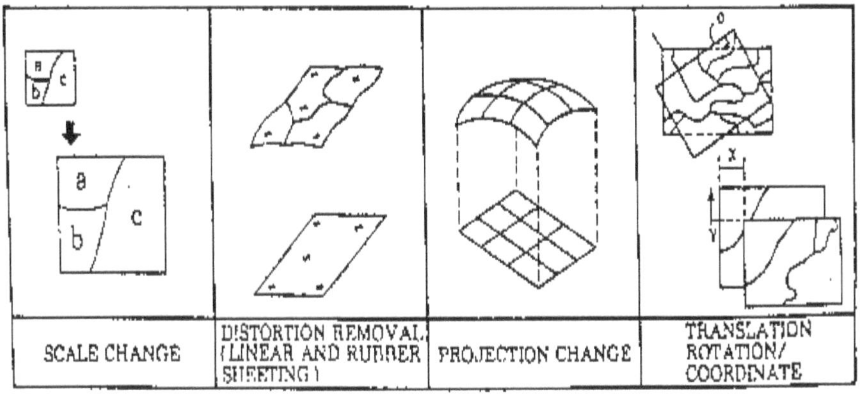

Some Geometric Conversions Which a Geographical Information System May Perform (after Dangermond, 1983)

Frequently it will be necessary to form a new map by accurately joining two mapped sheets (or many contiguous map sheets) along their edges, i.e. by edge matching, so that all linear and area features exactly coincide. A seamless dataset is then produced. When merging or integrating maps it will also be necessary to take into account possible variations in the data structure and format between any two maps, plus the ways that have been used to assign labels or to identify objects. So inevitably, integration must be performed with a great deal of consideration as to the integrity and make-up of the data being used.

Map Enhancement

These Geographical Information System functions simply consist of a series of operations which allow for the cartographic refinement of the finished map, i.e. at the manipulation stage factors concerned with map presentation can be improved. This may include adding a suitable border to the map, varying the width of mapped lines, altering chosen colours, varying the fonts or font size, altering the layout of the map or the position of textual features such as the key or title. If a 3-D image has been created this might be the stage at which a land use categorization could be draped over the image.

Illustration to Show How a Map Outline Can be Generalized

Interpolation

This is the procedure for estimating the values for any continuous (rather than discrete) "properties" at un-sampled sites along a line or within an area. This must be based on existing point observation data within the area (or along the line), which themselves should have been derived using valid measuring and sampling techniques. Figure shows how isolines (lines joining places of equal value) have been interpolated from known values at several points. The problem in interpolation is choosing the model which is best able to produce correct interpolations, i.e. a model which suits the data array and the way in which actual variability occurs. Many simple Geographical Information System interpolation procedures rely on the use of various weighting functions, i.e. this allows for the logical fact that near points used in an interpolation should count for more than distant points. There are more complex interpolation models which cover 2-D arrays of data points e.g. Thiessen polygons, the use of kriging or Fourier series, or models which can be applied to interpolating linear pathways, *e.g.* the fitting of spline functions. A special case of interpolation is centroid determination. In this function the Geographical Information System is able to calculate the co-ordinate location of the centre of a polygon. Extrapolation is simply the using of interpolation techniques to extend calculated trends beyond the area of specific study or interest, or beyond the range of the data held. Burrough (1986) and Martin (1991) provide comprehensive coverage on interpolation and Laurini and Thompson (1992) provide a detailed description of the various models for interpolation.

Buffer Generation

Since a fundamental concern of Geographical Information System is with spatial distances, it is frequently useful to determine what are known as buffer zones (zones of equal distance) around a point or an area, or along a line. Figure

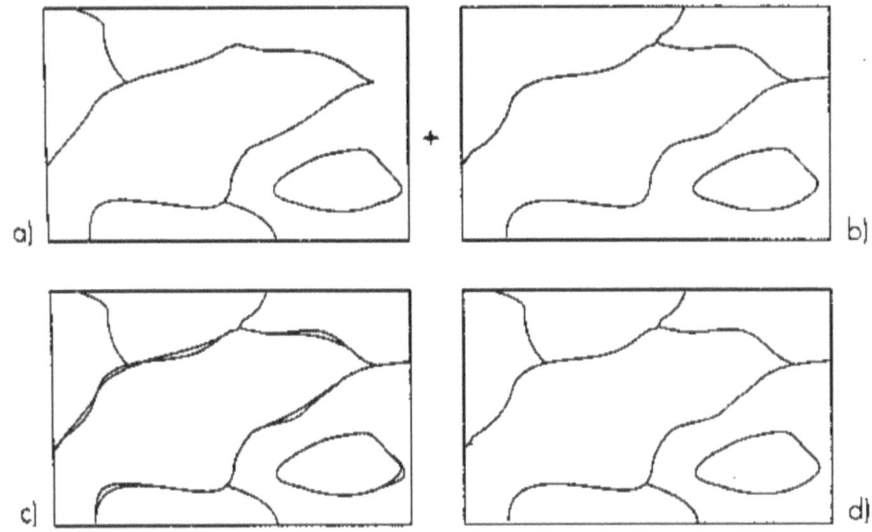

Illustration Showing How Overlaying Techniques Can Lead to Poor Integration (from Bernhardsen, 1992)

shows how simple buffer zones are created around a point, line and polygon in the raster structure. Buffers may be generated by the Geographical Information System at any preset distance and they might represent features such a maximum market range around a town, a legal exclusion zone, a zone of noise disturbance or a zone in which some economic rights exist. Buffer zones are clearly very useful in the compilation of many geographic models, though in reality their use may be restricted, mainly to isotropic surfaces.

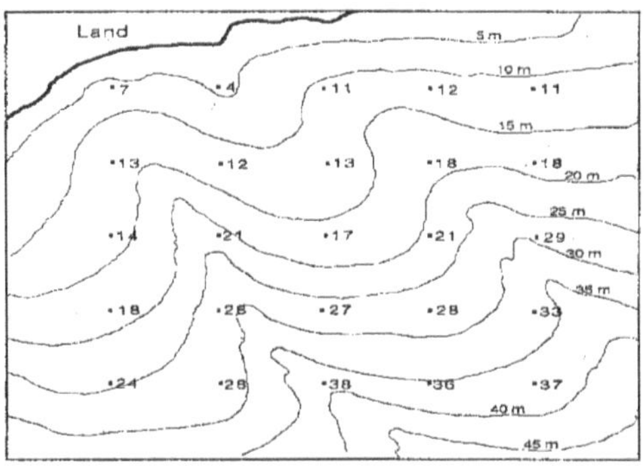

A Simple Example to Show Interpolated Isolines Data Searching (Query) and Retrieval

For discussion purposes the two processes of searching (querying) and retrieving can be viewed together, *i.e.* as a single process. In order that a Geographical Information System can perform any analytical procedure, it is essential that the software is able to selectively search and retrieve requisite data by as many criteria as possible. These criteria will include mapped data (lines, points and polygons) as well as attribute, numeric or textual data. The searching for data is performed using a dedicated Standard Query Language (SQL) and any search may be confined to a certain mapped area or to a specific theme. It is further possible to selectively retrieve data in various ways. For instance data can be classified by any theme, region or class. It is common to retrieve data using the rules of Boolean logic. Here the simple operators of "AND", "OR", "XOR" or "NOT" are stated to show which sort of conditions need to be met before the data is retrieved. Commands for search and retrieval using Boolean logic can be relatively complex. So, for instance, the GIS could be instructed to "find all the marine areas having a mean water temperature of >20C, in combination with a depth of <50 metres, which are situated in the waters of both country "x" and "y" and in which quotas do not yet operate". Complex requests like this can involve any parameters for which the data is held.

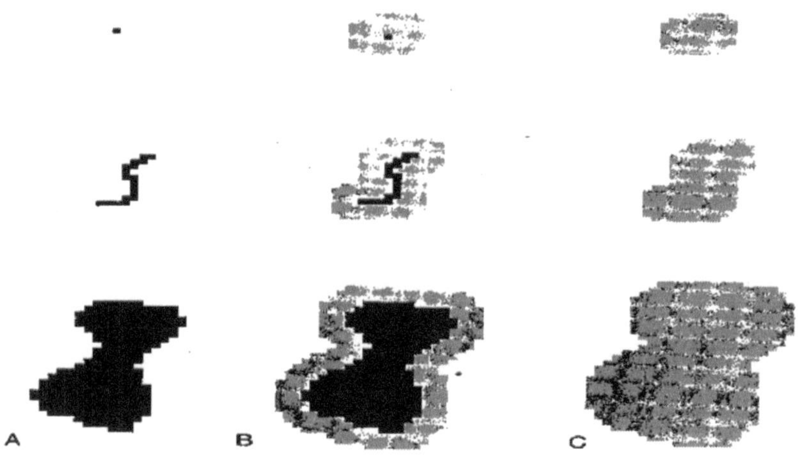

Buffers around a point, line and polygon

Data Analysis

It is the incorporation of analytical functionality which arguably distinguishes a true Geographical Information System from other forms of mapping packages. Recently there has been criticism that many Geographical Information System packages have lacked a sufficient range of analytical functions, but nowadays this would seem to be unfair in the sense that most packages have at least a limited range of such functions. Also, it is now usually a simple matter to link a Geographical Information System package to a specialist analytical software programme in which the analysis is carried out before reverting back to the Geographical Information System software for mapping. Additionally, it is not usually worth the time and financial effort involved for a software house to integrate lots of analytical functions, i.e. since most of these are only required for research purposes.

The analytical functions which most Geographical Information System software provides operate on both the spatial or the attribute data (or a combination of these). Most of the following analyses can be performed on both vector or raster structured data, though inevitably one or other of these is more efficient depending on the actual analysis being performed. Some authors have reviewed analytical capacities under headings which correspond to the types of data, i.e. point analysis, polygon analysis and linear analysis or vector analysis and raster analysis, though we have chosen to discuss the techniques under the headings of spatial, statistical and measurement analyses. Figure shows examples of raster analytical techniques. Laurini and Thompson (1992) or Bonham- Carter (1994) provides a detailed theoretical background on how the various analyses are performed.

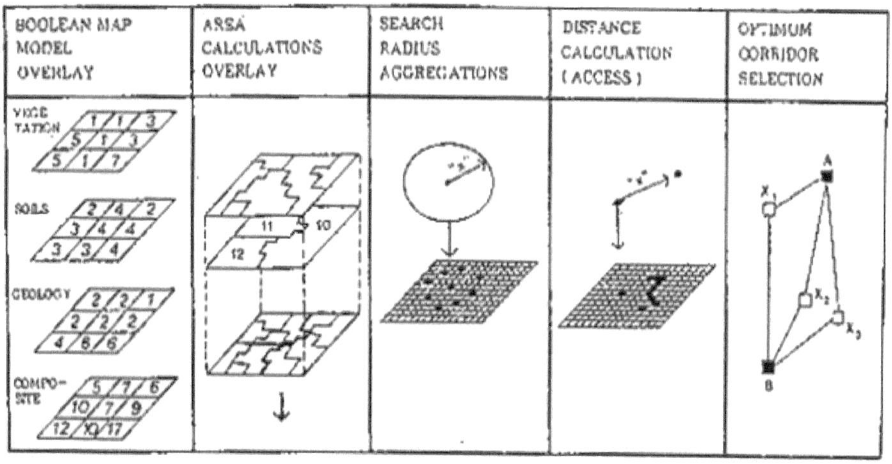

Some Raster Based Analytical Techniques (from Dangermond, 1983)

Spatial Analysis

In the above Table we gave examples of some of the important types of spatial analysis which may be performed. Here we can briefly expand upon these.

a. Connectivity (or network) analysis is useful for determining how well connected any particular site is via any method of communications. Thus a connectivity index can be worked out which shows, for example, for all towns (nodes) in any selected area, the relative number of road or other communications or pipeline connections (links) which exist between each town and all the other towns in that area. Connectivity can also be conceived in terms of distance, cost or time, and it is useful in optimizing route allocations. Figure illustrates how connectivity is recorded for a simple set of links and nodes.

b. Proximity and contiguity analyses are respectively simply methods of determining and indicating measures of distance between locations, or of showing a location's degree of adjacency to neighboring locations. Figure shows how contiguity is recorded for polygons in a small area. The creation of a buffer zone is an example of a proximity operation.

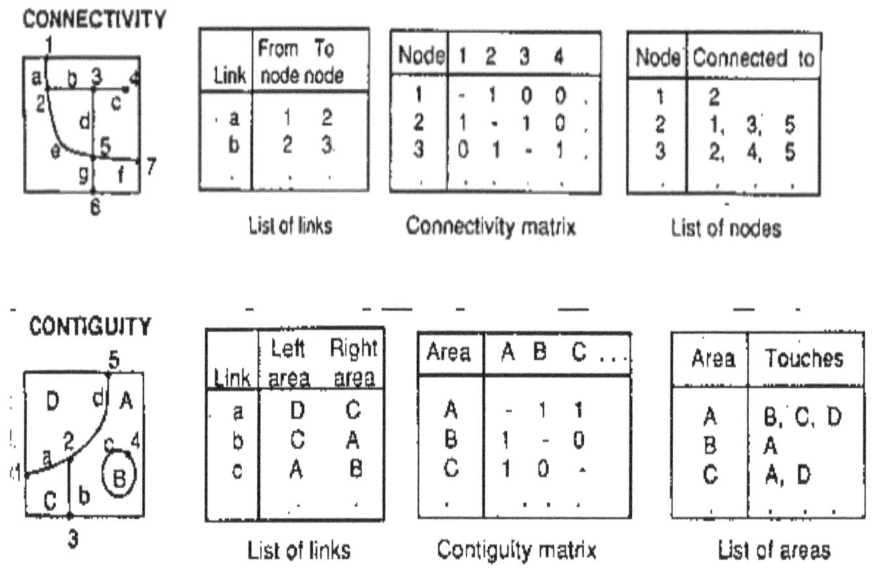

Method of Compiling a Contiguity Matrix (after Laurini and Thompson, 1992)

c. Inter-visibility defines, from map evidence, whether or not it is possible to have a direct line of sight between any two points on the map. Thus a calculation is made, bearing in mind the existence of high ground as shown by contours, whether or not hills or other high ground would obscure the line of vision.

d. Digital terrain modeling is the process whereby it is possible, using digitized height data, to build a 3-D model of any desired area. These models amy also be called 2.5-D since they only show surface heights and not true volumetric data. From the marine point of view it would be equally possible to use bathymetric data to construct visual models showing the physical appearance of selected areas of the sea floor.

e. Location optimization is now being widely used as a Geographical Information System based method which allows for the selection of optimum locations for the siting of any activity. This analysis is usually used by larger commercial companies when seeking, for instance, sites for new retailing outlets or for centralized distribution points. In these cases various spatially variable economic and social indicators, such as the social class structure of an area and the population density, would need to hold in a digital geo-referenced form. Similar analyses are also used by the forestry and agricultural sectors in seeking to optimize their operations, though here physical rather than economic criteria might be more important.

f. Trend surface analysis is a method for establishing whether a generalized spatial surface exists, i.e. one which may be obscured by a mass of detail in the real world. For instance, in any one country there may be an overall "wealth" surface which trends perhaps from east to west but which could well be obscured by numerous pockets of prosperity or poverty. From the marine viewpoint, it is quite likely that trend surfaces would exist with regard to the distribution of particular species, i.e. such that they would gradually decline outwards from a biologically optimum area but in an irregular, and thus perhaps obscured, way. The fitting of a trend surface therefore becomes a useful way of identifying spatial anomalies-points or areas which are above or below the general trend.

Statistical Analysis

As with the spatial analyses, there are a huge range of statistical functions which any individual Geographical Information System might be able to perform. Since many of these functions are not particular to Geographical Information System, i.e. they are commonly performed by statistical packages, or spreadsheet or database packages, then we need not elaborate on them here, except to mention that it is possible to link many existing statistical packages with Geographical Information System software as a means of executing statistical analyses. These include simple descriptive statistics showing measures of centrality, frequency analyses or measures of dispersion, plus more complex correlations and multi-variate analyses. Many GIS's are now capable of performing some complex spatial statistical analyses such as spatial autocorrelation and nearest neighbor analysis. Spatial autocorrelation is used to provide a measure of contiguity between areas, e.g. Figure shows the theoretical range of possible autocorrelations. If a marine species were shown to be distributed in any of these ways, then we might need to seek explanations.

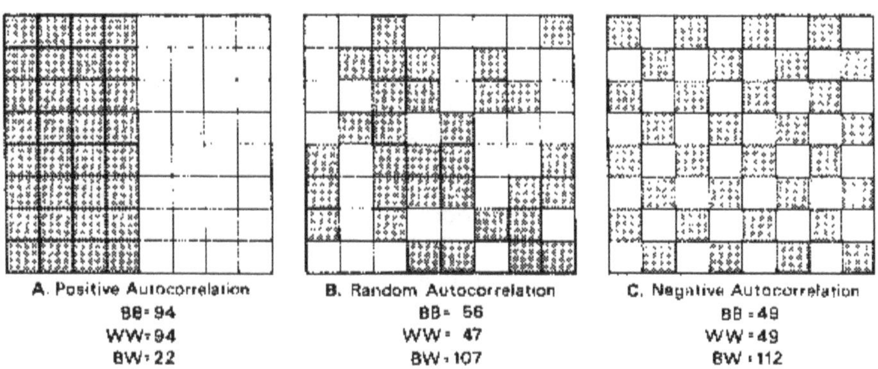

The possible range of spatial autocorrelation

Nearest neighbor analysis provides a relative measure of the dispersion of points in a given area, i.e. they may tend towards clustering, randomness or a uniform spread. There is now some recognition of the importance of geo-statistical capabilities as a functional tool within GIS's (Thomas, 1991).

Measurement

Under this heading, a Geographical Information System will be capable of performing a large number of operations on one or more data layers. Measurement will vary from simple counts (enumeration), to measuring linear or curvilinear distances, to calculating areas, perimeters or volumes or to recording directional or angular measurements. The derived measurements can then form the bases of further work using the Geographical Information System software, e.g. tabular or graphical displays. Clearly certain data structures will lend themselves more easily to different types of measurement. Thus it will be a simple task to compute area if a raster format is used, whilst distance measurements can be more accurate using vector data, i.e. if only because the central point of a pixel (in the raster structure), from where distances are measured, may not be the true starting point.

Data Display

A major anticipated use for any Geographical Information System will be to display the data, i.e. the display capacity will represent the output from the system as presented initially on the VDU. A fundamental usefulness of the concept of Geographical Information System is that it can display output at any stage in the processing of the data. So, the Geographical Information System provides the facility for maps to be incrementally built up, with desired modifications being possible at any stage. Modifications might be in terms of changes to the data inputs to the map, or in terms of the visual representation of the map. So the Geographical Information System user can control, review or experiment at any stage in order to achieve a meaningful final output. All good Geographical Information System software will have a range of graphic display features to control factors such as label size, fonts, colour or shading ranges, line widths, symbolism, map feature positions, etc. And the format of the display is not confined to maps - it may be in graphical, tabular or textual forms. The early 1990's has witnessed a huge emphasis in the Geographical Information System field on the perceptual science of "visualization" - how we look at maps, what information is being conveyed, how people may each view the same mapped scene differently, different ways of communicating information, etc. For those interested further in visualization we would recommend Earnshaw and Wiseman (1992), Bonham-Carter (1994) or Hearnshaw and Unwin (1994).

The data display itself can be temporary or permanent. Temporary display is that which is captured on the VDU. It is the functional user interface in that the VDU shows the results of any commands which have been given via the keyboard, and interactive experimentation can then take place at no cost, at great speed and in an almost infinite variety of ways. Only when the user is satisfied with any temporary screen view need permanent output is obtained. Permanent output is usually by means of hardcopy display, as obtained by the use of any of the variety of device, though it may also be permanently saved to an internal hard disk, or to some form of transferable disk or tape, or it may be sent to an alternative location via networking facilities. Hardcopy display is usually output to paper or film and may be in black and white or multi-colored. The display will vary in quality as a function of the Geographical Information System capability, the detail of the data, the scale of mapping, the quality of paper, the use of vector or raster structuring or, most importantly, with the quality of the output hardware being used and the printing resolution (in dpi) to which it is set. The best quality digital output is now superior to that which can be achieved by manual methods.

In the near future it is likely that Geographical Information System output, and indeed its full range of functionality, will be capable of display and/or use aboard suitably equipped fisheries vessels. Many vessels now have quite sophisticated navigation systems which utilize electronic charts in conjunction with radar and plotting facilities. These can display a variety of static information covering bathymetry, navigational features, land masses, restrictive areas, etc, plus the tracks of moving vessels. It will be a simple progression to extend this functionality so that other desirable layers will be capable of being integrated and displayed in an interactive mode, i.e. such that the vessel has on-board ability to perform a range of required Geographical Information System functions.

Database Management

Space prohibits a detailed look at databases or database management systems (DBMS, and we remind readers here that a database may be defined as a large collection of related data which has been structured in an ordered way but which is independent of any particular application. For Geographical Information System purposes this collection of data may be stored externally in digital form in a purpose created computer software database package, e.g. Oracle or dBASE, or internally in a database which forms part of the Geographical Information System software. Attribute data is frequently stored externally whilst the geographic data are more commonly stored within the Geographical Information System software.

Distributed databases are also commonly used in Geographical Information System and becoming more so. In this case the data may be held in disparate sources within or outside an organisation. Obviously, if a Geographical Information System has access to such databases then a vast range of extra data becomes available to the system. In the fisheries context, it is easy to envisage that a Geographical Information System could offer greater functionality if it could have direct access to oceanographic, meteorological and perhaps environmental databases, all of which were likely to be situated separately from the fisheries Geographical Information System. Laurini (1994) provides a useful overview of distributed databases, plus the many problems to be overcome before their use can become more universal. Anon (1993b) provides detailed information on how a major database, the Regional Maritime Database (BDRM), is being built to cover the marine areas along the West African coastline. This database will be capable of being accessed by any or all of the 10 participating countries, and all information is being specifically geo-referenced so that Geographical Information System functionality can be ensured.

Database Management Systems (DBMS)

A DBMS is a computer programme for creating, maintaining and accessing digital databases. There are a large number of commercial packages available for doing this. The DBMS provides the essential link between the Geographical Information System software, external data sources or graphics enhancing packages and any operations which the user might wish to perform. DBMS can work with different data types such as characters, numerals or dates; they have languages for describing or manipulating the data or for querying the database for particular pieces of information; they provide programming tools and they have particular file structures.

The Main Features Required of a Database Management System

* To create data bases which are in a carefully structured and consistently logical format.
* To create new data bases.
* To extract data from the database in a variety of ways.
* To persistently and constantly execute any commands.
* To display data as required.
* To edit data in any requisite way.
* To sort data.
* To allow for the transfer of data between various software packages.
* To protect data against loss, unauthorised entry, copying and destruction.
* To protect against any inconsistencies which might result from multiple simultaneous use of the database.
* To be independent of particular hardware needs

The reasons for having a DBMS are fairly basic. Thus firstly, it is obvious that the data needs to be maintained in the sense of being kept up-to-date, correct, properly ordered, properly structured, etc. It also has to be managed so that it is properly understood, and that it is available how, when, where and to whom it is required. Storage structures must be constantly monitored so as to minimize storage space and to maximize searching efficiency. And since most databases are constantly growing, new fields might need adding or indeed old fields might need deleting. A database manager is usually assigned who can not only manage all of the above, but who can also regulate access to the database, cope with systems failures when they occurred and who is able to link databases with external databases if required. He or she may also be required to cope with the legal sides of data management, i.e. making certain that inaccurate decisions are unlikely to be made as a consequence of using his data, and making sure that copyright infringements are not made.

For a DBMS to be most effective it is important that the data is stored in an orderly way. There are basically four types of DBMS structural/storage methods: - hierarchical, networked, relational and object oriented.

Hierarchical: In the hierarchical model each record can have a number of links to lower "levels", but only one link to a higher level. The highest link is the "root", lower levels are called "children" and levels above are called "parents". Figure shows a typical hierarchical data model for a hypothetical mapped area. Hierarchical data structures are easy to understand and to update or expand and they are useful where up and down searching is required, but they are not very good in circumstances where horizontal searching is carried out, i.e. where it might be necessary to locate all records which are at one level, since there are no connections at the same level.

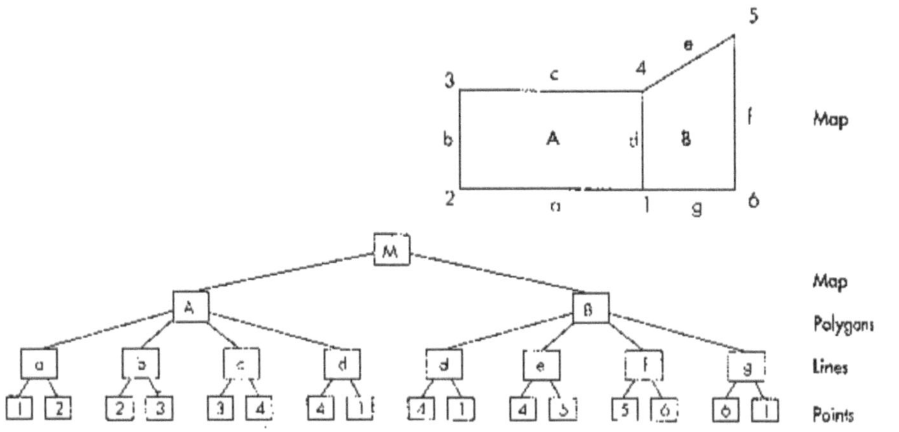

Hierarchical Database Structure Based on a Simple Map (after Bernhardsen, 1992)

Networked: This is similar to the hierarchical data model but here it is possible to have more than one parent, and thus many-to-many relationships can be found. Figure shows that a networked database structure can be analogous to a communications network where there may be many linkages between any combinations of centers. This type of data structure makes good use of the available data, with rapid connections being possible, but it is difficult to create and maintain. Both hierarchical and networked structures are now seldom used in GIS's.

Relational: Here data is organised in a series of tables, each of which contains one type of record. The rows of the tables correspond to records and the columns to fields of the records. Each table in the database will be linked by a common field, otherwise called a unique identifier (or a key attribute). Data is extracted from the database by defining the relationship which is appropriate to the query being asked. This could well involve the use of relational algorithms in order to construct new tables if required. Figure shows an example of the use of a relational database as it might apply in a fisheries situation.

Relational databases are very flexible and are easy to establish, use and maintain. Almost any relationship is possible to work out by the use of Boolean logic and mathematical operations, and additional data can easily be added to the database. The main drawback lies in the fact that searches can be very time consuming, i.e. given the calculations and other specifications involved and the huge numbers of tables which might be needed. The Standard Query Language (SQL) has been developed as the standard language for use with the relational DBMS, and this kind of database is the one most frequently used by GIS's. Floen et al (1993) describe in detail how a relational database has been set up in the Institute of Marine Research in Bergen, Norway, to access, query and manipulate all of their data holdings.

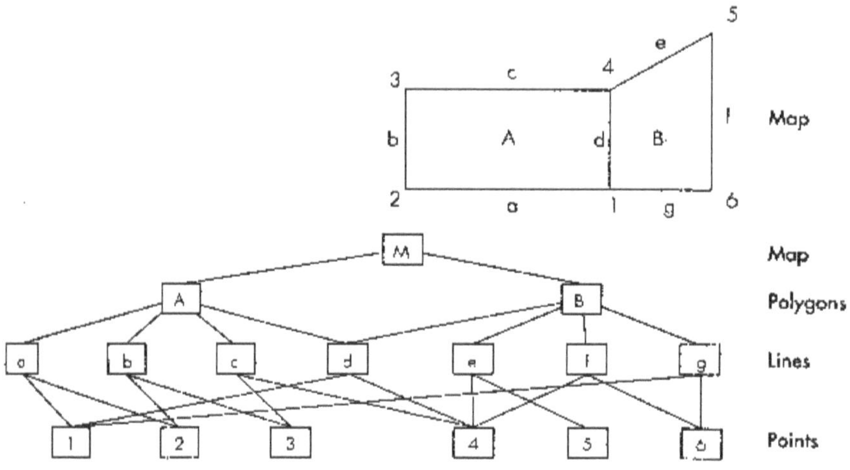

Networked Database Structure Based on a Simple Map (after Bernhardsen, 1992)

Object Oriented: This is a new type of DBMS which is slowly being developed, on a world-wide basis, mainly under the guidance of the Object Management Group (Leach, 1993). This group intends to make object oriented DBMS the norm for the future by ensuring that they are extremely efficient and that there is a world-wide common framework for their development so that they can work across all environments, on all hardware platforms and under any operating system. These databases take a sophisticated view of geographical entities and as such object oriented DBMS are conceptually difficult to explain and to understand. Basically all entities are considered as having the three fundamental concepts of: object, class and inheritance. For instance, geographic objects might be "road", "port", "sea", etc. Each object can then in turn have a class such as "Secondary (road)", "Fishing (port)"or "Shallow (sea)", and they may also have sub-classes. Each object can also be defined as having certain properties, e.g. a fishing vessel may have an owner, a value, a size, etc., and each object may have certain functions (operations or methods) which can be performed on them, e.g. select measure, locate, modify or draw. Inheritance means that when a class is added to an object, then all aspects of "class" can potentially be included in the new description, e.g. all aspects pertaining to "Fishing" can potentially be linked to the class of "Fishing port".

Object oriented databases provide a vehicle for representing data in a form similar to that of the real world. The fact that data is realistically modeled makes user interaction easy and the DBMS is also easy to update and therefore to maintain. There are currently many problems which must be overcome with these databases before they become widely adopted (Batty, 1992), the main one being the lack of a standard object-oriented query language (Kufoniyi, 1995). However, whilst in the past relational databases have been most favored by GIS users, the move is now towards the benefits of object orientation and this is likely to continue.

Global Positioning System (GPS)

The GPS (Global Positioning System) is a "constellation" of 24 well-spaced satellites that orbit the Earth and make it possible for people with ground receivers to pinpoint their geographic location. The location accuracy is anywhere from 100 to 10 meters for most equipment. Accuracy can be pinpointed to within one (1) meter with special military-approved equipment. Global Positioning System equipment is widely used in science and has now become sufficiently low-cost so that almost anyone can own a Global Positioning System receiver.

The Global Positioning System is owned and operated by the U.S. Department of Defense but is available for general use around the world. Briefly, here's how it works:

a. 21 Global Positioning System satellites and three spare satellites are in orbit at 10,600 miles above the Earth. The satellites are spaced so that from any point on Earth, four satellites will be above the horizon.

b. Each satellite contains a computer, an atomic clock, and a radio. With an understanding of its own orbit and the clock, it continually broadcasts its changing position and time. (Once a day, each satellite checks its own sense of time and position with a ground station and makes any minor correction.)

c. On the ground, any Global Positioning System receiver contains a computer that "triangulates" its own position by getting bearings from three of the four satellites. The result is provided in the form of a geographic position - longitude and latitude - to, for most receivers, within 100 meters.

d. If the receiver is also equipped with a display screen that shows a map, the position can be shown on the map.

e. If a fourth satellite can be received, the receiver/computer can figure out the altitude as well as the geographic position.

f. If you are moving, your receiver may also be able to calculate your speed and direction of travel and give you estimated times of arrival to specified destinations.

The Global Positioning System is being used in science to provide data that has never been available before in the quantity and degree of accuracy that the Global Positioning System makes possible. Scientists are using the Global Positioning

System to measure the movement of the arctic ice sheets, the Earth's tectonic plates, and volcanic activity.

Global Positioning System receivers are becoming consumer products. In addition to their outdoor use (hiking, cross-country skiing, ballooning, flying, and sailing), receivers can be used in cars to relate the driver's location with traffic and weather information. Here are some Web locations that describe Global Positioning System receiver products:

Definition

The Global Positioning System (GPS) is a technical marvel made possible by a group of satellites in earth orbit that transmit precise signals, allowing GPS receivers to calculate and display accurate location, speed, and time information to the user. By capturing the signals from three or more satellites (among a constellation of 31 satellites available), Global Positioning System receivers are able to triangulate data and pinpoint your location. With the addition of computing power, and data stored in memory such as road maps, points of interest, topographic information, and much more, Global Positioning System receivers are able to convert location, speed, and time information into a useful display format.

Global Positioning System was originally created by the United States Department of Defense (DOD) as a military application. The system has been active since the early 1980s, but began to become useful to civilians in the late 1990s. Consumer GPS has since become a multi-billion dollar industry with a wide array of products, services, and Internet-based utilities.

Global Positioning System works accurately in all weather conditions, day or night, around the clock, and around the globe. There is no subscription fee for use of GPS signals. Global Positioning System signals may be blocked by dense forest, canyon walls, or skyscrapers, and they don't penetrate indoor spaces well, so some locations may not permit accurate Global Positioning System navigation. Global Positioning System receivers are generally accurate within 15 meters, and newer models that use Wide Area Augmentation System (WAAS) signals are accurate within three meters. While the U.S. owned and operated Global Positioning System is currently the only active system, five other satellite-based global navigation systems are being developed by individual nations and by multi-nation consortiums.

How Global Positioning System works

Global Positioning System: A Satellite-Based Navigation System

The Global Positioning System (GPS) is a technical marvel made possible by a group of satellites in earth orbit that transmit precise signals, allowing Global Positioning System receivers to calculate and display accurate location, speed, and time information to the user. By capturing the signals from three or more satellites (among a constellation of 31 satellites available), Global Positioning System receivers are able to use the mathematical principle of trilateration to pinpoint your location.

With the addition of computing power, and data stored in memory such as road maps, points of interest, topographic information, and much more, GPS receivers are able to convert location, speed, and time information into a useful display format.

Global Positioning System was originally created by the United States Department of Defense (DOD) as a military application. The system has been active since the early 1980s, but began to become useful to civilians in the late 1990s. Consumer Global Positioning System has since become a multi-billion dollar industry with a wide array of products, services, and Internet-based utilities.

Global Positioning System works accurately in all weather conditions, day or night, around the clock, and around the globe. There is no subscription fee for use of Global Positioning System signals. Global Positioning System signals may be blocked by dense forest, canyon walls, or skyscrapers, and they don't penetrate indoor spaces well, so some locations may not permit accurate Global Positioning System navigation.

Global Positioning System receivers are generally accurate within 15 meters, and newer models that use Wide Area Augmentation System (WAAS) signals are accurate within three meters. While the U.S. owned and operated GPS is currently the only active system, five other satellite-based global navigation systems are being developed by individual nations and by multi-nation consortiums.

Global Positioning System for Driving, Outdoors, Sports, and More

More than Just Directions

a. GPS basics

b. car GPS

c. outdoor navigation

Imagine never being lost or needing to ask for directions as you drive. Imagine going out for a run or bike ride and capturing all of your speed, distance, elevation change and heart rate data and uploading it to a training log or an online map you can share. Imagine going hiking and always knowing the way back to camp. Imagine playing golf and always knowing the exact distance to the pin. These scenarios and many more are reality with the use of Global Positioning System (GPS) receivers.

1. In Your Car

Sales of aftermarket in-car Global Positioning System devices have more than doubled over the past two years, and automotive is the fastest-growing segment of the Global Positioning System market. Why? Better and brighter color screens, improved accuracy and usability of maps and directions, added features such as traffic alerts, the ability to wirelessly link a mobile phone to an in-car unit to get hands-free mobile speaker-phone use; all of these reasons and more.

Photo © TomTom

2. On the Trail

Handheld GPS units have dramatically changed outdoors travel, replacing map and compass with moving digital maps, and precise location, elevation, topographic and other data. These handhelds let you enter "waypoints" -- precise locations -- for a favourite fishing spot, safe passage across a stream, a campsite; anything of importance to the user. Handhelds can enhance outdoor safety, as well, making getting lost much less likely, and providing precise position information that can be relayed to rescuers in case of an emergency.

Photo © Magellan

3. Sports & Fitness

Some manufacturers offer GPS products specifically for cycling or running. In cycling, for example, a GPS unit can replace a conventional cycle computer, and add another layer of features, such as route mapping, and trip, heart rate, and cadence data that is uploadable to a computerized training log or to a Web site. Specialized golf GPS receivers provide precise yardage and other useful information and can be preloaded with your favourite courses.

Photo © Garmin

4. On the Water

Global Positioning System has been a boon to recreational and commercial boaters. Moving-map displays and navigation overlays are incredibly convenient, easy to use, and dramatically enhance all-weather, all-light-condition safety. The ability to provide precise position information in case of an emergency is another key safety feature. Built-to-purpose, portable chartplotters often come loaded with detailed coastal maps, and many additional maps for the earth's oceans and lakes are available.

Photo © Lowrance

5. In the Air

Portable aviation units offer intuitive, moving-map views with navigation overlays, providing a great supplement to a plane's instrumentation. Aviation features include: map page, terrain page, route page, position data page, "direct-to" navigation, airport information, and more.

Photo © Lowrance

Navigation

The Ten Essentials for Climbing Safety

Photograph © Stewart M. Green

Use all your navigation skills and tools to keep from getting lost in the Colorado Rockies.

Accessing Sponsored Links:

Mount Aconcagua Climbing International Expedition & Climbing Guide Services. Request a Brochure!www.patagonicas.com

PNI Digital Compasses Low power navigation solutions for Marine, Military & OEM applicationswww.pnicorp.com

Caving & Rope Climbing Caving, Rescue & Vertical Gear Training, Rope, Harnesseswww.onrope1.com

Climbing Ads

a. GPS Navigation

b. Tree Climbing

c. Climbing Knots

d. Indoor Climbing

e. Rock Climbing

f. Sponsored Links

Rock Climbing Summer Camp Ages 8 - 18, Intro to Advanced AMGA and PCIA Accreditedclimbacadia.org/ Trimble Resolution TEmbedded GPS Timing Receiver Motorola FS Encore replacementwww.dpie.com/gps

The first survival system on the Mountaineers Ten Essentials list is **Navigation**.

Get Lost and Found

You're out in the big wide world roaming around and it's easy to get lost, maybe in fog or dense forest or darkness. If you do get lost, the best thing is to get found. If you have the proper navigation tools, then you're the best one who can find yourself and you won't have to rely on having a search and rescue outfit called to track you down or not.

Map and compass are essential

The old navigation items on the classic Ten Essentials list were a map and compass. Both are excellent and mandatory items to have in your pack whenever you are trekking or climbing out in the wilderness. Of course, you need to know how to orient the map and to use your compass, accounting for declination or the difference between true and magnetic north, which varies depending on where you are in the world. Your compass points to magnetic north, which can be as much as 20 degrees off from true north.

Carry a Topo Map

Your map should be a topographic map published by the United States Geological Survey (USGS) and preferably a 7.5-minute quadrangle for sufficient close-up detail of the topography. Carry it in a waterproof container or get it laminated. I always carry my folded in a quart-size zip-lock baggie. The bottom of your map, if it's of a place in the United States, will have the angle of declination so that you can easily figure where true north is.

Global Positioning System Units

There are other high-tech navigation tools available that can help you stay found. The best is a GPS or Global Positioning System unit. These units use information from satellites to determine precisely where you are. If your unit has built in maps, it's a cinch to locate yourself. If it doesn't, then you'll have to translate your position, using latitude and longitude, onto your map to find out your location. Does a GPS unit take the place of a compass? No. A compass is lightweight, doesn't use batteries, and is an ideal back-up for GPS. Some cell phones also have a built-in GPS locator so if you are able to make a phone call, then the rescuers can figure out where you're at.

Use an Altimeter

Mountaineers often carry an altimeter to determine their elevation. An altimeter on your wrist watch will do the trick. Altimeters use a barometric sensor to measure air pressure, providing an estimate of your elevation.

Wands and Tape

A couple low-tech navigation tools are wands that can be used to mark your trail across glaciers or across snow-covered terrain and colored plastic surveyor's tape which you can tie around convenient trees to make a bread-crumb trail to follow back in dense woods. Just remember to remove the tape from the trees as you backtrack.

Use Your Brain

Lastly, use your brain the most important essential. Tell someone where you're going and when you'll be back. Learn navigation skills like map and compass reading. Carry extra batteries. And charge your cell phone before leaving—it just might work out there in the wild.

Part: III
Technical writing and Communication Skills

Chapter 16

Basics of Technical Writing

What is technical writing? Is there a difference between technical writing and technical communication?

Technical writing falls under the broader category of technical communication, which encompasses technical writing, editing, and design. Technical communicators organize, interpret, write, and edit scientific and technical material, gather information of a technical nature and present it to those who need it. Many different professionals fall under the broad category of "technical communicator." Technical Writer is the most commonly used job title for this occupation. Other titles include:

- Communication Specialist
- Information Designer
- Policy and Procedure Writer
- Science Writer
- Documentation Specialist
- Information Developer
- Proposal Writer
- Technical Editor
- Health Writer

- Medical Writer
- Publications Specialist
- Web Editor

Technical writers compose communication from product developers for users of the products. Users include consumers as well as scientists, engineers, plant executives, line workers, and production managers.

What kinds of skills we need?

- **Communication skills:** These include writing, self-communication, interviewing, research, and interpersonal communication skills. When writing technical documents, professional technical communicators must be both clear and concise. When planning and managing projects, they need to be able to anticipate obstacles, prioritize tasks, and manage their time. When gathering information for assignments, they must be able to efficiently collect it from the busy subject matter experts (SMEs) they are working with, or be able to find it in reference materials. Project meetings can be a great place to find useful information, as long as the technical communicator knows how to listen for it. Because technical communicators are almost always part of a larger project team, they must possess tact, gentle assertiveness, willingness to help, and the ability to adapt to rapidly changing circumstances.
- **Technical skills:** These include knowledge of hardware and software, general computer literacy, and familiarity with the tools of the profession like Microsoft Word, Adobe Frame Maker, and Robo Help. Technical communicators must be familiar with the appropriate networking and telecommunications equipment as well as specific database, telecommunication, and programming software. Technical communicators also need to be familiar with industry processes, such as ISO requirements.

What kind of education we need?

Many employers prefer applicants with a four-year college degree in communications, engineering, or journalism, or those possessing a degree or certificate in technical writing. A college degree is recommended for people interested in careers in technical communication, although a major in technical communication is not necessary. Individuals interested in becoming technical writers should take courses in

- Communication
- English
- Journalism
- Chemistry
- Computer-related subjects
- Graphic arts

- Mathematics
- Mechanical drawing
- Physics

A 2002 survey of Society for Technical Communication members showed the seven academic backgrounds most common among technical communicators:

English	32%
Technical Communication	22%
Business Administration	10%
Computer Science	9%
Science	9%
Engineering	8%
Journalism	7%
Other	28%

Note: The question was presented as multiple-choice and some respondents checked off more than one area of study, so the percentages listed add up to more than 100 percent.

What to do?

- **Apply directly to employers:** Direct application to employers remains one of the most effective job search methods, but it should by no means be the only method that you should use.
- **Join a professional organization:** Many professional associations, such as the Society for Technical Communication, maintain job listings on their Internet sites. Professional association meetings can be a good place to make contacts and get advice from those experienced in the field.
- **Meet others in the field:** Networking with other professionals is also an effective way of finding jobs, including those that are not advertised outside of the company.

How to get a job?

Technical writers should prepare a portfolio of technical writing samples that may include projects completed as a student or a volunteer as well as professional work. This portfolio is over and above the normal résumé or curriculum vitae that a technical writer must have, and will show any prospective employer the quality of work that you have produced in previous assignments. This work can include Web pages you've designed, technical manuals you've edited, or brochures you've designed, among others.

What are employers looking for?

Only a few years ago, computer literacy was considered merely an asset whereas today it is a requirement, and some technical writers may be required to know and be able to read programming languages. Applicants must have good computer skills and may need a working knowledge of specific industry operations and procedures. Applicants should also have good communication skills and be able to convey scientific and technical information clearly and accurately. Increasingly, companies require technical writers to be knowledgeable about computer graphics and desktop publishing, including multimedia production. Some employers will only hire experienced writers who specialize in one field of technology. Some employers require a strong background of technical knowledge and experience, combined with writing skills. Employers also select trainees from among technicians who have backgrounds in science, military equipment, and communications.

What kinds of project working on?

Writers must write in a concise and easy-to-read manner for consumer publications or in highly specialized language for experts. With the increased use of desktop publishing, technical writers are increasingly responsible for the publication process including graphics, layout, and document design.

Technical communicators work in every medium, including print, illustrations, video, multimedia, online help systems, Web sites, and training materials. They work in almost every industry, including:

- Banking, finance, and insurance
- Legal, government and defense
- Computing, engineering, and telecommunications
- Healthcare, biotechnology, pharmaceutical, and environmental
- Manufacturing and aviation

Within their chosen industry, many technical writers will specialize further. For example, technical writers in the computer industry might specialize in software documentation, tutorials, or user manuals. Technical writers design and develop a broad range of information and produce a variety of communication products. User's guides, reference manuals, policies and procedure guide, quick references, online help, online wizards, guided tours, and online cue cards explain how to use products, services, and policies. Web sites provide support and documentation. Technical reports, articles, and books exchange "basic" scientific information. Proposals, catalogs, brochures, videotapes, audio tapes, and demonstrations market products and services. Workbooks, tutorials, quick references, and online coaches train users. Newsletters, magazines, and e-zines have a combination of purposes.

What kinds of working conditions can we expect?

- **Work week:** The normal work week is 40 hours; however, this is often exceeded to meet publication deadlines, and tight deadlines can create stressful working conditions.
- **Environment:** Most technical writing is done at a desk in an office. However, during planning and production of publications, writers may be

required to travel to discuss a project with others. Some technical writers are able to do much of their work from a home-office, using teleconference technology and the Internet to communicate with others.

- **Pay:** According to the Society for Technical Communication's 2003 Salary Survey, the mean salary for entry-level technical writers/editors in the United States was $43,260. For these professionals in Canada, it was $41,030. Some companies pay time-and-a-half for overtime, while others offer compensatory time off.
- **Benefits:** Benefits available in some companies include vacation, holidays, flexible hours, sick leave, bonuses, health and life insurance, and retirement plans.
- **Tools:** Technical writers use personal computers and word processing or desktop publishing packages as typical tools for text, graphic, and multimedia production.
- **Supervision:** Writers may work alone or together under the supervision of a publications chief or editor, a product or procedure specialist, or a marketing manager.

What's the difference between freelancing and working directly for a company?

About one fourth of technical writers are independent freelancers and consultants. Choosing to work as a freelance technical writer can be both liberating and frustrating. You can choose when you want to work, and you have a wider variety of experience available to you than the technical writer that works for a single company. However, finding freelance work requires self-discipline and motivation. Jobs can sometimes be hard to come by. Self-employed technical writers must carefully estimate the number of hours needed to complete a project under contract. Technical writers who work under contract or freelance may work from their home or at the employer's site. Self-employed technical writers must pay for their own insurance.

Is this a secure profession?

Technical writing has become an increasingly essential occupation in business and government, and jobs can be found in almost any industry sector because of the need for users' guides, instruction manuals, and training manuals. Demand for technical writers is expected to grow because of the need to communicate new scientific and technical information to others. The U. S. Bureau of Labor Statistics' *Occupation Outlook Handbook* (2004-5 editions) predicts that over the next several years, among the different areas of writing, the most job opportunities will be for technical writers and writers with specialized training. It goes on to say that demand for technical writers and writers with expertise in specialty areas, such as law, medicine, or economics, is expected to increase because of the continuing expansion of scientific and technical information and the need to communicate it to others. The full entry can be found at the US Department of Labor, Bureau of Labor Statistics.

Six Principles of Technical Writing

If you take a technical writing workshop, the materials will address numerous areas; some are very specific while others are more general in nature. A good starting point is to look at six principles of technical writing. Reviewing these basics can provide a great platform from which we can launch forays into specific areas of interest and documentation.

- Use Good Grammar
- Write Concisely
- Use the Active Voice
- Use Positive Statements
- Avoid Long Sentences
- Punctuate Correctly

The fundamental purpose of scientific discourse is not the mere presentation of information and thought but rather its actual communication. It does not matter how pleased an author might be to have converted all the right data into sentences and paragraphs. What matters is, whether a large majority of the reading audience accurately perceives what the author had in mind.

Basics of Technical Report Writing –a case study

It was refreshing to spend a Friday evening listening to Leah Guren sharing her insights, acquired from a 30 year old career in Technical communication. She talked about the fundamentals that lay at the core of Technical Writing. The focus was on what process is followed, what are the right questions to be raised, how well the information is analyzed rather than any particular tools or domain. The talk was titled "The Golden Rules (Technical communication theory and application)".

She discussed 10 Golden rules that work for all types of TC (Technical communication) projects and provide a practical methodology. These golden rules are a result of Leah's own experience and research.

1. Paper is Permanent

It's about owning what I am writing. It says "I am responsible for what I am writing". So it's important to write correctly, proof read and edit the documents properly and provide a proper structure to the content.

Let's see why it is so important.....

- Users judge the quality of product, s/w based on the quality of documentation.
- *Quality* states the measure of accuracy. Thereby demanding absence of errors, like typos, spellings, punctuations, incorrect data.
- *Usability* defines clear and accurate information.
- Is the content standalone?
- Can someone use it?

- ❖ Do you understand what you are writing?
- ❖ Is the information misleading or ambiguous?

Get your grammar and punctuation rules correct so as to give logical explanations to developers.

2. Know your Audience

Who is the user, it is essential to know who is going to use the document. Based on that knowledge, one include demographics, take into consideration the technical and product knowledge of the users, analyze the language skills of the users as well as make safe assumptions.

Ask yourself

- Are you communicating to existing users, experienced users or novice users.
- How users are consuming the content, online, printed, and so on.
- Does the documentation pay attention to the special needs of a certain section of the population for instance for those who are color blind), such as color blindness?
- Are you adapting yourself to the type of audience.

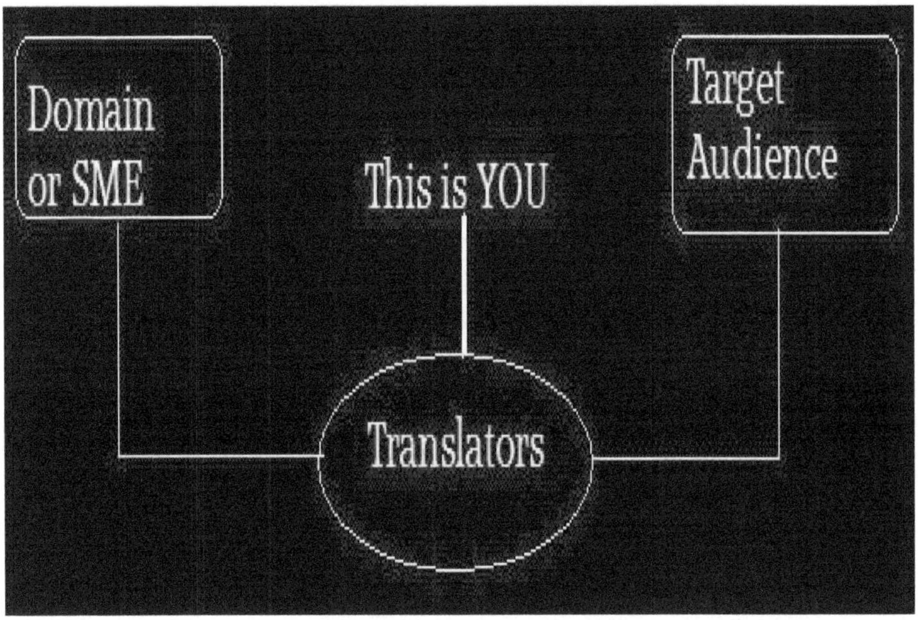

Have you heard about FOGG Index? It tells you how good or bad is your writing and does it match the learning level of your audience. Check out FOGG Index

- Do you engage with tech support to find out the pain points, how good is the content, what is working for the user what is not working.
- Are you taping other internal resources like marketing team to get feedback on the user documents.
- Engage with the customer.
- Create Personas- an artificial or fictitious person/s who represents the user/s - and put it in front of you to check at every point, whether you are writing for the right person.

3. Highlight Hazards

Watch out for things that need special attention of the user. Figure out the pain points, high risk things and use proper means to clearly specify the hazard.

As a Technical Writer your job includes:

- Finding hidden hazards through scenarios.
- Use a ranking mechanism to differentiate types of severities
- Danger
- Warning or Caution or Attention
- Note
- Make hazard visually clear by placing it at the right place.
- Break it out of body text
- Highlight it, Use icons, Write imperative sentence, Use a simple do/do not statement
- Explain the ramifications and workarounds

4. Break it out

Do you know 7 years ago it was known that users spend 30 seconds on a page to look for information? What is crucial for a 'Technical Writer' is giving proper structure to the content so that its easier for the user to find the information.

What can you do?

- Keep the sentences short.
- Think visually and reduce the text, use graphical illustrations if useful
- Provide structure to documents, use headings, bullets, Include tables, Steps, Flowcharts for decision making and branching options in a process.
- Think about the way people need information (text, non-text).

5. Don't write blind

If you don't understand do not write, it's your job to know what you are writing. Important to know the product you are writing about. Do not rely on second hand information. Explore the product, see it, use it, touch it because if you cannot perform the action you are writing about you cannot explain it correctly. Always ask open ended questions to get the right information from the SME.

6. Be Consistent

When you are writing the same thing in three different places in the document do not write it differently each time. This is exactly the opposite of what we are taught in school. But that's the rule here. Take care of technical terminology and pick one term. It's even important for localization. So,

- Decide how to refer to the product.
- Consider interface elements and actions
- Don't forgot styles, fonts, and layouts
- Use a style sheet.

7. Signpost

Use signposts to direct the user to the correct information at the appropriate place in the document. Provide proper references wherever needed. Use layout and typography to indicate relationship of elements. As part of a documentation suite make all parts aware of each other.

8. Don't violate (legitimate) standards

- Do not use non-standard terminology.
- Recognize the difference between a legitimate (de jure) rule and a bad de facto custom. Thus, if you find certain things to be customary as they have been passed from generations of writers, which are incorrect and does not have a proper usage. Go ahead, make yourself heard and change it.
- Develop knowledge about issues relating to, regulatory and compliance such as ISO compliance, EU directives or any certifications or legal issues.

9. Contemplate before you illustrate

Make a rational decision regarding, when you must have graphics. Sometimes you may not need it at all or maybe you may need only a part of it.

What to do ?

- Use appropriate graphic types
- Use callouts or annotations to focus attention.
- Use various image views like exploded view, isometric view, and so on.
- Place the graphic after the content.

10. Cut the fluff

Fluff is a serious issue with usability. Flowery, vague language can also be seen as fluff. So Keep it simple and short (KISS).

Common fluffs include:

- formal and pompous writing

- hidden strong verbs
- unnecessary passive voice

Thus, these golden rules are intended to provide a user centric approach to documentation, help to justify editorial choices, and provide the analytic decision tree for all projects. The topic was not only interesting but her approach to presenting the topic and her style of engaging with the audience made the experience even more enriching.

Note: Main points are taken from Lean Guren presentation handout notes shared with the audience.

With an increase in the field of technical writing, technical documentation translation, and the technology industry in general, there has also been an increase in the demand for workers within these industries. Having already written about what it takes to be a translator, it seemed logical to take a look at the step that comes before the translation process: technical writing.

What is technical writing?

Although technical writing covers a large range of documents and fields, technical writing still has its own specific style that technical writers must follow. The basic definition of technical writing is communication that involves imparting information about a technical or specialized field, such as computers, engineering, or biotechnology. A more specific definition of technical writing includes three parts:

- Communicating about a topic that is specialized or technical in nature.
- Communicating through technology like the internet or social media.
- Giving instructions for using or doing something specific (even if it's complicated).

The most important aspect of technical writing is that it is thoroughly and clearly understood by readers. Without clarity, there is little point in producing any technical documents.

What is technical writing used for?

As mentioned previously, technical writing is meant to communicate instructions or help for something that is technical. This applies to pretty much anything within the technical field. Some technical writers focus on writing grants, while others write helps materials for new software. In any case, technical writing involves taking something complicated and creating a simple explanation for it.

What is the process for technical writing?

There are some basic steps that apply to almost any piece of technical writing. You'll probably notice that these steps are extremely similar to any other method of writing. That's because, when you get down to it, technical writing is simply a more refined style of writing.

- **Plan**: Gather information and analyze the audience you are targeting. Make sure you understand the writing task completely, as well as the end goal. Plan what needs to be conveyed to the reader or audience.
- **Outline**: Make an outline of the essentials that you will write. Organization is essential for the content of the piece to be clear. Some companies may require the outline to be approved before you can even move on to writing the actual piece.
- **Write**: Write clear, concise sentences that communicate whatever the audience needs to know. That's where the first two steps will come in handy. Depending on the piece you're writing, you may have to include statistical information to support the point you're making.
- **Revise**: Chances are that, your writing is not always perfect on the first go-round. Revision gives you a chance to make obvious changes before submitting it to peer or expert editors.
- **Edit**: Most companies will have superiors who will edit your writing and then send it to the subject matter experts for feedback. This way, your writing will be as correct as possible, both grammatically as well as subject wise.
- **Publish**: Once the piece has been edited and received approval from all the essential people, it will be ready to be published.

What does it take to be good at technical writing?

While many employers of technical writers are looking for a college degree and some sort of technical experience, there are certain skills that are considered important for any technical writer.

- **Forward Thinking** – The ability to plan ahead is incredibly important to the technical writing process. If you can think ahead in terms of what you're writing, that skill will serve you well in the field.
- **Detail-Oriented** – Every word counts in technical writing. Noticing the details in a technical piece is incredibly important to almost any industry or company that you may work for. One word can make the difference between success and failure.
- **Excellent Writer** – This should go without saying, but technical writing is a demanding brand of writing that may tax your writing skills. Make sure that's what you want to do before you get into it.
- **Succinctness** – Being able to say as much as possible with as few possible words. That is the key to technical writing. If you are a little long-winded, you may want to do some brevity exercises before applying for a technical writing position.
- **Large Vocabulary** – Having a large vocabulary is important, but having a specialized vocabulary is even more necessary. It depends on the company and the field, but this is often something you may have to learn as you go.

Just know that the larger your command is over the language, the better your writing will be.

- **Impartiality** – Technical writing does not leave any room for opinions or biases. If you can't write something without sounding promotional, you may want to try a different brand of writing.

Chapter 17
Proposal Writing

A Guide to Proposal Planning and Writing

Overview

Grants Marketplace, grant seeking tips represent a condensed version of information presented in another Greenwood Press publication, Proposal Planning and Writing, by Lynn E. Miner in depth presentation of grants processes and products, including paragraph by paragraph analyses of proposals funded by private and public sponsors, refer to Models of Proposal Planning and Writing (Greenwood Press, 2005) by Jeremy T. Miner and Lynn E. Miner. Grant seekers usually enter the grants arena with many questions. "How do I find grant money?"

"What are my chances of getting a grant?" "Is it easier to get public or private grants?" "Do I have to know the 'right people' in order to get a grant?" "How big a grant can I get?" "What can I do with my grant money?" "How hard is it to get my first grant?" "Should I talk to grant makers before submitting a proposal?" "What do grant makers look for in a proposal?"

Grant seeking is a multibillion dollar a year business. If it were a single company, it would rank at the top of the Fortune 500 list. The GRANTS database identifies approximately 10,000 public and private programs that disperse grant dollars. The prefatory comments in this Guide offer time-tested suggestions on how you can plan and write successful grant proposals so that you can get your share of those dollars. These and many others often translate into one fundamental question, "Is it all really worth it?" For those organizations that received part of the over 250 billion dollars given last year in public and private philanthropy, the answer is clearly "Yes, it really is worth it."

Motivations of Grant Makers or inequities

Why do grant makers "give away" money? Grant makers (sponsors) are vitally concerned about social problems and injustices. In fact they are so concerned, that they are willing to invest their money to address these problems. In essence, they see a gap between what is and what ought to be. Another name for the "gap" in grant parlance is the "need." The gap represents their view of the world. Grant makers exist because gaps exist; their goal is to close these gaps. Successful grant writers understand the sponsor's view of the world and express that view in the grant proposal. Successful grant writers are able to reflect the "priorities" of the sponsor. Too often, grant applicants focus on their own need for funds instead of matching their projects with the sponsor's priorities. You should select sponsors that share your values and tailor your proposals to them. Sponsors view grants as investments in an improved future. Proposals are funded when they express the same priorities shared by the sponsor. Projects are rejected when they do not precisely reflect the priorities of the sponsor.

Getting Started

There are **three main steps** to follow in successful grant seeking. **First,** you must identify potential grant makers who would be interested in supporting your project. You should use the entries in this Directory as a starting point to select those prospects with a high probability of financing your needs. You can use one of the four indexes Subject, Sponsoring Organizations, Grants by Program Type, or Geographic--to locate the appropriate grant opportunities for you. If you need additional reference materials, you may wish to consult other publications in the Greenwood Grants Collection or contact the nearest Foundation Center Library, which will contain reference books on private foundations in your region as well as basic information on government and corporate grants. **Second**, after you have identified your list of potential prospects, you should contact key people who can help you plan your proposal before you start writing. In essence, you must do your homework if you are going to be successful. A sure way to fail in grant seeking is to write a proposal without talking to key people who can maximize your possibility of success. The Systems and Procedures section below offers a few of the basic proposal-planning strategies. **Third,** after you have qualified your prospects and planned an effective approach, you must produce a carefully written, well-reasoned proposal. Some grant proposals are rejected because they contain bad ideas; most grant proposals are rejected because they contain good ideas poorly written. There are basically two types of grant proposals: (1) long proposals to government agencies, and (2) shorter letter proposals for private sponsors. The final section of this Guide offers proposal planning and writing tips for both types.

Proposal Planning

Overview

Getting fund

This section of the Guide identifies a few of the other basic reference sources for finding public and private grants. While identifying possible funding sources is not particularly difficult, the greater challenge lies in knowing what to do with that information once you have it. As a result, it is important to understand the pre proposal contact process: what to do after you identify a potential sponsor but before you mail your proposal. Following these steps will significantly improve your chances of finding out about Public Grants

Catalog of Federal Domestic Assistance

The federal government remains a major provider of grant dollars, despite its decline in funding for social services and health and welfare programs during the last two decades. Most federal agencies have some type of grant making program. Many of the federal grant programs listed in the Directory of Research Grants originate from the most current CFDA. The CFDA is published every spring and updated each fall. The program number entry for the government grants in this Directory refers to the Grant program number in the CFDA.

Federal Register

In order to keep current with the continually changing federal grant scene, you may wish to periodically check the Federal Register, the government's "daily newspaper." It is also available from the Superintendent of Documents and lists notices of legal rules and regulations, and application deadlines for new grant programs from federal agencies.

FedBizOpps

Besides awarding grants, the federal government also awards contracts. A grant is a mechanism to support a project whereas a contract is an instrument to procure a project. FedBizOpps is the single source for federal government procurement opportunities that exceed $25,000. All federal agencies must use FedBizOpps to tell the public about contract opportunities over $25,000.

Finding Out About Private Grants

Private grants come from both foundations and corporations. Brief overviews are presented for both sponsor categories.

Overview of Private Foundations

There are more than 65,000 private foundations in the United States. Annually, they award more than $20 billion. While the figures vary slightly from year to year, the 10,000 largest foundations have 90 percent of the assets and make 80 percent of the awards. By federal law, foundations must give away 5 percent of their market value assets or interest income each year, whichever is greater. This means, for example, that the W. K. Kellogg Foundation with $6 billion in market assets must award at least $300 million annually. Foundations must follow the

5 percent rule or they risk losing their tax-exempt status. At present, roughly 15 percent of the private foundations maintain Web sites that disseminate grant-related information, and that number is slowly growing. There are two major Web sites that maintain links to foundations.

Appealing to Foundations

Foundations award grants to those organizations presenting a convincing case that they will help the foundation reach its long-term goals. The grant appeals can assume several different forms. Some foundations make their money available for specific purposes, e.g., building funds, operating support, equipment, or seed capital. Some foundations make their money available to serve specific populations, e.g., frail elderly, minorities, homeless. Some foundations make their money available to specific types of organizations, e.g., hospitals, universities, boys' and girls' clubs. Some foundations make their money available to specific geographic areas, e.g., a city, a county, a state, a region. Some foundations have their own specific priorities and interests, which determine the types of programs they support. The program descriptions in this Directory often list sponsor preferences and any program restrictions or requirements for applying. With these considerations in mind, cast your project in a way that appeals to the foundation's self-defined mission.

Analyzing Foundation Tax Returns

To gain additional information about all foundations, large or small, review their tax records. By law, foundations must submit IRS 990-AR (Annual Reports) or 990-PF (Private Foundation) returns. The 990s are the private foundation's equivalent to your individual 1040 income tax records. While you would not want anybody looking at your personal tax returns, you can examine the tax returns of private foundations online free at the Guide Star Web site. Generally, the information is quite useful in identifying foundation personnel, grant recipients, and award amounts. In some invest their dollars. Use the pre proposal contact process described in the Systems and Procedures section below to investigate further these apparent inconsistencies instances you may even notice a mismatch between what sponsors say they fund and how they actually.

Overview of Corporate Philanthropy

Most corporations follow a concept of "profitable philanthropy." They often fund projects that will bring those better products, happier or healthier employees, lower costs, or an improved public image--all things from which they benefit. Your challenge is to describe your project in terms that will benefit them. If your organization doesn't have a history of attracting corporate donations, start small and request larger grants as you establish credibility. You may wish to request no monetary support as a first grant. Corporations are very cost conscious; challenge grants, dollars awarded to match other grants, have special appeal because corporations feel they are getting the most for their money. While there are nearly 2.5 million corporations, only about one-third of them make contributions to nonprofit organizations.

Appealing to Corporations

Corporations exist to make a profit. When you are seeking a corporate grant, you are asking for the stockholders' income. When profit increase, corporate giving increases slightly. When profit decrease, corporate giving decreases dramatically. When corporations make grants, they look for something in return. What can you offer them?

1. An improved corporate image? Will they have a better community reputation by funding you? Will funding your project make the local residents more productive or satisfied?
2. An improved environment around the corporation? Will your project offer improved transportation, communication, or ecology?
3. An improved benefits package? Will your project offer new or better health programs, cultural activities, or recreational facilities?
4. An improved pathway to attaining corporate goals? Will your project offer new personnel, personnel training, or availability of resources?

Your corporate proposal must emphasize what they are "buying" with their grant prestige, employee satisfaction, or increased profits. As a result, your request should involve a project that is related in some way to their business. For instance, corporations often feel they are unfairly taxed to provide the public services required to deal with many social problems, such as illiteracy and high dropout rates. You can argue that their support will reduce long term tax liabilities for such problems. Corporations typically support those organizations with which they already have a relationship. Don't give up if you don't make it on the first try. Use any business contacts your board, staff members, or volunteers have to help advocate for your project. Most corporations have an informal application process. Personal contact, crucial to success in any grant solicitation, is especially important here. Again, show in concrete terms how their grant to you will benefit them. Try your answers out on your own corporate board members before you present them to the corporate funding officials you are soliciting.

"One-Stop Shopping" for Public and Private Grants

There are many times when grant seekers want one comprehensive source that provides current information on both public and private grant makers. The GRANTS database is one such tool. It offers descriptive information on approximately 10,000 grant opportunities covering the range from curriculum development and teacher training grants to fellowship and conference opportunities. Multiple indexes provide handy cross references. The data for this Directory come from the GRANTS Database, available electronically on the Web through Grant Select, as well as online through The Dialog Corporation and Knowledge Express Data Systems (KEDS), and on CD-ROM with bimonthly updates (through The Dialog Corporation).

Systems and Procedures

After reviewing the large body of proposal funding information, the beginning proposal writer is often left with two reactions: "I had no idea so much information is available," and "How do I possibly organize and manage so much information?" The following well-established systems and procedures will help you organize and process your grant information in order to increase your chances of getting funded. With a systematic approach, you can significantly reduce your proposal development time, an important advantage, especially when you have a pressing grant deadline.

As you begin the process of prospect research with the entries in this Directory, look for those sponsors who share your view of the world. As you review your initial prospect list, sort them into two categories: (1) Maybe: on the basis of the program description, it appears that this sponsor might be interested in my project; and (2) No: it seems unlikely, based on the program description, that this sponsor would be interested in my project. Said differently, initial prospecting will not identify the final list of sponsors to whom you will submit proposals. Rather, it will identify the point from which you must gather additional prospect information before you can say "Yes, I definitely should submit a proposal." The gathering of additional information may be the most critical phase of proposal development. Successful grant seekers follow a four-step process in conducting pre-proposal contacts so they can fine-tune their proposal planning and gain a competitive funding edge.

Step One

Write to the Program Officer

Write the program officer who was identified from your initial prospect research and request a copy of the sponsor's application forms and guidelines. In addition, request a list of past grantees and reviewers, if appropriate. Some sponsor Web sites list reviewer names. If you are unable to get specific names of past reviewers, ask the program officers for general information on the types of reviewers they use their age, background, and training; how they are selected; how they are used in the review process; and how points are allocated to a proposal. This information will allow you to match your proposal writing style to the level of sophistication of your reviewers.

Step Two

Call a Past Grantee

From the information you gathered in step one, contact a past grant winner from this sponsor. Ask to speak with the project directors or the individuals who wrote the proposal. Indicate where you got their names and raise questions that will assist you in learning about the funding source. Questioning is the foundation of pre-proposal contacts, getting essential background information to decide (1) if you should submit a proposal, and (2) if you submit, how best to frame the proposal so it matches the "values glasses" of the sponsor. The list of questions that you could pose is theoretically endless. Nevertheless, if you want to write a successful grant, you must PREP first, where PREP is an acronym to distinguish between four basic types of questions.

1. Position: what are the baseline situations, present circumstances, and basic facts?
2. Rationale: what are the problems, needs, and injustices that exist today?
3. Expectation: what are the implications for addressing these problems?
4. Priority: what approaches are most likely to lead to an improved situation now?
5. Collectively, PREP questions span a continuum of time from past actions to future intentions.

Position Questions

Did you call or visit the sponsor before writing the proposal? This will give you a clue about the extent to which the grantee engaged in pre-proposal contact. Who did you find most helpful on the funding source staff? This query will help identify an "in-house hero," an agency staff person who may be the best source of inside information for you. How close was your initial budget to the awarded amount? The interest here is to identify the extent to which budget negotiations took place.

Rationale Questions

You got funded because the sponsor was convinced you could solve some big problems they were concerned about. What were those big problems? Look for "big picture" problems that really trouble the sponsor. Are there problems or difficulties in this area that are particularly challenging now? Take note of priorities among complex problems: what are the top issues? Generally speaking, what are the disadvantages of the way these problems are being handled now? Pay attention to what's wrong today and will be worse tomorrow.

Expectation Questions

Given the problems you identified, what are the implications of those difficulties? Get the informant talking about the consequences of existing problems. What's the desired impact on these problems, balancing project breadth, depth, and financial resources available? The answer to this question helps you find a proper balance within budgetary constraints. What would you do differently next time? Invariably, people learn from the positive experience of getting a grant and have a number of suggestions about things they would do next time to strengthen a proposal.

Priority Questions

Why did the sponsor think it important to solve the problem you identified? This will give you insight to the sponsor's motivation for solving the problem. What are the benefits you see of your approach? Look for reasons why did the sponsor found this solution so useful. Would this approach be useful for cost reasons or something else? Narrow the general benefits to that major motivator: money.

Step Three

Call a Past Reviewer

Contact some past reviewers and indicate that you understand they were reviewers for the grant program you intend to approach. Your goal is to learn about the actual process to be followed as your proposal is reviewed. For example, if a reviewer has only three minutes to review your proposal, you will write differently than if the reviewer has three hours to review your proposal.

Ask Position Questions

How did you get to be a reviewer? Usually you just submit a resume and express an interest, showing how your background and expertise meshes with agency interests. Consider becoming a reviewer yourself because it is an easy way to get "inside information" and improve your success rate. How many proposals were you given to read and how much time did you have to read them? This answer will give you an idea of what your immediate competition will be like and the conditions under which proposals will be evaluated. Did you follow a particular point or scoring system? Invariably, some portions of a proposal carry greater weight than other portions. This information will enable you to concentrate your greatest efforts on the highest-scoring portions.

Rationale Questions

What were you told to look for? Sometimes, reviewers are instructed to look for elements of proposals that are not requested in the application guidelines. Your proposal should respond to all items on the application guidelines and the reviewer's evaluation form. What were the disadvantages of the way the targeted problem areas were being approached in the proposals you read? This answer highlights reviewers' insights into problems not being addressed in the proposals they reviewed, problems that you may be able to incorporate in your proposals. What difficulties linger that still is not being addressed? The answers are errors that you want to be sure to avoid, such as failing to number the pages, omitting the resumes of project directors or consultants, or miscalculating budgets differently now that you have been a reviewer? People invariably learn from the positive experience of seeing the inside process of awarding grants and have a number of suggestions about things they would do if they were asked to write a proposal again.

Priority Questions

What's not happening in this area that should? This answer highlights areas that need attention. What would be the key features of an ideal solution? Among various options, this answer indicates the components of an ideal solution. What might be the benefits of this approach? The response compiles the benefits list for the intervention strategy.

Step Four

Contact the Program Officer

Tell your program officer you have studied the program guidelines carefully and you have some additional questions. Realize, however, that your credibility will decrease if you ask questions that are answered in their written guidelines. If there were no budget limits, what should have been proposed that wasn't? Playing "what if" invites creative solutions that are not normally considered.

Position Questions

How much of the money distributed this year will be available for new awards as opposed to no competing continuation awards? This answer will tell you how much money is actually available for new applicants like you. How are proposals being evaluated? This response will help you to identify the yardstick against which proposals are measured. Would you review our draft proposal if we got it to you early? A favorable response will help you better cast your proposal to meet the sponsor's expectations.

Rationale Questions

Why have you targeted your program dollars towards this problem? This answer explains why money will solve the problem. What are the biggest hurdles in this area now? The response points out the biggest challenges people in the field now face. Which dimensions of this problem need to be addressed next? This answer points to the "big impact" needs.

Expectation Questions

Does my project fall within your current priorities? If it does, begin writing. If it doesn't, explore different activities that might yield a better fit or ask for suggestions of other grant programs that might be interested in your project. What would you like to see addressed in a proposal that others may have overlooked? This question provides program officers with an opportunity to articulate their pet ideas. Can you provide me with a copy of the reviewer's evaluation form? Use this form to organize your proposal, using the same headings and subheadings, even if they differ from those in the application guidelines.

Priority Questions

Would this approach produce what is needed? This answer helps to map out a successful action plan. What are the long-term benefits of this solution? This answer describes the long-range implications of your solution. What outcomes do you expect from grantees? This answer clarifies what the program officer will expect from you. Successful grant seekers who follow this four-step proposal planning process can use that information to write winning proposals.

Proposal Writing

Overview

Public sponsors usually require full proposals that range from 15 to 100 pages and contain such sections as a cover letter, title page, abstract, introduction, need/

problem, objectives, methods, evaluation, dissemination, budget, and appendixes. In contrast, private sponsors often require a letter proposal, a brief two- to five-page document in letter form that concentrates on the problem and solution portions. The remaining section of this Guide offers suggestions and tips on the major components of public and private proposals. It walks you step-by-step, from beginning to end, through an integrated process of planning and writing persuasive proposals.

Purpose of Introduction Statement

The introduction is a credibility statement that describes your professional and organizational qualifications and establishes the significance of your idea. Your qualifications, or credibility, may have more to do with your being funded than anything else. In a government proposal, the application guidelines may or may not ask for an introductory section. The introduction section establishes the tone of the whole proposal. Novice proposal writers focus on their own need for funds instead of using the introductory section to link their project with the sponsor's priorities. Successful grant seekers capitalize on the partnerships they build with sponsors during their pre proposal contact and cast their projects in ways that mesh with sponsors' values. For novice writers, the psychological orientation is "I-I, Me-Me" while successful writers take a "You-You" perspective.

Key Questions to Answer

- Clearly establish who you are.
- Describe your organizational goals.
- Establish your credibility in the project topic area.
- Lead logically to the problem statement.

Writing Tips for Introduction Section

The introduction section of a proposal represents a credibility statement about you and your environment. While your resume is an important credibility statement, particularly in government proposals, it may not communicate the fact that you work in an environment that will support your project. Weave this point into your introduction. Tell the reviewer about your track record in projects of this kind and how this project fits into your overall organizational goals. If you don't have a strong track record in your proposed project area, borrow credibility from other field experts through the use of project consultants, letters of endorsement, and supporting statistics.

Statement of Problem or Need

Purpose of Problem Statement

Your statement of the problem your need represents the reason behind your proposal. Experienced grant seekers know it is the single most important part of the proposal that influences funding success. The problem statement specifies the conditions you wish to change. It should be supported by evidence drawn from your experience, from statistics provided by authoritative sources, and from appropriate literature reviews. Your need statement should quickly summarize the problem, show your familiarity with prior research or work on the topic, reinforce your credibility for investigating the problem, and justify why this problem should be investigated. Do not assume that everyone sees the problem as clearly as you do. Even if the problem is obvious, your reviewers want to know how clearly you can state it.

Key Questions to Answer

- ❖ Demonstrate a precise understanding of the problem or need that you are attempting to solve?
- ❖ Clearly convey the focus of your project early in the narrative?
- ❖ Indicate the relationship of your project to a larger set of problems or issues and justify why your particular focus has been chosen?
- ❖ Establish the importance and significance of the problem?
- ❖ Justify why your problem should be of special interest to the sponsor?
- ❖ Demonstrate that your problem is feasible to solve?
- ❖ Make the reviewer want to read further?
- ❖ Indicate how the problem relates to your organizational goals?
- ❖ State the problem and outputs in terms of human needs and societal benefits?

Writing Tips for the Problem Section

A common error is to paint the problem in grand or general terms. Don't say "little is known about...", "there is a lack of information about...", or "no research has dealt with..." this problem. Arguing for something that isn't makes for a weak need statement. Instead, go one step further. Explain the consequences of the information void. Describe the need in human terms. For example, if you want to buy computers for your school, talk about the computer literate students who will benefit in the future. Beyond discussing the importance of the project's topic, demonstrate the need for your methodology; the reviewers should be able to anticipate your solution based upon your analysis of the problem. This important transition paragraph is frequently left out of proposals written by beginning proposal writers.

Objectives

Purpose of Objectives Statement

Your objectives specify the outcome of your project, the end product(s). When sponsors fund your projects, they are literally "buying" your objectives. That's why it is extremely important to state your objectives clearly. When you write your objectives, follow the acronymic advice: "Keep them **S-I-M-P-L-E.**" Your objectives should be,

Specific: indicate precisely what you intend to change through your project.

Immediate: indicate the time frame during which a current problem will be addressed.

Measurable: indicate what you would accept as proof of project success.

Practical: indicate how each objective is a real solution to a real problem.

Logical: indicate how each objective systematically contributes to achieving your overall goal(s).

Evaluable: indicate how much change has to occur for the project to be effective.

Although these categories are not mutually exclusive, each of your objectives should meet at least two or three of these six criteria.

For instance, given the goal of "improving the quality of life for homeless individuals in our city," a proposal objective might be for the "Midwest Home Shelter Agency to reduce the number of homeless [Specific] [Practical] [Logical] during the next 24 months [Immediate] by 15 percent [Evaluable] as noted in the Department of Social Welfare Homeless Survey Report [Measurable]." Your objectives section indicates precisely what you intend to change through your project and what you would accept as proof of project success for your target population. Your objectives provide the yardstick you use to conduct your evaluation section. In other words, if you write your objectives in precise, measurable terms, it is easy to write your proposal evaluation section because you know exactly what will be evaluated.

Key Questions to Answer

- Clearly describe your project's objectives, hypotheses, and/or research questions?
- Signal the project's objectives without burying them in a morass of narrative?
- Demonstrate that your objectives are important, significant, and timely?
- Include objectives that comprehensively describe the intended outcomes of the project?
- State your objectives, hypotheses, or questions in a way that they can be evaluated or tested later?
- Demonstrate why your project's outcome is appropriate and important to the sponsor?

Writing Tips for Objectives Section

List your specific objectives in not more than one or two sentences each in approximate order of importance. Don't confuse your objectives (ends) with your methods (means). A good objective emphasizes what will be done and when it will be done, whereas a method will explain why or how it will be done. Include goals (ultimate) and objectives (immediate) statements.

Methods

Purpose of Methods Section

The methods section describes your project activities in detail, indicating how your objectives will be accomplished. The description should include the sequence, flow, and interrelationship of activities as well as planned staffing for the project. It should present a clear picture of the client population, if any. It should discuss the risks of your method, and indicate why your success is probable. Finally, tell what is unique about your approach.

Key Questions to Answer

- Explain why you chose one methodological approach and not another.
- Describe major activities for reaching each objective.
- Indicate the key project personnel who will carry out each activity.
- Show the interrelationship among project activities.
- Identify all project data that will be collected for use in evaluating proposal outcomes.

Data Collection

You will probably need to collect some data as a part of your project. Common data collection methods include achievement tests; psychological tests; role-playing exercises; clinical examinations; personal diaries; ratings by program staff, management participants, or experts; interviews; observations by program staff or evaluators; daily program records (telephone logs, tracking slips, referral forms); historical program records and archives; government records; searches of news media; questionnaires; and surveys. You can either make up your own data-gathering instruments or use existing ones.

Writing Tips for Methods Section

Begin with your objectives. Describe what precise steps you will follow to carry out each objective, including what will be done, who will do it, and when it will be done. If you have trouble writing this section, assume the sponsor's check just arrived in the mail. What is the first thing you will do? Hire additional staff? Order equipment? What will you do next? Keep asking and answering the "What's next?" question and you will lead yourself through the methodology section sometimes called procedures in other proposal guidelines. Once you have determined the sequence of events you will follow in completing your project, cast the major milestones into a time-and-task chart. In graphic form, it segments your total project into manageable steps and lets your reviewers know exactly what you will be doing and when. It says to the reviewers that you are organized and have thought out the major steps of your project. It lets them know you have done significant planning and are not just proposing on a whim. It gives them a road map of the territory you plan to cover. Finally, the time-and-task chart represents a clear, one-page, visual summary of the entire methodology section.

Evaluation

Purpose of Evaluation

Evaluations pinpoint what is really happening in your project so you can improve your project efficiency. Based on evaluation information, you can better allocate resources, improve your services, and strengthen your overall project performance. Beyond these immediate benefits, a project evaluation can uncover needs to be served in your next proposal and make it easier to get and sustain funding. If you want to include an evaluation component in your proposal but know nothing about the subject, consider borrowing ideas from the evaluation plans developed for similar programs or ask a colleague or consultant to review the rest of the proposal and develop an appropriate evaluation strategy. Too frequently, proposals don't explain how the project will be evaluated. At best, they mention some vague process, such as holding a discussion meeting or assigning the evaluation to an expert, with no specifics on how the evaluation will be conducted or what will be learned from the evaluation.

Key Questions to Answer

- Describe why evaluation is needed in the project.
- Provide a definition of what is meant by evaluation.
- Clearly identify the purpose of your evaluation and the audiences to be served by its results.
- Demonstrate that an appropriate evaluation procedure is included for every project objective.
- Provide a general organizational plan or model for your evaluation.
- Demonstrate that the scope of the evaluation is appropriate to the project
- Demonstrate the extent to which the project is practical, relevant, and generalized.
- Describe what information will be needed to complete the evaluation, the potential sources for this information, and the instruments that will be used for its collection.
- Clearly summarize any reports to be provided to the funding source based on the evaluation, and generally describe their content and timing.

Using Evaluators Effectively

Whether you use an internal or an external evaluator, or both, is sure to include them in the proposal development process. A common proposal-writing mistake is to budget an amount for evaluation costs and worry later about the evaluation procedure. Instead, involve evaluators in the proposal writing. Be sure to give them a copy of your project objectives. Remember that pointed objectives will simplify the evaluation process. An evaluator should provide you with important proposal information. Specifically, ask your evaluators to identify precisely what will be evaluated, what information they will need to conduct the evaluation, where that information will be obtained, how often data will be collected, what data collection instruments will be used to get that information, what evaluation design will be used, what analyses will be completed, and what questions you will be able to answer as a result of the evaluation.

How to Evaluate

Evaluation is essentially a four step process. As you will see, if the objectives and methodology sections of your proposal are precise, you are well on your way to completing the evaluation protocol.

1. Identify precisely what will be evaluated. If you wrote measurable objectives, you already know what to evaluate.
2. Determine the methods you will use to evaluate each objective. More precisely, you will need to describe the information you will need and how you propose to collect it.
3. Complete your evaluation design. Specify the analysis you plan to make and then carry out your evaluation by collecting and interpreting the data needed for each objective. Your evaluation design may be simply to observe the behavior of a particular population or something more complex like a rigorous experimental and multiple control group design.
4. Summarize the resulting data analyses and indicate its use. Consider including mock data tables that show what your resulting data might look like.

Note that of these four steps, the first two are completed as you write the objectives and methods sections of your proposal. In other words, you are half-done with the evaluation section before you start it.

Writing Tips for Evaluation Section

Include a separate evaluation component for each project objective. Strengthen your evaluation section by including examples of surveys, questionnaires, data collection instruments, data analysis forms, and other evaluation methodologies in order to demonstrate the credibility of your evaluation section. If you use outside evaluators, identify costs, credentials, and experience. Evaluation sections are less likely to be included in basic research than training grants. Replicability is the primary evaluation criterion in most basic science research proposals.

Dissemination

Purpose of Dissemination

Dissemination is the means by which you let others know about your project: its purpose, methods, and accomplishments. Among other things, it generates publicity for your sponsor and you. As grants become more competitive, dissemination of results is increasingly important. No longer is it sufficient to say you will submit a journal article or present a paper at a professional society meeting. Instead, specify the tentative titles, target journals, and submission dates. Likewise, indicate which meetings will be attended, including dates and locations for presenting papers.

Key Questions to Answer

- Indicate why dissemination activities are important to your project.
- Clearly identify the intended outcome of the dissemination effort.
- Include a feasible and Does appropriate plan for dissemination.
- Succinctly describe any products resulting from the dissemination effort.
- Demonstrate that you are well grounded in theory and research on the dissemination and utilization of knowledge.
- Provide sufficient detail on proposed dissemination procedures to justify the budget request.
- Specify clearly who will be responsible for dissemination and why they are capable.
- Indicate why the dissemination will get the necessary information to the appropriate audiences in a form they can use when needed.

Writing Tips for Dissemination Section

Project results can be disseminated verbally or visually. The visual/verbal distinction emphasizes how you as project director choose to communicate your project results to your target audiences. Additionally, your project dissemination strategies can be active or passive. The active/passive distinction refers to your target audiences and the role they play in processing the visual or verbal information you present. If you choose to write up a report of your project results, you have chosen a visual channel of communication to which the reader responds passively, since reading is a passive process. On the other hand, you may involve the target audience in a hands-on demonstration of project results. In this illustration, the results are presented verbally and the target audience is actively involved. While the visual/verbal and active/passive distinctions are not wholly discrete, the following table illustrates how the more common dissemination strategies might be classified.

Process Verbal Visual

a. Active Conferences and Seminars
b. Demonstrations
c. Site Visits
d. Web Casts and Chat Rooms
e. Teleconferences
f. Courses/Seminars
g. Displays/Poster Sessions
h. Commercial Distributors
i. Instructional Materials
j. Web Sites

k. Video Conferences
l. Passive Convention Papers
m. Staff Presentations
n. Books and Manuals
o. Computer Disks
p. Executive Summaries
q. Interim Working Papers
r. Journal Articles
s. National Information Sources
t. Newsletters
u. Pamphlets
v. Press Releases

In most proposals, you will want to use an appropriate mix of active/passive and verbal/visual dissemination strategies.

Budgets

Purpose of the Budget

A project budget is more than just a statement of proposed expenditures; it is an alternate way of expressing your project. Programs officers will look at your budget to see how well it fits your proposed activities. Incomplete budgets are examples of sloppy preparation. Inflated budgets are signals of waste. Budgets that are too low cast doubt on your planning ability. In essence, your budget is as much a credibility statement as your project narrative.

Key Questions to Answer

- Does your budget provide sufficient resources to carry out your project?
- Include a budget narrative that justifies major budget categories.
- Present the budget in the format desired by the sponsor.
- Provide sufficient detail so the reviewer can understand how various budget items were calculated.
- Separate direct costs from indirect costs and describe what is covered in the latter.
- Relate budget items to project objectives.
- Include any attachments or special appendixes to justify unusual requests.
- Identify evaluation and dissemination costs.

Allowable Budget Categories

Unless the sponsor guidelines dictate otherwise, you can include in your budget request such things as accounting, advertising, animals, audiovisual instruction, auditing, binding, books, computer time, consultants, dues, equipment, fringe benefits, indirect costs, instruments, insurance, legal services, maintenance, periodicals, postage, publication, recruitment, registration fees, relocation, renovation, rent, repairs, salaries and wages, security, subcontracts, supplies, telephone, travel, and tuition.

Direct Costs

Those costs that are line items listed in the budget as explicit project expenditure are called direct costs. The direct costs are usually categorized into personnel (people) and non personnel (things) components. Personnel costs include such items as salaries, wages, consultant fees, and fringe benefits. Non personnel costs include such items as equipment, supplies, travel, and publication charges. Space and utilities may be reflected as direct costs or included as a part of your indirect cost rate.

Indirect Costs

Those costs that are not directly listed in the budget and yet are costs incurred in the project are called indirect costs. Indirect costs are real costs that are hard to pin down, such as payroll and accounting, library usage, space and equipment, and general project administration. Do you include in your proposal budget the costs associated with preparing payrolls or the time your boss spends talking with you about the project? While you could cost out those factors, and others, they become more difficult to quantify. At the same time, they are real project costs someone has to write your payroll checks. Rather than calculating a strict cost accounting of these nebulous factors, many sponsors allow you to calculate a percentage of your direct costs and add it to your budget request. Semantically, the federal government uses the term indirect costs to refer to these extra project operating costs. These costs are usually figured as a percentage of the grant, either of the total direct costs or the total project salaries and wages. Organizations regularly receiving federal grants have an approved federal indirect cost rate that is included in the budgets of federal proposals. If you plan to submit federal proposals periodically but do not have a federal indirect cost rate, ask your federal program officer to refer you to the appropriate federal agency so you can negotiate a federal indirect cost rate for your organization.

Foundations usually use the term administrative costs rather than indirect costs when referring to extra project operating costs, though the terms are interchangeable. Foundations vary considerably in their policies regarding the allow ability of administrative costs. Some will pay administrative costs on grants, and their application guidelines specify the allowable percentage of total direct costs. Others say explicitly in their application materials that they do not allow administrative costs.

In contrast to governments and foundations, corporations use the term overhead to mean administrative or indirect costs. As business professionals, they are accustomed to the concept of overhead and are apt to have a fairly high overhead rate. In most instances, corporate application materials do not specify a policy regarding the payment of overhead. You have three options for determining a reasonable overhead rate to include in your proposal. First, if they are a publicly held company, you may be able to determine their overhead rate by studying the latest annual report. Second, some corporate officials might be willing to tell you what their overhead rate is as a percentage of their total costs. Third, if you have no basis for calculating their corporate overhead, you can use your federally negotiated indirect cost rate since it is an audited figure. If none of these three options is acceptable, your fallback position is to include all costs as direct cost items.

Cost Sharing

Those costs that your organization will contribute to the total project costs are called shared costs. You may contribute partial personnel costs, space, volunteer time, or other costs towards the total project expenses. Your cost sharing may be in the form of "hard" dollars (cash) or "soft" dollars (in kind contributions). In kind contributions do not require a cash outlay from your organization yet would cost real dollars if you had to pay for services rendered.

Cost sharing may be mandatory or voluntary. Mandatory cost sharing is often referred to as "matching funds;" it is required whereas voluntary cost sharing is optional. Your promise of cost sharing in a proposal budget may be a key factor in a sponsor's funding decision.

Mandatory

As one of the eligibility requirements of the grant, the sponsor requires you to share a certain percent of the total project costs. For example, "Local organizations are required to provide a local match totaling 75 percent of the requested grant funds."

Voluntary

You offer cost sharing in your proposal as an incentive to get the grant award. For instance, a sponsor may indicate, "Consideration will be given to organizations with in kind contributions." In response, you may offer 20 percent cost sharing of personnel time toward the total project cost. However, you can cost share too much: for some agencies, higher levels of cost sharing require more administrative monitoring on their part, something program officers may wish to avoid. Accordingly, check with your project officers to see if they have a "preferred level" of voluntary cost sharing. Assuming you've decided to cost share on your proposed budget, the funds may come either from internal or external sources-or both.

Internally

You may allocate a portion of your direct or indirect costs to your proposed project. These shared costs may take on the form of cash or in-kind contributions. Consider this internal cost sharing example: assume you decide to cost share 20 percent of the project director's salary towards your proposed project. This means that instead of your project director receiving 100 percent of her salary from your agency personnel budget, she will now receive 80 percent from that source and the remaining 20 percent from the cost sharing account on the grant; you merely reallocate a portion of her salary; her income remains the same; the source(s) of income are changed on the bookkeeping records.

Externally

You may allocate extramural dollars from other sources to the project. For instance: (a) you have a matching grant from another sponsor; (b) a wealthy philanthropist has given you unrestricted dollars that can be earmarked to this project; or, (c) revenue is generated from another fund raising activity, e.g., golf outing income, can be directed to this project. In each case, you can redirect dollars from those sources to help support the total costs of your proposed project, thereby showing your sponsor you are financially committed to supporting your proposal.

Writing Tips for Budget Section

Here are some tips for planning your budget. Make sure your calculations are as clear as possible. Show the basis of your calculations. If the project is to occur in phases, identify the costs associated with each phase. Don't overlook budget support for such things as service or maintenance contracts, insurance, shipping, or installation. If you anticipate training costs associated with the purchase of new equipment, include those costs in your budget as well. Include a budget narrative immediately following your budget to explain or justify any unusual expenditure items, even if one is not required by the sponsor.

Some sponsors expect you to continue funding your project after the grant expires. If you have a financing plan for future funding, briefly outline it. Other fund-raising options include membership fees, user charges, local organizations, other granting agencies, wealthy individuals, product sales, publications, service fees, direct mail, bequests, memorial gifts, phone-a-thons, and capital campaigns.

Abstract

Purpose of Abstract

The abstract is usually the last written and first read section of your proposal. It should be carefully constructed, providing a cogent summary of your proposed project. It should provide a quick overview of what you propose to do and clear understanding of the project's significance, generalizability, and potential contribution. Project end-products should be clearly identified. Often, proposal reviewers must write up a summary of your project for presentation to a larger review panel. If you write a quality abstract, you make your reviewer's job easier. If the abstract is poorly written, the reviewer's job is more difficult and your funding chances diminish.

Key Questions to Answer

- Does my abstract effectively summarize the project?
- Does it place appropriate emphasis on the various proposal components?
- Does it enumerate project outcomes?
- Does it comply with length or word requirements of the sponsor?
- Does it use key headings and subheadings to highlight proposal sections?

Writing Tips for Proposal Abstracts

Do not write the abstract until you have completed the proposal. Generally, the abstract section contains 250 to 500 words. Include at least one sentence each on problem, objectives, and methods, using the major headings and subheadings you used in the proposal.

Appendixes

Purpose of Appendixes

Appendixes contain information peripheral to your proposal, such as reprints of articles, definitions of terms, subcontract data, consortia agreements, tabular data, certifications, lists of board members and officers with titles, recent annual reports, organizational fiscal reports, organizational charts, resumes, past success stories, significant case histories, agency publications, publicity, and letters of support and commitment. Some grant making agencies do not circulate copies of appendixes when transmitting proposals to reviewers. (Ask your program officer about this, and if materials are not circulated, include essential proposal information in the narrative.) Nevertheless, the use of appendixes is recommended, especially when page limits are sponsor-imposed.

Key Questions to Answer

- Could reviewers evaluate the proposal without any appendix information?
- Have strong letters of support and commitment been included?
- Are assurances of cooperation provided in instances of inter-agency support?
- Are the resumes included for all key project personnel and consultants?

Writing Tips for Appendixes

After your proposal is written, reread it to make sure your reviewers could make an informed funding decision without any appendix information. Include strong letters of support and endorsement. Attach assurances of cooperation in instances of interagency proposals. Be sure to include the resumes of all key project personnel, including consultants.

Proposal Appearance

While you will obviously spend much time working on the content of your proposal, you should also pay attention to the appearance or design of your proposal. Just as clothing is important in the business world for establishing initial impressions, so, too, is the appearance of your proposal as it reaches the reviewer's hands. The proposal should "look" familiar to the reader. A familiar proposal is a friendly proposal. Look at the printed materials issued by the sponsor. Note their use of type size and style, white space, and headings. Structure your proposals to match the private foundation's publication preferences; when appropriate, use the same type size, style, layout, and headings as they do in their publications. Your proposal will look more credible when you consider these factors.

As you learn about your audience and consider proposal appearance, try to anticipate which reading styles the reviewers are likely to use: skim reading, search reading, or critical reading. Recall that your earlier prospect research from a past reviewer or program officer identified the likely manner in which your proposal would be reviewed. Reviewers skim proposals when they have many pages to read in a very short time. Reviewers search proposals when they are following an evaluation sheet that assigns points to specific proposal sections. Reviewers always critically read proposals, especially when the reading occurs in the time luxury of a mail review. The following writing and editing tips are particularly appropriate for all three reading styles.

Proposal Writing Tips

Read and reread proposal guidelines--and believe them! Use the guidelines and the reviewer's evaluation form, if available, as a basis for creating your proposal outline. Strengthen your proposal by following the format tips that experienced proposal writers use.

Bold Type

Bold type is easier to read than underlining, italics, or all capital letters as a means of creating emphasis. Use bold type to emphasize key words and ideas, but avoid overemphasis.

Lists

Lists help to get the message to the reader with a sense of immediacy, without being wordy. Furthermore, because lists are easy for readers to skim, they convey chunks of information quickly. Use a numbered list when items need to be examined in a specific sequence. Use a bulleted list when items are all equally important.

Ragged Right Margins

A ragged right margin is easier to read than one that is right justified because the proportional spacing slows readability. It is easier for the reader's eye to track from the end of one line to the beginning of the next line when the right-hand margins are jagged.

Type Size

Adjust your word processing program for a 12-point type style. Text will be smaller than 12 point becomes difficult to read and makes the reviewer's job harder. Rather than reducing type size to make all your ideas fit on the page, try tightening sentences and editing wordy phrases.

Type Style

If a type style is not specified in the application guidelines, consider using serif typefaces for the text of your proposal and sans serif typefaces for titles and headings. Serif type styles such as Times Roman and Courier have small strokes that finish off the main stroke of a letter and make it easier to read. Sans serif type styles such as Universal and Arial, which do not have the small finishing strokes, are ideal for titles and headings because they stand off from the body of the text. Avoid using ornate type styles, such as Cloister Black and Freestyle Script, which are hard to read and distract from the content of the proposal.

Proposal Editing Tips

Successful grant seekers estimate that they spend 25 percent of their time writing the first draft of their proposal and 75 percent of their time editing it. Editing is a multistage process: edit for only one feature at a time. The multiple editing loops through the proposal ensure that all elements are presented with punch and persuasion.

Headings

Headings and subheadings act like a table of contents placed directly in your proposal text; that is, at a glance they reveal the main ideas and the organization of your proposal to the reader. Ask your program officer for a copy of the reviewer's evaluation form, and use those same headings and subheadings in your proposal. If a reviewer's evaluation form is not available, use headings and subheadings that are specific to your proposal.

Page Numbering

Place page numbers in the top right or bottom center of the pages of your proposal. Do not number the first page.

Proofreading

Proofread and proofread your proposal. Proofread your proposals in multiple readings, looking for different features on each reading. As you proofread, look at

1. Content and Organization: Does your proposal have enough substance? Are your ideas complete? Is your organization logical?
2. Clarity: Have you included appropriate transitions? Are ideas expressed clearly? Are all acronyms defined?
3. Mechanics: Are words spelled correctly, especially proper names? Are all numbers and computations accurate? Are sentences grammatically correct, including subject-verb
4. Agreement? Are sentences punctuated properly?
5. Design: Is the proposal design visually appealing? Did you include ample white space? Are headings specific to your project?

Transitions

Transitional expressions are words and phrases that signal connections among ideas; these connectors can help you achieve coherence in your writing. Common transitional words can indicate: addition (also, and then, further, moreover), example (as an illustration, for instance, in fact, that is), result (accordingly, consequently, hence, in short), and summary (finally, in conclusion, on the whole, to sum up).

White Space

Use white space to break up long copy. Ample white space makes your proposal appear inviting and user-friendly. White space can indicate that one section is ending and another is beginning, or that an idea is so central to the proposal that it needs to be set off by itself. Judicious use of white space breaks your proposal into smaller, manageable chunks of information. Even a simple use of white space between paragraphs helps the mind to see the information in that paragraph as a unit. One creative use of white space is the making of lists.

Letter Proposal

A letter proposal is a short grant proposal, usually two to four pages long. Written in letter form, it is primarily targeted to private sponsors, such as foundations and corporations, though it can be used as a pre-proposal for federal sponsors. Most federal program officers like to receive a letter proposal because it presents them with a "concept paper," or a "conceptual shell" of what you propose. However, some private sponsors use the letter proposal as a screening device and request an expanded proposal if your idea captures their interest. In either case, you face the challenge of clear, concise writing. In certain respects, a short proposal is more challenging to write than a long proposal. In seven brief sections, you must anticipate and answer the major questions that the sponsor will be asking as your letter proposal is read. Each sentence must carry a heavy load of information. To aid in the writing process, the components of a letter proposal are identified and discussed below.

Part One: Summary

Your objective is to summarize the entire proposal in one sentence. The critical elements of the sentence include: (1) self-identification (your organizational name); (2) uniqueness (your claim to fame); (3) sponsor expectation (what you want them to do); (4) budget request (how much money you want); and (5) project benefit (major project outcomes).

Part Two: Sponsor Appeal

Your objective is to explain why you are approaching this sponsor. Conduct background research on the sponsor to determine prior funding patterns, usually available in annual reports and tax records. Identify values that the sponsor seems to cherish as evidenced by their funding patterns, e.g., high-risk projects not normally funded by the government, cutting-edge research, demonstration projects with a national impact, or low cost/high benefit projects.

Part Three: Problem.

Your objective is to briefly summarize the current problem. Focus the problem or need statement from the sponsor's perspective, not yours. Funding your project is not their end goal. You must show how funding your project can be a means for them to reach their end goal their mission. Remember that a need is really a gap between what is and what ought to be. Document that gap with statistics, quotations, reasoning, or surveys and express it in human terms. Limit your documentation to brief but clear statements. Beware of the excessive use of statistics, which only confuses the reader.

Part Four: Solution

Your objective is to describe your approach to the problem. Summarize the objectives that you will meet with your approach. Convey confidence that you can close the gap between what is and what ought to be. You can detail your precise methodology in a one-page attachment by use of a time-and-task chart. This visual device lets reviewers know exactly who will be doing what and when. Do not include extensive methodological detail in the letter proposal.

Part Five: Capabilities

Your objective is to establish your credentials to do the project. More precisely, your job is to establish three types of credibility: you have a (1) credible organization proposing a (2) credible idea to be directed by a (3) credible project director. You must demonstrate what is unique about your group in order to show that you can solve this problem.

Part Six: Budget

Your objective is to request a specific amount (in dollars) in the proposal. Ask for a precise amount. Base your request on the review of tax records or other giving references so you are asking for a reasonable amount as viewed by the sponsor. Express your request in meaningful units, e.g., hours of instruction, numbers of students, or healthy patients. If you plan to submit this or a similar proposal to other sponsors as well, mention this.

Part Seven: Conclusion

Your objective is to identify the desired action you wish the sponsor to take. Avoid the hackneyed "We'd be happy to talk with you further about this. Please call if you want more information." Identify a contact person for more details if requested. Have a "heavyweight" (?? Influential? Resourceful?) sign the letter.

Grant Review and Funding Decisions

Once your proposal is submitted, it will undergo review. Whether your proposal is funded or declined, you should plan to deal with either outcome option. Further, since your proposal represents a valuable piece of intellectual property, consider submitting it to other sponsors as well while you are awaiting a funding decision.

Review Criteria

Proposal review usually covers five basic areas: scope of work, personnel, facilities, track record, and budget information. Experienced proposal writers often conduct proposal review sessions within their organizations prior to formal submission using the same evaluation form that their actual reviewers will use.

Dealing with Grant Decisions

Planned reactions become planned options. How do you plan to behave if your proposal is funded? Rejected? What are your options? When you have a powerful itch, it is almost unbearable waiting to get it scratched! Having to wait to get what you want demands patience and tolerance unless you have planned options. Patient people turn to other activities to meet other needs while they are waiting for grant decisions. This keeps them strong and in control. Strong people wait a lot. It may take many months before the decision on your proposal is made.

At some point, you will whether your proposal was successful. Besides getting started, there are certain things you should do. For example, if you were successful, request a copy of the reviewer comments, if allowed by the sponsor. Ask the program officer about common mistakes other grantees make so you don't fall into the same trap. Ask how you can be a good steward of their money. Clarify the submission deadlines for technical and financial reports. You can keep your program officer very happy if you submit your reports on time. Invite your program officer to come and visit you. Add your program officer to your organizational public relations list for information about your agency. If you were turned down by the sponsor, thank the source for considering the proposal. Ask what can be done to improve the proposal. If it is their policy, request reviewer comments, particularly verbatim comments; otherwise you may only receive summary comments, which are less specific. Ask if you should reapply next year. Use this as an opportunity to build a relationship with the sponsor for the next submission cycle. Periodically send a photocopy of articles or publicity with a note: "Thought you might be interested in this." Invite them to your agency to get to know you better. Avoid making them feel as if you only need them at submission time.

Multiple Submissions

One of the things you should do while waiting to hear from your first sponsor is submit your proposal to other sponsors. This is commonly done and, indeed, expected by sponsors. However, you are ethically obligated to tell a sponsor that you have submitted a similar proposal to a different sponsor. This will not jeopardize the likelihood of getting your proposal funded. In fact, it could help as there is a close communications network among sponsors with similar interests. Co-funding is not uncommon in some cases; that is, several sponsors may contribute to the total project cost. Finally, engaging in multiple submissions communicates to sponsors that you are seriously committed to your project and are willing to exert considerable effort to secure funding.

The Problem: Spiraling Tensions

Despite proactive community relations programs, an unchecked tension exists between municipal police and minority community members. Many factors contribute to the growing minority community police tensions: increasing complexity of urban life, unemployment discrimination, and housing problems. Although these nationwide social problems were not created by the police, the police must cope with the consequences of these problems. This vast social dislocation spawns minority attitudes of prejudice and contempt. To counterbalance these problems, many police communities have adopted public relations programs to "sell" their departments to the minority communities without the concomitant need to be ready to work with those communities. As a result, there is an ever-widening gap between present and potential minority community acceptance of police behavior.

The Solution: Evaluating Police Community Relations Bureaus

Successful claims regarding the effectiveness of police community relations bureaus remain undocumented. Police departments are latching on to a new fad without understanding the key components of a police community relations program. Some features of the bureau approach work; others don't. The goal of this project is to identify the successful features of existing bureaus, so that success can be delivered more quickly to police departments serving substantial numbers of minority citizens.

Steps to Writing a Successful Project Proposal – an example

Master of Counselling (MC) Program

University of Lethbridgei

Office of Graduate Studies & Research in Education Requirements

The Office of Graduate Studies and Research in Education requires a written proposal (5 – 7 pages for MC students, excluding references, double spaced with headings as noted below) in APA format. For steps in the approval process students and supervisors should refer to the Guidelines for Completing the Project.

Purposes of a Proposal

The project proposal

- allows the writer to clarify what it is he/she wants to do, why and how he/she wants to do it,
- presents what she/he wants to do in the manner and timeframe proposed, and
- once approved, provides a written contract between the student and the project supervisor

Components of a Project Proposal

1. Title

The title of the project is very significant. It will go on the spine of the published document when it is bound and becomes accessible in the University of Lethbridge Library. The title is one of the ways that people using the University of Lethbridge electronic search engine will locate the project in the holdings. The title must be clear, appropriate for the topic and less than 45 characters, including spaces and punctuation.

2. Introduction (WHAT is this about?)

The Introduction to the project provides a general introduction to the phenomena or issue of interest, and is usually contained in 2 pages. The issue or problem under investigation is described, and background and/or context for understanding the nature of the issue is provided. In writing this section, students should provide answers to two main questions:

- What is the project all about?
- Why is the project important or worthwhile?

The Introduction will also typically conclude with a brief description of the structure of the remainder of the document.

3. Research Question (WHAT am I trying to find out? WHY?)

Every project must state a research question or a statement on what is the intent of the project. It is not a "null hypothesis statement" but rather you are stating the big, overarching question that is guiding your study. Several smaller questions may even be nested in the larger one. The research question and the title are two required elements of all proposals. The connections between the two must be obvious. (Hopefully, they are obvious to the writer, but they may not be to the audience.) Where the research question appears in the proposal is something each writer must decide. It could appear in the introduction; it could follow the introduction or the background or the literature review or it could appear within any of the above sections. Just don't leave it until the reference section!

4. Method (HOW? WHO? WHERE?)

(This section must make sense within the context of the document and be linked with the sections preceding it.)

In this section provide a clear, explicit and thorough description of how you will complete your project and the timetable for completing each step. For example, what databases will you be using during your literature review? What search terms and exclusion criteria will you be using? When will you be starting your literature review and when will it be finished? It is the writer's responsibility to ensure that the proposal is clear about what is being proposed, with whom, where and when. (WHY should already have been explained.) Approximately 1-2 paragraphs are suitable for this section. A project does not involve interviewing people or collecting raw data as this type of collection of data would require ethics approval and a host of other documents would need to be created such as informed consent. This section needs to include a statement that you will adhere to an approved code of ethics of.

5. Project (What do I do when I'm done doing what I said I would do?)

Describe what the final product will look like. For example, if it is a manual, provide details about the manual's length, formatting style, number of lessons, etc. If the product is to be something creative and other than text, provide examples of what the final product might look like. If the proposal includes writing a series of fictionalized autobiographical accounts, a sample should be included in the proposal. If the proposal includes a website describe it.

Title Page

Acknowledgements Page

Abstract

Table of Contents

Chapter 1: Overview (proposed number of pages :)

Sub headings:

a.

b.

c.

Chapter 2: Review of Literature (proposed number of pages:)

Sub headings:

a.

b.

c.

Chapter 3: Methods (proposed number of pages:)

Chapter 4: Outcome

a.

b.

c.

Chapter 5: Synthesis

a. Strengths

b. Limitations

c. Recommended Future Research

Timeline

Include a draft timeline from start to finish. Start with the proposed convocation date and work backwards to revising the final draft of project (expect at least two to three rewrites). Availability of the supervisor is also very important factor to be considered in to your timeline. References to anything cited in the text of the proposal must adhere to APA guidelines. (As a matter of fact the entire proposal needs to adhere to APA guidelines). APA formatting requires only those materials cited or referred to in the text be listed in the References. A separate section entitled "Bibliography" lists other materials (books, journal articles, etc.) related to the project but not specifically referred to in the document. Buy an APA manual. The $40.00 is worth the time and frustration it will save or find an editor who knows APA and will edit/proofread your text for APA formatting. A Few Final Words of Encouragement and Wisdom .

For most projects, one third of the entire time allotted to doing the project can be taken up with writing the proposal. It is a creative, thinking, clarifying, explorative process, at the end of which you will have created/produced a persuasive and realistic document which is as definitive as possible without being a straight jacket.

Chapter 18
Technical Report Writing

Technical Report

The major focus of this technical writing course is the technical report. Just about everything you study and everything you write is geared toward preparing you to write this final report. The early, short assignment involving instructions or descriptions and the like give you practice using headings, lists, notices, and graphics; in handling numbers and abbreviations; and of course in producing good, clear, well-organized writing.

For many students, the technical report is the longest document they've ever written. It normally involves some research; often the information comes not only from published sources in the library, but also from sources outside the library, including un-published things such as interviews, correspondence, and video tapes. It may also be the fanciest document: it uses binding and covers and has special elements such as a table contents, title page, and graphics.

As you think about what you want to write about for this project, don't shy away from topics you are curious about or interested in, but don't know much about. You don't need to do exhaustive research; normally, you can pull together information for an excellent report from several books and half-dozen articles. Our real focus is the writing: how well adapted to a specific audience it is, how clear and readable it is, how it flows, how it's organized, how much detail it provides. We are also focused on format: how well you use headings, lists, notices; how well you incorporate graphics; how well you handle the front- and back-matter elements; and how nice a job you do of turning out the final copy of the report.

You don't need a fancy laser printer and you don't need to be a trained graphic artist to produce a fine-looking report. A simple typewriter or dot-matrix printer, scissors, tape, whiteout, a good-quality photocopier, and access to nice (but inexpensive) binding are all you need. Plan on doing a first-rate job on the report; remember that past students have shown prospective employers their reports and have benefited by doing so.

Your job in this unit then is defining the following:

- **Report topic**: Decide what subject you are going to write on; narrow it as much as possible.
- **Report audience**: Define a specific person or group of people for whom you are going to write the report. Define the circumstances in which this report is needed.
- **Report purpose**: Define what the report will accomplish what needs of the audience it is going to fulfill.
- **Report type**: Decide on the type of report for example, technical background report, feasibility report, instructions, or some other.

You can do these in any order: for some people, it helps to start by defining an audience or a report type. For others, beginning by picking a topic is more stimulating. Once you have defined these elements, you can start testing your report-project ideas by asking yourself these questions:

- Is there hard, specific, factual data for this topic?
- Will there be at least one or two graphics?
- Is there some realistic need for this report?

Types of Technical Reports

Technical-background report

The background report is the hardest to define but the most commonly written. This type of technical report provides background on a topic for example, solar energy, global warming, CD-ROM technology, a medical problem, or U.S. recycling activity. However, the information on the topic is not just for anybody who might be interested in the topic, but for some individual or group that has specific needs for it and is even willing to pay for that information. For example, imagine an engineering firm bidding on a portion of the work to build a hemodialysis clinic. The engineers need to know general knowledge about renal disease and the technologies used to treat it, but they don't want to have to go digging in the library to find it. What they need is a technical background report on the subject.

Instructions

These are probably the most familiar of all the types of reports. Students often write backup procedures for the jobs they do at their work. Others write short user manuals for an appliance, equipment, or program. If there is too much to write about, they write about some smaller segment--for example, instead of instructions on using all of WordPerfect, just a guide on writing macros in WordPerfect.

Feasibility, recommendation, and evaluation reports

Another useful type of report is one that studies a problem or opportunity and then makes a recommendation. A feasibility report tells whether a project is "feasible" that is, whether it is practical and technologically possible. A recommendation report compares two or more alternatives and recommends one (or, if necessary, none). An evaluation or assessment report studies something in terms of its worth or value for example, a college might investigate the feasibility of giving every student an e-mail address and putting many of the college functions online. The same college might also seek recommendations on the best hardware and software to use (after the feasibility report had determined it was a good idea). In practice, however, it's hard to keep these two kinds of reports distinct. Elements of the feasibility and recommendation report intermingle in specific reports but the main thing is to get the job done.

Primary research report

Primary research refers to the actual work someone does in a laboratory or in the field in other words, experiments and surveys. You may have written a "lab report," as they are commonly called, for one of your previous courses. This is a perfectly good possibility for the technical report as well. In this type of report, you not only present your data and draw conclusions about it, but also explain your methodology, describe the equipment and facilities you used, and give some background on the problem. You can modify this type by summarizing other primary research reports.

Technical specifications

In this report type, you discuss some new product design in terms of its construction, materials, functions, features, operation, and market potential. True specifications are not much on writing the text is dense, fragmented; tables, lists, and graphics replace regular sentences and paragraphs whenever possible. Thus, specifications are not a good exercise of your writing abilities. However, you can write a more high-level version one that might be read by marketing and planning executives.

Report length proposal

As you may be aware, proposals can be monster documents of hundreds or even thousands of pages. Most of the elements are the same, just bigger. Plus elements from other kinds of reports get imported such as feasibility discussion, review of literature, and qualifications; these become much more elaborate. The problem with writing a proposal in our technical writing class is coordinating it with the proposal you write at the beginning of the semester. Several students have set up scenarios in which they proposed internally to write an external proposal, in which they went after some contract or grant.

Business prospectus

If you are ambitious to run your own business, you can write a business prospectus, which is a plan or proposal to start a new business or to expand an existing one. It is aimed primarily at potential investors. Therefore, it describes the proposed business, explores the marketplace and the competition, projects revenues, and describes the operation and output of the proposed business.

Audience and Situation in Technical Reports

A critical step in your early report planning is to define a specific audience and situation in which to write the report. For example, if you wanted to write about CD audio players, the audience cannot be an ambiguous entity like "anybody who is considering purchasing a CD player." You have to define the audience in terms of knowledge, background, and need for the information.

- Why does the audience need this information?
- How will readers get access to this information?

Just as critical to the planning process is defining the situation. When you define audience, you define who the readers are, what they know or don't know in relation to the topic, what experience or background they have in relation to the topic, and why they want or might need the information. Sometimes this leaves out a critical element: just what are the circumstances that bring about the need for the information.

Topics for Technical Reports

Just about any topic can be worked into a good technical report project. Some are a little more difficult than others and that's where your instructor can help. This indeed is the purpose of the proposal assignment that gives your instructor a chance to see what you want to do and to guide you away from problems such as the following:

Editorializing

For the report project, avoid editorial topics. For example, don't attempt to write a technical report on the *pros and cons* of gun control, abortion, marijuana, and the like. You can, however, develop these topics: for example, describe the chemical, physiological aspects of marijuana or the medical techniques for abortion or the developmental stages of the fetus. These get into substantial technical areas. However avoid editorializing there are other courses where you can do this.

Fuzzy topics

Some topics just don't work, for some reason. For example, dream analysis can be very fuzzy and nebulous. So can UFOs. You want your report to have hard factual data in it. The preceding topics are difficult to pin down this way. However, good reports have been written on the apparatus used in dream research laboratories. Maybe somebody can even figure out a good way to handle UFOs.

Tough technical topics

As mentioned earlier, don't shy away from interesting topics that you don't feel you know enough about. No one expects a doctoral thesis. Use the report project as a chance to learn something new. Of course, its common sense that we often write better about things we know about. If this is a concern for you, look around you in your work, hobbies, or academic studies. The word technical refers to anybody of specialized knowledge.

General Characteristics of Technical Reports

You're probably wondering what this technical report is supposed to look like. Here is a brief review of some of the chief characteristics of the technical report:

- **Graphics:** The report should have graphics. Graphics include all kinds of possibilities, as a later chapter in this book will show. If you can't think of any graphics for your report project, you may not have a good topic. Get in touch with your instructor, who can help you brainstorm for graphics.
- **Factual detail**: The report should be very detailed and factual. The point of the report is to go into details, the kind of details your specific audience needs.
- **Information sources:** Your report should make use of information sources. These may include not only books and articles that can be found in libraries but also technical brochures, interviews or correspondence with experts, as well as first-hand inspections. If you don't believe any information sources are necessary for your report project, contact your instructor.
- **Documentation:** When you use borrowed information in your technical report, be sure to cite your sources. The style of citing your sources (also called "documenting" your sources) used in this course is called the number system.
- **Realistic audience and situation:** The report must be defined for a real or realistic group of readers who exist in a real or realistic situation. Most students invent an audience and situation. And the audience can't merely be something like "anybody who might be interested in global warming." Instead, it has to be real, realistic, and specific: for example, "Texas Coastal Real Estate Developers Association, interested in reliable information on global warming, to be used to aid in long-range investment planning."
- **Headings and lists**: The report should use the format for headings that is required for the course, as well as various kinds of lists as appropriate.
- **Special format**: The technical report uses a rather involved format including covers, binding, title page, table of contents, list of figures, transmittal letter, and appendixes. These have to be prepared according to a set standard, which will be presented in a later chapter.
- **Production:** The technical report should be typed or printed out neatly. If graphics are taped in, the whole report must be photocopied, and the

photocopy handed in (not the original with the taped in graphics). The report must be bound in some way.

- **Length**: The report should be at least 8 double spaced typed or printed pages (using 1-inch margins), counting from introduction to conclusion. This is the minimum required length. However a report of this length is rather skimpy. There is no real maximum length, other than what your time, energy, and stamina can handle. But remember that sheer weight does not equal quality (or better grade). If you get into a bind with a report project that would take too many pages, consult your instructor there are numerous tricks we can use to cut it down to size.
- **Technical content:** You must design your report project in such a way that your poor technical-writing instructor has a chance to understand it in other words, you must write for the non-specialist. Also, at some point, you may get concerned about the technical accuracy of your information. Remember that this is a writing course, not a course in engineering, nursing, science, electronics, or the like. Make a good faith effort to get the facts right, but don't go overboard.

Checklist for the Technical Report

- Use the following questions to ensure that your technical report is structured properly according to our specifications:
- Do you include all the required components in the required order, for example, transmittal letter, followed by title page, followed by figure list, and so on?
- Do you address your report to a real or realistic audience that has a genuine need for your report? Do you identify in the introduction what background the audience needs to read and understand your report?
- Does your report contain specific, factual detail, focused on the purpose of the report and the needs of the audience and aimed at their level of understanding?
- Does your report accomplish its purpose? Is that purpose clearly stated in the introduction?
- Does your report use information sources and do you properly document them?
- Does your report use the format for headings that is standard for this course?
- Does your report use the format for lists that is standard for this course?
- Does your report use graphics and tables? Does your report use the format for graphics and tables that is standard for this course? Specifically, are your figure titles (captions) to our class specifications?
- Is page 1 of your introduction designed according to the standard for this course?

- Does every new section (which starts with a first-level heading) start on a new page? Have you checked for widowed headings (headings that start at the very bottom of a page)? Stacked headings (two or more consecutive headings without intervening text)? Lone headings (a single heading within a section)? Parallelism in the phrasing of headings?
- Does the title page of your report include a descriptive abstract, and is it written according to the specifications of the chapter mentioned in the abstracts?
- Do you include an informative abstract in your report; is it positioned properly in relation to the other report components; and is it written according to the specifications in the chapter on abstracts? Specifically, does your informative abstract summarize the key facts and conclusions of your report rather than act as just another introduction or descriptive abstract?
- Does the introduction of your report include the elements necessary in good introductions, such as audience, overview, and purpose? Do you avoid the problem of having too much background in the introduction, or having an introduction that is all background?
- Return to the table of contents for the Online Technical Writing Course Guide.

A good report is easy to recognize. It has a precise and informative title, a clear and well organized layout, is easy to handle, and opens flat to reveal both text and diagrams. It is written in a fluent and concise style; headings clearly indicate the content of each section; and diagrams, tables and graphs are clear and labeled. Reading a well written report is a pleasure. We cannot give you a detailed absolute, immutable set of rules for producing reports because the report's purpose should determine what it contains and how it is constructed.

Engineers and scientists write formal reports for many reasons, including the documentation of experiments and designs. As an engineer or scientist working on the design of an airplane seat, you might write several formal reports. One formal report might propose a new design for the seat. A second formal report might update the progress on the construction of a test seat. Yet another formal report might document tests performed on the design. Still another formal report would assess whether the new design should replace the existing design.

What distinguishes a formal report from an informal reporting of information? The answer lies not in the topics of formal reports, but in the expectations of the audiences for formal reports. In a formal report, the audience expects a methodical presentation of the subject that includes summaries of important points as well as appendices on tangential and secondary points. The readers for a formal report are often two or more distinct audiences. These distinct audiences include professionals specializing in the report's subject matter, professionals not specializing in the report's subject matter, and managers overseeing the report's subject matter.

Some guidelines

1. The READER is the most important person.
2. Keep the report as short as possible.
3. Organize the report for the reader's convenience.
4. Give complete and correct references.
5. Write in a concise and fluent style.
6. Put diagrams in the right place for the reader with the right titles.
7. Give the whole picture in miniature form in the summary.
8. Check the report thoroughly for technical errors, typing errors and inconsistencies.
9. Make the report look as good as it surely is.
10. The READER is the most important person.

Yes, 1 and 10 are the same. This is not a mistake, we have repeated 1 because it is one guideline that overrides all the others. A report is written once but are (you hope) read many times.

Aims You need to have a clear set of aims before starting to write a report. In formulating the aims you need to consider:

- What is the intended content of the report;
- For whom is it being written; and
- Why is it being written?

You will find it helpful to list a set of objectives under each of these headings before starting to write a report.

What is the intended content?

This depends on what you were asked to do: build a mathematical model, design a controller; compare design techniques; or write some software. For example, if you were asked to design a controller you might list the intended contents as:

1. Description of system to be controlled.
2. Choice of design method.
3. Details of design procedures and justification for decisions.
4. Assessment of performance of the design.

For whom is it being written?

This is clearly important in terms of setting aims. You are writing a dissertation so that we can assess your ability. Your aim is to obtain a high mark. This you can subdivide into several.

Objectives

1. To demonstrate your technical competence.
2. To show your organizational skills.
3. To demonstrate that you work hard and have initiative.
4. To demonstrate that you can write clearly and concisely.
5. To demonstrate that you can produce a good report.

Producing the Report

Having established the aims of the report you can begin collecting material: you probably have some already. You can begin organizing the report before you have collected all the material. A good way of doing this is to separate the material into three categories:

1. Obviously the most important information which must go into the report.
2. Borderline information which might be of use to some readers, or which might amplify or substantiate other more important material.
3. Information that you find interesting (or cannot bear to throw away) but which is not relevant to the report.

Beginning and End

A report has to start and has to end but you do not necessarily start writing at the beginning and stop writing at the end. Before elaborating further let us look at the general structure:

- Preliminaries.
- Body of the report.
- References (or bibliography).
- Appendices.

Frequently a good starting point is to decide what is to go in appendices and to assemble or write each appendix. The appendices should be used to remove information from the body of the report that is not essential to the majority of the readers. For example, details of how to use a particular computer program to obtain a controller design. This is valuable information for anyone who subsequently wishes to use the program but is not relevant to the reader who is interested in how the controller performs. Another use of appendices is to hold program listings or detailed tables of results. An appendix can also be used to contain background information which most readers will know but a few need to be told. For example, detailed derivations of formulae or theorems.

Format distinguishes formal reports from an informal reporting of information

A well crafted formal report is formatted such that the report's information is readily accessible to all the audiences. For that reason, formal reports are split into different sections. One way to group these sections is in terms of the front matter, main text, and back matter. The front matter, which presents preliminary information for the report, serves to orient all intended audiences to what the report contains. The text portion of the formal report is the report's "story" and contains the introduction, discussion, and conclusion of the report. The text delivers a methodical explanation of the report's work to the report's primary audience. The portion of the report, which contains the appendices, glossary, and references, serves to provide secondary information to all readers as well as primary information to secondary readers.

Front Cover

The front cover of a formal report is important. The front cover is what people see first. When the report sits flat on a desk, the front cover is in view. Therefore, the front cover should contain the report's title and the author's name. Because reports are often revised and republished, the front cover should also contain the date of publication. The front cover has no page number. Space the title, name, and date to achieve a nice balance on the page. If possible, type the title in a larger font size than the name and date. Use initial capitals for the title.

Title Page

The title page for a formal report often contains the same information as is on the cover. In some formats, there is a summary included. Most often, because of space restrictions, that summary is descriptive (more like a table of contents in paragraph form). Sometimes, though, this initial summary is informative and geared toward the technical audience of the report. In such situations, that summary is often named an "Abstract."

Contents Page

The table of contents includes the names of all the headings and subheadings for the main text. In addition, the table of contents includes names of all headings (but not subheadings) in the front matter and back matter. For instance, the contents page includes listings for the appendices (including appendix titles), the glossary, and the references.

Summary

Perhaps no term in engineering writing is as confusing as the term "summary." In general there are two types of summaries: descriptive summaries and informative summaries. A descriptive summary describes what kind of information is in the report; it is a table of contents in paragraph form. An informative summary is a synopsis of the text portion of the report; it is analogous to a baseball box score. Unfortunately, few people use these terms to name the summaries in reports. The names you're likely to run into are "abstract," "executive summary," and plain old "summary."

Abstract

An "abstract" usually, but not always, refers to a summary written to a technical audience, and depending on its length can be either descriptive, informative, or a combination of both. As you might imagine, short abstracts are typically descriptive and longer abstracts are typically informative. Abstracts generally do not include illustrations. Sometimes the word "abstract" is preceded by the word "descriptive," which is usually a clue that you should write a descriptive summary written to a technical audience. Other times the word "abstract" is proceeded by the word "technical," which is usually a clue that an informative summary written to a technical audience is called for.

An "executive summary "is the most consistently defined term it refers to an informative summary written to a management audience. Since it is informative, it includes the most important results and conclusions of the document. Since it is written to a management audience, it includes enough background for the manager to understand those results and conclusions. Stylistically, it is tailored so that a manager can read it quickly and garner what happened in the report. Whether it contains illustrations or not depends on the format.

The 'catch all' term "summary" can be used to refer to anything, either a descriptive summary, or an informative summary, a summary with illustrations, or a summary without any such images and illustrations. So how do you proceed if a company, laboratory, or professor asks you to write a "summary" for a formal report? Well, the best thing to do is to look at examples of summaries in previous reports for that company, laboratory, or professor. In formatting the main summary of your report, treat the name ("Abstract," "Executive Summary," or whatever your instructor prescribes) as a major heading. If illustrations are allowed, number them using the abbreviation of the summary's title. For instance, if the summary is named an "Executive Summary," number the illustrations ES-1, ES-2, and so on. Equations should be numbered in the same way.

Main Text

The text portion of your formal report contains the introduction, discussion, and conclusion of your report. Begin all major headings ("Introduction," for example) on a new page. Use Arabic numerals for numbering pages of the text and begin the first page of your text.

Introduction

The introduction of a report prepares readers for understanding the discussion of the report. Like the title and summary, the introduction is written for the widest audience possible.

Discussion

The discussion or middle part of the report is the story of your work. You do not necessarily present results in the order that you understood them, but in the order that is easiest for your readers to understand them. In your discussion, you not only present results, but you also evaluate those results. Note that you do not generally use the word "Discussion" as the title for the major headings in this part of the report. Rather, you choose titles that reflect the content of the sections.

Conclusion

The conclusion section analyzes for the most important results from the discussion and evaluates those results in the context of the entire work. In your conclusion, you often make recommendations based on those evaluations. The conclusion is much like an informative summary except for one thing-in the conclusion, you are writing to an audience who has read your report. Note that you do not necessarily have to use the word "Conclusion" as the title for this section. Depending on the situation, you might for example choose "Conclusions and Recommendations." In still other situations, your conclusion might span two sections.

Back Matter

The back matter portion of your report contains your appendices, glossary, and references. The back matter portion usually begins from the page following the conclusion.

Appendices

Use appendices to present supplemental information for secondary readers. When the occasion arises in the text, refer readers to information in the appendix. This section compares three software pages to run tests on Homodyne's blood analyzer. Homodyne's blood analyzer performs test for such diseases as syphilis, tuberculosis, and the AIDS virus. The analyzer has a complex design, which is discussed in Appendix B. The three software packages considered in this report are...

Treat each appendix as a major heading. If you have only appendix, call it the "Appendix." If you have more than one appendix, number the appendices with letters: Appendix A, Appendix B, and so on. As with all major headings, skip three returns from the top margin and center the appendix name and title. Illustrations in appendices are numbered as follows. In both a single appendix and in an Appendix A, figures and tables are numbered A-1, A-2, and so on. Equations in Appendix A are numbered in the same way. In an Appendix B, illustrations and equations follow a B sequence.

Glossary

Use a glossary to define terms for secondary readers. Arrange terms in alphabetical order. Use italics or underlines to key readers to terms that the glossary will define. Footnote the first italicized or underlined term in the text and key readers to the location of glossary, where that term and all future underlined or italicized terms will be defined. Use a reverse indent for each definition and treat each definition as a separate paragraph.

References

Use a reference page to list alphabetically the references of your report. Also skip a space between each citation.

Structure of a report

The following should roughly be the structure of a report. Note that these are just guidelines, not rules. You have to use your intelligence in working out the details of your specific writing.

- **Title and abstract:** These are the most-read parts of a report. This is how you attract attention to your writing. The title should reflect what you have done and should bring out any eye-catching factor of your work, for good impact.

  The abstract should be short, generally within about 2 paragraphs (250 words or so total). The abstract should contain the essence of the report, based on which the reader decides whether to go ahead with reading the report or not. It can contain the following in varying amounts of detail as is appropriate: main motivation, main design point, essential difference from previous work, methodology, and some eye-catching results if any.

- **Introduction:** Most reports start with an introduction section. This section should answer the following questions (not necessarily in that order, but what is given below is a logical order). After title/abstract introduction and conclusions are the two mainly read parts of a report.

  - ❖ What is the setting of the problem? This is, in other words, the background. In some cases, this may be implicit, and in some cases, merged with the motivation below.
  - ❖ What exactly is the problem you are trying to solve? This is the problem statement.
  - ❖ Why is the problem important to solve? This is the motivation. In some cases, it may be implicit in the background, or the problem statement itself.
  - ❖ Is the problem still unsolved? This constitutes the statement of past/related work crisply.
  - ❖ Why is the problem difficult to solve? This is the statement of challenges. In some cases, it may be implicit in the problem statement. In others, you may have to say explicitly as to why the problem is worthy of a BTech/MTech/PhD, or a semester project, as the case may be.
  - ❖ How have you solved the problem? Here you state the essence of your approach. This is of course expanded upon later, but it must be stated explicitly here.
  - ❖ What are the conditions under which your solution is applicable? This is a statement of assumptions.
  - ❖ What are the main results? You have to present the main summary of the results here.
  - ❖ What is the summary of your contributions? This in some cases may be implicit in the rest of the introduction. Sometimes it helps to state contributions explicitly.

- How is the rest of the report organized? Here you include a paragraph on the flow of ideas in the rest of the report. For any report beyond 4-5 pages, this is a must.

The introduction is nothing but a shorter version of the rest of the report, and in many cases the rest of the report can also have the same flow. Think of the rest of the report as an expansion of some of the points in the introduction.

- **Background:** This is expanded upon into a separate section if there is sufficient background which the general reader must understand before knowing the details of your work. It is usual to state that "the reader who knows this background can skip this section" while writing this section.
- **Past/related work:** It is common to have this as a separate section, explaining why what you have done is something novel. Here, you must try to think of dimensions of comparison of your work with other work. For instance, you may compare in terms of functionality, in terms of performance, and/or in terms of approach. Even within these, you may have multiple lines of comparison functionality-1, functionality-2, metric-1, metric-2, etc.

  Although not mandatory, it is good presentation style to give the above comparison in terms of a table; where the rows are the various dimensions of comparison and the columns are various pieces of related work, with your own work being the first/last column.

  While in general you try to play up your work with respect to others, it is also good to identify points where your solution is not so good compared to others. If you state these explicitly, the reader will feel better about them, than if you do not state and the reader figures out the flaws in your work anyway.
- **Technical sections:** The main body of the report may be divided into multiple sections as the case may be. You may have different sections which delve into different aspects of the problem. The organization of the report here is problem specific. You may also have a separate section for statement of design methodology, or experimental methodology, or proving some lemmas in a theoretical paper.

  The technical section is the most work-specific, and hence is the least described here. However, it makes sense to mention the following main points:

  - **Outlines/flow**: For sections which may be huge, with many subsections, it is appropriate to have a rough outline of the section at the beginning of that section. Make sure that the flow is maintained as the reader goes from one section to another. There should be no abrupt jumps in ideas.
  - **Use of figures**: The cliché "a picture is worth a thousand words" is appropriate here. Spend time thinking about pictures. Wherever

necessary, explain all aspects of a figure (ideally, this should be easy), and do not leave the reader wondering as to what the connection between the figure and the text is.

- **Terminology:** Define each term/symbol before you use it, or right after its first use. Stick to a common terminology throughout the report.

- **Results:** This is part of the set of technical sections, and is usually a separate section for experimental/design papers. You have to answer the following questions in this section:

  - What aspects of your system or algorithm are you trying to evaluate? That is, what are the questions you will seek to answer through the evaluations?
  - Why are you trying to evaluate the above aspects?
  - What are the cases of comparison? If you have proposed an algorithm or a design, what do you compare it with?
  - What are the performance metrics? And, why?
  - What are the parameters under study?
  - What is the experimental setup? Explain the choice of every parameter value (range) carefully.
  - What are the results?
  - Finally, why do the results look the way they do?

  The results are usually presented as tables and graphs. In explaining tables and graphs, you have to explain them as completely as possible. Identify trends in the data. Does the data prove what you want to establish? In what cases are the results explainable, and in what cases unexplainable if any?

  While describing a table, you have to describe every row/column. And similarly while describing a graph; you have to describe the x/y axes. If necessary, you have to consider the use of log-axes. If you are presenting a lot of results, it may be useful to summarize the main take-away points from all the data in a separate sub-section at the end (or sometimes even at the beginning) of the results section.

- **Future work:** This section in some cases is combined along with the "conclusions" section. Here you state aspects of the problem you have not considered and possibilities for further extensions.

- **Conclusions:** Readers usually read the title, abstract, introduction, and conclusions. In that sense, this section is quite important. You have to crisply state the main take-away points from your work. How has the reader become smarter, or how has the world become a better place because of your work?

A technical report is a formal report designed to convey technical information in a clear and easily accessible format. It is divided into sections which allow different readers to access different levels of information. This guide explains the commonly accepted format for a technical report; explains the purposes of the individual sections; and gives hints on how to go about drafting and refining a report in order to produce an accurate, professional document.

A technical report should contain the following sections

Section	*Details*
Title page	Must include the title of the report. Reports for assessment, where the word length has been specified, will often also require the summary word count and the main text word count
Summary	A summary of the whole report including important features, results and conclusions
Contents	Numbers and lists all section and subsection headings with page numbers
Introduction	States the objectives of the report and comments on the way the topic of the report is to be treated. Leads straight into the report itself. Must not be a copy of the introduction in a lab handout.
The sections which make up the body of the report the	Divided into numbered and headed sections. These sections separate different main ideas in a logical order
Conclusions	A short, logical summing up of the theme(s) developed in the main text
References	Details of published sources of material referred to or quoted in the text (including any lecture notes and URL addresses of any websites used.
Bibliography	Other published sources of material, including websites, not referred to in the text but useful for background or further reading.
Acknowledgements	List of people who helped you research or prepare the report, including your proofreaders
Appendices (if appropriate)	Any further material which is essential for full understanding of your report (e.g. large scale diagrams, computer code, raw data, specifications) but not required by a casual reader

Presentation

For technical reports required as part of an assessment, the following presentation guidelines are recommended;

Script	The report must be printed single sided on white A4 paper. Hand written or dot-matrix printed reports are not acceptable.
Margins	All four margins must be at least 2.54 cm
Page numbers	Do not number the title, summary or contents pages. Number all other pages consecutively starting at 1
Binding	A single staple in the top left corner or 3 staples spaced down the left hand margin. For longer reports (e.g. year 3 project report) binders may be used.

Planning the report

There are some excellent textbooks contain advice about the writing process and how to begin. Here is a checklist of the main stages;

- Collect your information. Sources include laboratory handouts and lecture notes, the University Library, the reference books and journals in the Department office. Keep an accurate record of all the published references which you intend to use in your report, by noting down the following information;

Journal article

author(s)

title of article

name of journal (italic or underlined)

year of publication

volume number (bold)

issue number, if provided (in brackets)

page numbers

Book

author(s)

title of book (italic or underlined)

edition, if appropriate

publisher

year of publication

Creative phase of planning

Write down topics and ideas from your researched material in random order. Next arrange them into logical groups. Keep note of topics that do not fit into groups in case they come in useful later. Put the groups into a logical sequence which covers the topic of your report.

- Structuring the report. Using your logical sequence of grouped ideas, write out a rough outline of the report with headings and subheadings.

Writing the first draft

Who is going to read the report? For coursework assignments, the readers might be fellow students and/or faculty markers. In professional contexts, the readers might be managers, clients, project team members. The answer will affect the content and technical level, and is a major consideration in the level of detail required in the introduction.

Begin writing with the main text, not the introduction. Follow your outline in terms of headings and subheadings. Let the ideas flow; do not worry at this stage about style, spelling or word processing. If you get stuck, go back to your outline plan and make more detailed preparatory notes to get the writing flowing again.

Make rough sketches of diagrams or graphs. Keep a numbered list of references as they are included in your writing and put any quoted material inside quotation marks.

Write the Conclusion next, followed by the Introduction. Do not write the Summary at this stage.

Revising the first draft

This is the stage at which your report will start to take shape as a professional, technical document. In revising what you have drafted you must bear in mind the following, important principle;

- The essence of a successful technical report lies in how accurately and concisely it conveys the intended information to the intended reader.

During year 1, term 1 you will be learning how to write formal English for technical communication. This includes examples of the most common pitfalls in the use of English and how to avoid them. Use what you learn and the recommended books to guide you. Most importantly, when you read through what you have written, you must ask yourself these questions;

- Does that sentence/paragraph/section say what I want and mean it to say? If not, write it in a different way.
- Are there any words/sentences/paragraphs which could be removed without affecting the information which I am trying to convey? If so, remove them.

Diagrams, graphs, tables and mathematics

It is often the case that technical information is most concisely and clearly conveyed by means other than words. Imagine how you would describe an electrical circuit layout using words rather than a circuit diagram. Here are some simple guidelines;

Diagrams	Keep them simple. Draw them specifically for the report. Put small diagrams after the text reference and as close as possible to it. Think about where to place large diagrams.
Graphs	For detailed guidance on graph plotting.
Tables	Is a table the best way to present your information? Consider graphs, bar charts or pie charts. Dependent tables (small) can be placed within the text, even as part of a sentence. Independent tables (larger) are separated from the text with table numbers and captions. Position them as close as possible to the text reference. Complicated tables should go in an appendix.
Mathematics	Only use mathematics where it is the most efficient way to convey the information. Longer mathematical arguments, if they are really necessary, should go into an appendix. You will be provided with lecture handouts on the correct layout for mathematics.

The report layout

The appearance of a report is no less important than its content. An attractive, clearly organised report stands a better chance of being read. Use a standard, 12pt, font, such as Times New Roman, for the main text. Use different font sizes, bold, italic and underline where appropriate but not to excess. Too many changes of type style can look very fussy.

Headings

Use heading and sub-headings to break up the text and to guide the reader. They should be based on the logical sequence which you identified at the planning stage but with enough sub-headings to break up the material into manageable chunks. The use of numbering and type size and style can clarify the structure as follows;

Originality and plagiarism

Whenever you make use of other people's facts or ideas, you must indicate this in the text with a number which refers to an item in the list of references. Any phrases, sentences or paragraphs which are copied unaltered must be enclosed in quotation marks and referenced by a number. Material which is not reproduced unaltered should not be in quotation marks but must still be referenced. It is not sufficient to list the sources of information at the end of the report; you must indicate the sources of information individually within the report using the reference numbering system.

Information that is not referenced is assumed to be either common knowledge or your own work or ideas; if it is not, then it is assumed to be plagiarized i.e. you have knowingly copied someone else's words, facts or ideas without reference, passing them off as your own. This is a **serious offence**. If the person copied from is a fellow student, then this offence is known as collusion and is equally serious. Examination boards can, and do, impose penalties for these offences ranging from loss of marks to disqualification from the award of a degree.

This warning applies equally to information obtained from the Internet. It is very easy for markers to identify words and images that have been copied directly from web sites. If you do this without acknowledging the source of your information and putting the words in quotation marks then your report will be sent to the Investigating Officer and you may be called before a disciplinary panel.

Finalizing the report and proofreading

Your report should now be nearly complete with an introduction, main text in sections, conclusions, properly formatted references and bibliography and any appendices. Now you must add the page numbers, contents and title pages and write the summary.

The Summary

The summary, with the title, should indicate the scope of the report and give the main results and conclusions. It must be intelligible without the rest of the report. Many people may read, and refer to, a report summary but only a few may read the full report, as often happens in a professional organization.

- Purpose - a short version of the report and a guide to the report.
- Length - short, typically not more than 100-300 words
- Content - provide information, not just a description of the report.

Proof Reading

This refers to the checking of every aspect of a piece of written work from the content to the layout and is an absolutely necessary part of the writing process. You should acquire the habit of never sending or submitting any piece of written work, from email to course work, without at least one and preferably several processes of proofreading. In addition, it is not possible for you, as the author of a long piece of writing, to proofread accurately yourself; you are too familiar with what you have written and will not spot all the mistakes. When you have finished your report, and before you staple it, you must check it very carefully, yourself. You should then give it to someone else, e.g. one of your fellow students, to read carefully and check for any errors in content, style, structure and layout. You should record the name of this person in your acknowledgements.

Word processing / desktop publishing

Advantages	*Disadvantages*
Word processing and desktop publishing packages offer great scope for endless revision of a document. This includes words, word order, style and layout.	Word processing and desktop publishing packages never make up for poor or inaccurate conten
They allow for the incremental production of a long document in portions which are stored and combined later	They can waste a lot of time by slowing down writing and distracting the writer with the mechanics of text and graphics manipulation.
They can be used to make a document look stylish and professional.	Excessive use of 'cut and paste' leads to tedious repetition and sloppy writing.
They make the process of proofreading and revision extremely straightforward	If the first draft is word processed, it can look so stylish that the writer is fooled into thinking that it does not need proofreading and revision!

Two useful tips

- Do not bother with style and formatting of a document until the penultimate or final draft.
- Do not try to get graphics finalized until the text content is complete.

Report Writing Rules

- ➢ Use The Most Common Words.
- ➢ Do not use jargon or slang. The only exception is when Directly and necessarily quoting.

- Don't be conclusive. Write what you know, not what you think you know.
- Short sentences are better than long sentences. If a sentence could be made into two sentences, writing it as one sentence is wrong.
- Refer to yourself (and only to yourself) with pronouns such as I, me, my and mine (write in first person). the only allowable exception to this is when necessarily making a direct quote.
- Write who did something and then what they did, not the other way around (use active voice instead of passive voice).
- Be clear. If a reader must guess at what you meant, you're wrong. If anything you write, has more than one meaning, then you're wrong.
- Only use pronouns when it is clear to whom the pronouns refer. When in doubt, use names.
- A report should present the facts in the order in which they occurred to you (chronological order).
- One "said" per paragraph is sufficient unless there's a chance of confusion regarding who said what.
- Use specific nouns of which you have information. Use last names for suspects, victims and witnesses. Those last names are generally more specific than first names.
- Statements of witnesses, victims and suspects should refer to them by their names or with pronouns such as he, him, his, she, her and hers.
- Write in the past tense.
- Write in complete sentences.
- Don't use comparative modifiers unless a basis for comparison is given.
- Don't spell out a number and then immediately repeat it as a numeral in parentheses.
- Use correct punctuation.
- Within the statement of a suspect, victim or witness, put events in the order in which things occurred to that person
- The overall structure of a report is paragraphs arranged in the order in which the events depicted occurred to you, the writer.
- Indicate the beginning of each paragraph by skipping a line before the new paragraph. Do not skip a line until going from one type of paragraph to another.
- Do not write a report in a question-and-answer format. There should be no "I asked..." followed by "he said...." in fact, leave out every sentence beginning with "I asked..." no one cares what you asked.
- Report facts, not opinions. Inferences are acceptable, but before making an inference, give the supporting facts.

- Only quote a person if: necessarily quoting slang, what the person says is an admission or confession, there are words that help prove the crime, it is a denial, or the suspect uses embarrassing language.
- A report is a summary of the facts that appeared to be important at the time of the writing. If a fact doesn't appear to be important to the criminal prosecution, leave it out.
- Use correct spelling.
- Use correct grammar.
- Do not use labels or titles in reports. Do not call someone victim smith or suspect brown. Do not use Mr., Mrs. or Ms. in reports.
- Make sure the reader knows the source of information for any information in a report.
- Leave out the phrase "described as." it never adds information and fails to tell who did the describing.
- Use 24-hour time. 24-hour time is always four numerals without colons. 0001 hours is one minute after midnight. 1201 hours is one minute after noon.
- Report all the elements of the crime in question.
- Establish a clear time frame for the crime based on facts.
- Make sure the reader understands the perspectives of any witnesses. Were they 10 feet away? Across the street?

Chapter 19

Introduction to Technical Communication

Technical writing courses introduce you to some of the most important aspects of writing in the world of science, technology, and business in other words, the kind of writing that scientists, nurses, doctors, computer specialists, government officials, engineers, and other such people do as a part of their regular work.

To learn how to write effectively for the world of work, you'll study common types of reports, special format items such as lists and headings, simple techniques for putting graphics into reports, and some techniques for producing professional looking final copy.

Technical-writing courses build on what you've learned in other writing courses. But there's lots that is new to learn! If you currently have a job in which you do some writing, you'll discover that you can put what you learn in your technical-writing course for immediate use.

About Technical Writing

You're probably wondering what is "technical writing ". Someone might even have told you, "It's this course where they make you write about rocket science and brain surgery." Well, not really . . . . Actually, the field of technical communications is a fully professional field with degree programs, certifications, and yes! even theory. It's a good field with a lot of growth and income potential.

An introductory technical-writing course for which this book has been developed is a good way to start if you are interested in a career in this field.

However, the focus for technical-writing courses is not necessarily career as a technical writer but an introduction to the kinds of writing skills you need in practically any technically oriented professional job. No matter what sort of professional work you do, you're likely to do lots of writing and much of it technical in nature. The more you know about some basic technical-writing skills, which are covered in this guide and in technical-writing courses, the better job of writing you're likely to do. And that will be good for the projects you work on, for the organizations you work in, and most of all good for you and your career.

Technical communications or technical writing, as the course is often called is not writing about a specific technical topic such as computers, but about any technical topic. The term "technical" refers to knowledge that is not widespread, that is more the territory of experts and specialists. Whatever you're major is, you are developing an expertise you are becoming a specialist in a particular technical area. And whenever you try to write or say anything about your field, you are engaged in technical communications.

Another key part of the definition of technical communications is the receiver of the information the audience. Technical communications is the delivery of technical information to readers (or listeners or viewers) in a manner that is adapted to their needs, level of understanding, and background. In fact, this audience element is so important that it is one of the cornerstones of this course: you are challenged to write about highly technical subjects but in a way that a beginner a non specialist could understand.

This ability to "translate" technical information to non specialists is a key skill to any technical communicator. In a world of rapid technological development, people are constantly falling behind and becoming technological illiterates. Technology companies are constantly struggling to find effective ways to help customers or potential customers understand the advantages or the operation of their new products.

So relax! You don't have to write about computers or rocket science write about the area of technical specialization you know or are learning about.

About Technical Writing Courses

In technical writing courses, the primary focus is on writing the technical report, due toward the end of the semester. Just about everything you do in the course is aimed at developing skills needed to produce that report. Of course, some technical-writing courses begin with a resume and application letter (often known as the cover letter), but after that you plan the technical report, then write a proposal in which you propose to write that report. Then you write short technical papers where you get accustomed to using things like headings, lists, graphics, and special notices not to mention writing about technical subject matter in a clear, concise, understandable way that is appropriate for a specific audience.

Technical communication

Technical communication is the use of semiotic means to document or convey information. Individuals in a variety of contexts and with varied professional credentials engage in technical communication. Some individuals are designated as technical communicators. These individuals use a set of methods to research and document technical processes or products. Technical communicators may put the information they capture into paper documents, web pages, and digitally stored text, audio, video, and other media. The Society for Technical Communication defines the field as any form of communication that focuses on technical or specialized topics, communicates specifically by using technology or provides instructions on how to do something. More succinctly, the Institute of Scientific and Technical Communicators defines technical communication as factual communication, usually about products and services. Whatever the definition of technical communication, the overarching goal of the practice is to create easily accessible information for a specific audience.

Technical communication is a professional task performed by specialized employees or consultants. For example, a professional writer may work with a company to produce a user manual. Some companies give considerable technical communication responsibility to other technical professionals such as programmers, engineers, and scientists. Often, a professional technical writer edits such work to bring it up to modern technical communication standards.

To begin the documentation process, technical communicators identify the audience and their information needs. The technical communicator researches and structures the content into a framework that can guide detailed development. As the body of information comes together, the technical communicator ensures that the intended audience can understand the content and retrieve the information they need. This process, known as the 'Writing Process', has been a central focus of writing theory since the 1970s, and some contemporary textbook authors apply it to technical communication. Technical communication is important to most professions, as a way to contain and organize information and maintain accuracy.

The technical writing process can be divided into five steps:

- Determine purpose and audience
- Collect information
- Organize and outline information
- Write the first draft
- Revise and edit

Determining purpose and audience

All technical communication serves a particular purpose typically to communicate ideas and concepts to an audience, or direct an audience in a particular task. Technical communication professionals use various techniques to understand the audience and, when possible, test content on the target audience. For example, if bank workers don't properly post deposits, a technical communicator would:

- Review existing documentation (or lack thereof)
- Interview bank workers to identify conceptual errors
- Interview subject matter experts to learn the correct procedures
- Author new material that describes the correct procedures
- Test the new material on the bank workers

Similarly, a sales manager who wonders which of two sites is better for a new store might ask a marketing professional to study the sites and write a report with recommendations. The marketing professional hands the report off to a technical communicator (in this case, a technical editor or technical writer), who edits and formats the document, and adds graphics as needed. In each of these instances, a technical communicator transfers knowledge from someone who has it to someone who needs it. This is the basic definition of technical communication.

Audience type affects many aspects of communication, from word selection and graphics use to style and organization. A non-technical audience might misunderstand or not even read a document that is heavy with jargon while a technical audience might crave detail critical to their work. Busy audiences often don't have time to read entire documents, so content must be organized for ease of searching for example by frequent headings, white space, and other cues that guide attention. Other requirements vary according to particular audience's needs.

Collecting information

Technical communicators must collect all information that each document requires. They may collect information through primary (first-hand) research or secondary research, using information from existing work by other authors. Technical communicators must acknowledge all sources they use to produce their work. To this end, technical communicators typically distinguish quotations, paraphrases, and summaries when taking notes.

Organizing and outlining information

Before writing the initial draft, the technical communicator organizes ideas in a way that makes the document flow well. Once each idea is organized, the writer organizes the document as a whole accomplishing this task in various ways:

- Chronological: used for documents that involve a linear process, such as a step-by-step guide that describes how to accomplish something
- Parts of an object: Used for documents that describe the parts of an object, such as a graphic showing the parts of a computer (keyboard, monitor, mouse, etc.)
- Simple to complex (or vice versa): starts with easy ideas and gradually goes into complex ideas
- Specific to general: starts with many ideas, then organizes the ideas into sub-categories
- General to specific: starts with a few categories of ideas, then goes deeper

After organizing the whole document, the writer typically creates a final outline that shows the document structure. Outlines make the writing process easier and save the author time.

Writing the first draft

After the outline is complete, the writer begins the first draft, following the outline's structure. Setting aside blocks of an hour or more, in a place free of distractions, helps the writer maintain a flow. Most writers prefer to wait until the draft is complete before any revising so they don't break their flow. Typically, the writer should start with the easiest section, and write the summary only after the body is drafted.

The ABC (abstract, body and conclusion) format can be used when writing a first draft of some document. The abstract describes the subject, so that the reader knows what the document covers. The body is the majority of the document and covers topics in depth. Lastly, the conclusion section restates the document's main topics. The ABC format can also apply to individual paragraphs beginning with a topic sentence that states the paragraph's topic, followed by the topic, and finally, a concluding sentence.

Revising and editing

Once the initial draft is laid out, editing and revising can be done to fine-tune the draft into a final copy. Four tasks transform the early draft into its final form.

Adjusting and reorganizing content

In this step, the writer revises the draft to elaborate on topics that need more attention, shorten other sections and relocate certain paragraphs, sentences, or entire topics.

Editing for style

Good style makes writing more interesting, appealing, and readable. In general, the personal writing style of the writer is not evident in technical writing. Modern technical writing style relies on attributes that contribute to clarity:

- Headings, lists, graphics
- Generous white space
- Short sentences
- Present tense
- Active voice (though some scientific applications still use the passive voice)
- Second and third person as required

Technical writing as a discipline usually requires that a technical writer use a style guide. These guides may relate to a specific project, product, company, or brand. They ensure that technical writing reflects formatting, punctuation, and general stylistic standards that the audience expects. In the United States, many consider the Chicago Manual of Style the bible for general technical communication.

Other style guides have their adherents, particularly for specific industries for instance, Microsoft Style Guide is used in some information technology settings.

Editing for grammar and punctuation

At this point, the writer performs a mechanical edit, checking the document for grammar, punctuation, common word confusions, passive voice, overly long sentences, etc.

Introduction to technical communication as a profession

If you are a newcomer or a beginner, or if you have been doing technical writing or editing for a while but want to know what kind of a profession you have gotten into, this workshop shows you what tasks are involved, what standards and values are worshipped, and where the money is.

First, you explore the profession. You learn about the top 10 indicators of success in this profession, take a quick look back at 2,000 years of technical communication and explore the roles of technical communicators today. You discover the role technical communicators can play in work settings such as government labs, start-ups, Fortune 500 industries, and as freelance consultants.

Then you explore the process by which technical communicators accomplish their work. You learn about information design and development processes, in particular user-centered processes. You learn how your work as a technical communicator fits into a product life cycle in a high-tech environment.

Should I take this workshop?

Yes....

- If you want to know whether your skills match those needed for technical communication
- If you wonder what kind of work is available in the field
- If you want to try your hand at some typical tasks performed by technical communicators
- If you want to find out how a technical communicator's work fits into an organization's process
- If you want to know how much money people make in this profession, and how to get promoted
- If you want to know what technical communication managers care about in your writing, interview, and work on the job

No...

- If you want basic instruction on writing English, you should take another class.
- If you heartily dislike writing, you should avoid this class. You're asked to do a lot of writing and editing here.

- If you are looking for a course in public relations or marketing writing, skip this one.
- If you are involved primarily in scientific writing, or responding to requests from regulatory agencies, this course is probably not for you. We focus on technical communication in the high-tech industry, where 90% of the jobs are, even now.

How it works

- You learn through in-class exercises, solving problems, writing drafts, reviewing and editing.
- You can ask lots of questions, so you can find out how your own background prepares you for technical communication.
- You get an insider's view of the profession, and lots of tips to help you when you are preparing your resume, looking for work, interviewing, and, once you have a job, getting promoted.
- You learn the way technical communication looks to a professional, so you can make the right moves when you are looking for work, and promotions.
- Benefits

When you complete this workshop, you will be able to:

- Identify key roles, tasks, tools, salaries, and career paths for a technical communicator
- Identify job types, locations and skills that fit your interests
- List at least a dozen tasks a technical communicator performs
- Write parts of typical documentation
- Identify critical aspects of editing
- Describe the way that technical communication fits into the life of the larger organization
- List the phases of a design and development process
- Describe various tools and techniques used in each phase of the process
- Use 100 key industry terms correctly

Agenda for the in-person workshop (3 days)

Day One

- What is this profession?
- Signs that you will do well in technical communication
- What does a technical communicator do?
- How do I write a procedure?

- How do I analyze and edit technical writing?
- What tools do I need?

Day Two

- What kind of jobs are there in technical communication?
- Organizations to explore.
- How do I get a job?
- How will I fit into an organization's process?
- How you work within the process of product management
- Creating an information plan
- Doing task analysis
- Creating a content specification
- How you work within a user-centred design process

Day Three

- How can I test a manual?
- How do I handle emotions in this job?
- Where did this profession come from?
- Review of key ideas and terms.

Agenda for the online workshop (6 weeks)

Week 1

- What is this profession?
- Signs that you will do well in technical communication
- How to write a procedure

Week 2

- What will I do as a technical communicator?
- What kind of jobs are there in technical communication?
- Organizations to explore.
- Revising and formatting your procedure for your portfolio.

Week 3

- Understanding the levels of editing
- Editing a procedure
- Editing sentences and paragraphs
- Using a style guide
- Editing with a style guide

Week 4

- How your work fits into the organization's life
- The idea of process
- Waves of process in industry
- The process of product management
- Creating an information plan
- Analyzing your audience

Week 5

- Doing task analysis
- Drawing up the task list
- Completing the information plan
- Creating a content specification
- Adapting your writing for your audience and their idea of the task

Week 6

- Putting your user at the center of the process
- Testing your manual
- Developing your emotional intelligence, to become a star performer
- Final exam

Comments from students

- A comprehensive perspective.
- Great instructor, very complete course materials
- The combination of lecture and team assignments made coursework interesting. Lots of topics-all very interesting and motivational.
- Jonathan Price is a very effective instructor. I was concerned before taking this course that it would be very repetitious and boring, since I've completed quite a few of the other certificate courses, but I still learned quite a bit, especially about resumes, job searches, and web sites.
- Great teacher! Really an expert in his field. Best person to teach this introduction to the profession. More concrete and hands-on than I expected, which was a wonderful surprise?
- It was an overview course, but specific examples were included that gave you a real feel of what the profession is really like.
- Very well organized. Informative, relevant, useful, intelligent, witty, fun.

Chapter 20

International Technical Communication

Society for Technical Communication

The Society for Technical Communication (STC) is a professional association dedicated to the advancement of the theory and practice of technical communication. With membership of more than 6,000 technical communicators, it's the largest organization of its type in North America. The society publishes a quarterly journal and a magazine ten times a year and hosts an annual international conference (STC Technical Communication Summit). STC also provides high quality online education in the form of live Web seminars, multi week online certificate courses, virtual conferences, recorded seminars, and more. Association News magazine lauded the STC education program as an "incredible e-learning system" that provides "highly credible peer-to-peer learning while minimizing costs. Its Headquartered in Fairfax, Virginia, USA, STC is the largest organization of its type in the world according to its website. It includes 87 chapters, 10 Special Interest Groups (SIGs), and over 6,000 members worldwide. STC members work in a wide range of roles.

Most STC members belong to one or more communities, which are either geographic chapters or special interest groups (SIGs). Most chapters are in the United States, but STC includes members from 14 countries. The largest group outside the U.S. is the chapter in Toronto, Ontario, Canada. STC publishes a quarterly journal, Technical Communication, and a monthly magazine, Intercom.

The organization traces its roots to the Society of Technical Writers (STW) in Boston and the Association of Technical Writers and Editors (ATWE) in New York. Both were founded in the United States in 1953. These organizations merged in 1957 to form the Society of Technical Writers and Editors (STWE). In 1960, this group merged with the Technical Publishing Society (TPS), based in Los Angeles, to become the Society of Technical Writers and Publishers. In 1971, the organization's name was changed to the Society for Technical Communication.

The organization's main journal developed from the TWE Journal to the STWE Review to the STWP Review to Technical Communications to Technical Communication. Editors of this journal have included Douglas E. Knight, Allan H. Lytel, A. Stanley Higgins, Frank R. Smith, George Hayhoe, and Menno de Jong. Other important leaders in the history of STC include Robert T. Hamlett (first president of ATWE), A. E. Tyler (first president of TPS), Samuel A. Miles (president of the Society of Technical Writers and Editors, which became ATWE's New York chapter in 1955), Vernon R. Root, Robert O. Shockney, and Stello Jordan. In 2011, Alan Houser was elected vice president of the organization; per their by laws, he became president in 2012, and was succeeded by his own vice president Nicky Bleiel in 2013. STC's annual publications competition for 2012-2013 was held in Washington, D.C. The organization also has branches internationally. On November 12, 2012, STC's branch in India held its 14th annual conference in Bangalore.

Membership

STC members are technical communicators who work across a wide range of fields. STC helps to support those Individuals? with access to best in class technical communication education and training programs, career development, and professional networking opportunities.

An individual can join STC as a Classic, Gold Value Package, New TC Professional, or Student member. Members can select the category that best fits their individual needs. STC also offers a Corporate Value Program that companies who employ five or more technical communicators. All STC memberships operate on a calendar-year basis (rather than anniversary).

- **Classic membership**: The member receives access to a wealth of benefits to support their career. This category includes access to online publications, free archived seminars, the salary database, and STC Job Bank. Classic members also receive discounts on STC events and education.
- **Gold Value Package membership**: This package combines the benefits of Classic membership and built-in education. Members receive Classic member benefits and membership in one chapter and one Special Interest Group (SIG). Gold Value Package members also receive five live webinars, 50% off an online course, and heavily discounted registration for the STC Summit as part of their membership.

- ❖ **New TC Professional membership:** This membership aims to aid in the transition from student to new professional. Individuals who have graduated in the past three years are eligible for this option. New TC Professional members receive Classic member benefits, as well as membership in a chapter and SIG.
- ❖ **Student membership**: Students actively seeking a degree can join STC as a student and receive Classic member benefits, membership in a professional and student chapter, the academic SIG, and a SIG of their choosing.
- ❖ **Corporate Value Program:** Companies with five or more employees interested in membership can join STC and receive discounted membership prices for each employee who joins. The individuals receive Classic membership benefits while the company receives added benefits as well.

Any STC member can choose to join a professional chapter, student chapter, or SIG. The professional and student chapters are primarily regional though many have a strong online presence that allows members around the world to participate. SIGs are composed of STC members with common experiences and interests who share their knowledge online

Technical communication summit (Annual Conference)

In May or June of each year, STC holds the 'Technical Communication Summit', the largest conference for technical communicators in the world. The Summit includes over 80 education session segregated into relevant subject areas or tracks; networking events such as the Opening General Session, Welcome Reception, Communities Reception, and Closing Lunch; an Honors Banquet; and an exhibit hall with dozens of companies offering technical communication products or services. The Summit also includes pre conference education for an additional price. Also included in the conference registration is post-conference access to STC's Summit@aClick. This unique feature provides audio/graphic synchronized recordings of almost all the sessions. Summit@aClick takes away the problem of only being able to attend one session per hour. You can benefit from all of the education and information available at the Summit by listening and watching the other sessions on Summit@aClick after you get home. This product is also available for purchase by non-attendees. With education, networking, and professional development opportunities, attendees regularly say that they love the STC Summit.

In addition to the Technical Communication Summit, STC offers online education both to its members and nonmembers, with members receiving discounted registration rates. STC offers both live and recorded online education:

- ❖ **Multi-week Online Courses:** STC offers multiple online courses for more in depth learning. The courses are typically divided into 90 minute sessions and presented live online over several weeks. Through STC's online courses, attendees can strengthen foundational skills and advance their knowledge of the profession.

- **Virtual Conferences:** STC offers two to four virtual conferences each year. Virtual conferences are usually three day events, three hours per day, with multiple sessions per day on the overall topic. Past topics have included technical editing, usability and accessibility, content strategy for mobile devices, and starting your own business; recordings of past virtual conferences are also available.
- **Live Web Seminars**: The Society produces multiple live, one-hour, Web seminars each month on technical communication topics ranging from accessibility to user experience design. Expertise levels range from beginner to advance with audio via the phone or Voice over Internet Protocol (VoIP) and the presentation via the Web.
- **Recorded Web Seminars**: STC's recorded Web seminars are recordings of our popular live Web seminars, allowing technical communicators to view them at any time and on any computer. The webinars are recorded within AdobeConnect and include the audio of the webinar plus the PowerPoint presentation. More recordings are constantly being added.
- **Free Archived Seminars** (members-only): This is an archive of STC seminars, free for members to watch at their leisure. The audio recording of the sessions has been synched up to the PowerPoint presentation and is available in an embedded video player. There are almost 150 free seminars available to members for viewing as of the end of 2013.

STC recognizes outstanding individuals by conferring the titles of Fellow, Associate Fellow, and Honorary Fellow. STC's Honorary Fellow for 2009 was "Jimmy Wales, the founder and creator of Wikipedia."

STC also sponsors honorary societies for technical communication students with a grade point average of 3.5 or above:

- Sigma Tau Chi recognizes students in four-year or graduate programs
- Alpha Sigma recognizes students in two-year or certificate programs

Chapter 21

Technical Presentation

Technical presentations serve engineering, scientific and high tech purposes, describing advances in technology, problem resolution, product design and project status. In general, technical presentations serve one of two purposes: (1) to inform (e.g., knowledge transfer, classroom instruction) or (2) to persuade (e.g., convincing others to adopt a design approach or accept the results of an evaluation process).

Subject Matter Experts (SMEs), individuals who are well-versed in the topic being presented, generally make technical presentations. Audiences may range from highly technical (fellow engineers and technologists) to non-technical, and part of the challenge of preparing and presenting technical material is gauging the knowledge level of the audience.

One reason for this is that, unlike other public speaking occasions, technical presenters are not usually expert presenters. Instead, they're usually experts in the material, but not in projecting, engaging the audience, demonstrating enthusiasm, using humor and other best practices successful presenters use. They may have little or no preparation for, or background in, public speaking, yet they are required to do so. Often they present material as if the material should speak for itself, no additional enthusiasm or expression required.

Tips for Designing Technical Presentations

The Do's

- Keep the fonts consistent in both style and size throughout the whole presentation.
- Use common fonts that are available on every computer, such as Arial, Times New Roman or Calibri. This way, there will not be any surprises if the computer used for the presentation does not have the unusual font you chose installed, and therefore substitutes another font.
- Include relevant photos and graphics such as simple charts or diagrams. Consider whether the audience can understand the information presented or if you need to simplify the chart / diagram for clarity.
- Make sure that graphics are of good quality so the information is easily deciphered at the back of the room.
- Make labels on charts large enough to be read at a distance.
- Use heightened contrast on your slides. Consider creating the same presentation in two formats, one presentation with dark text on a light background, and a second, duplicate presentation using light text on a dark background. This way, you are ready for either a very dark room or very light room to present in and can choose the suitable presentation accordingly.
    - Keep the number of slides to a minimum. Present only what is necessary and don't overwhelm the audience with too much information. Technical information is hard enough to digest.
    - Allow time for a question period at the end of your presentation
    - Know everything about your topic so that you are prepared for *any* question that arises, even if the question was not covered in the material you presented.
    - Have detailed handouts ready to give out *after* the presentation. This allows for the audience to later reflect on the presentation and the information is ready at hand for any necessary follow-up.
- ***The Don'ts***
    - Don't confuse the audience with disorganized slides so that the purpose of the presentation is not crystal clear.
    - Don't overwhelm your audience with busy slides. Think of that old cliché -- "less is more".
    - Don't use *small* images or *small* text on your slides. Think about those people at the back of the room.
    - Don't use script type fonts. They are notoriously difficult to read at the best of times, let alone on a screen.
    - Don't use more than three or four related points on each slide.

- Don't use a fancy background. It may be pretty or even on topic, but the text will be difficult to read. Keep to a subtle backdrop for the information.
- Don't add pictures for the sake of decoration. Make sure there is a point to be made and that information is obvious to the viewer.
- Don't use sounds or animations unless they are to emphasize a point. Even then, it is risky as they can detract from the main focus of the presentation.
- Don't use acronyms unless all members of the audience are familiar with them.
- Don't include more than four or five items on a chart. Even though Excel charts can be made to show great detail, a slide show is not the place for this information. Stick to important facts only.

A Guide to Technical Presentations

The Three Most Important Elements of a Technical Presentation

- Organization
- Visual Aids
- Delivery and Style: Your Presence and Preparation

Before You Present...

Identify the Purpose

- What is your general purpose?
- What is your specific purpose?
- What is your thesis statement?
- What do you want your audience to learn?
- What action should be taken next?

Perform an Audience Analysis

- Identify who your audience will be
- Identify your objectives for the audience
- Identify their understanding of the subject
- Determine how willing they will be to accept your ideas

Organization: Introduction

- Identify who you are establish your "presence /"
  - Why should they listen to YOU? Express your qualifications, passions, become a bit "human" to the audience

- Earn the audience's attention
  - What will they get from listening to you?
  - Why should they listen?
- Give a roadmap:
  - Tell your audiance what you're going to tell them
  - Tell them
  - Then tell them what you total them in
  - Explain where you plan to go, set up the story
  - Explain what the audience can anticipate

Organization: Transitions

- A word or phrase that signals when a speaker has finished one thought and is moving onto another

  These are the turn signals necessary for changing lanes.
- 2 parts to a transition: Transitions state the idea that the speaker is leaving (the review part) and the idea that the speaker is coming up to (the preview part)

Ex: Now that you know what the TPS report is (review), I will discuss its functions (preview).

Organization: Conclusion

- Purpose: Tell them what you told them
  - Offers audience a sense of closure
  - Reinforces thesis
- Tips
  - Signal the end verbally and non-verbally
  - Restate the thesis using a strong concluding statement
  - Make conclusions strong and brief.

Organization: Conclusion

- Don't drag out the conclusion
- Don't end on a weak or rambling note
- Don't introduce new points
- Don't say "so in conclusion"
- Don't end with "Any Questions?" (if the audience was engaged, then they will have questions)

Visual Aids

Fonts Are Important

- Use Microsoft sans serif fonts
- Use readable font sizes
- Use appropriate color combinations
- 40 point Title
- 28 point Heading
- 24 point Sub-headings
- 18 point References and Labels
- Avoid using 12 point font or smaller

The Abuses of Capital Letters

- Bullet points typically have one capital letter at the beginning
- Just because You think a word is Important does Not mean it should be Capitalized
- ALL CAPITAL LETTERS MAKE IT HARDER FOR YOUR AUDIENCE TO DECIPHER WORDS

The Exclamation Point

Don't over use this. Be careful of when and where you use it.

Avoid Using Too Many Words

- Use Bullet Points
- Use Condensed Sentences
- Practice with the laser pointer.
- Use sparingly: Only when necessary

Delivery

- Is clearly audible, fresh, and energetic sounding
- Is a polished version of yourself
- Looks and sounds natural
- Is human, not robotic or like a TV news anchor
- Has controlled and planned body movements
- Is extemporaneous (don't read slides!)
- Is always focused on the goal (speak so your audience will understand)

A Few Words on Practicing

- It is the most important aspect of a presentation... seriously
- It is in practice that you find out if you

- Make sense
- Sound knowledgeable
- Or sound like an idiot

The Best Way to Practice

- Highlight in your notes difficult pronunciations or phrases that need emphasis or a slower rate of speaking
- Practice in a room similar to where you will present
- Imagine the audience to whom you will present
- Always practice out loud
- Plan movement and gestures
- Practice with all of the equipment you will use .

Practice 7 to 11 Times

- 1st Practice: Use notes and try to get a sense of the rhythm of the speech
- 2nd to 4th practice: Make notes of difficult transitions, phrases, or words
- 5th to 7th practice: Focus on how to maintain freshness and energy without making the speech seemed "canned"
- Practice two more times beyond the plan of pain plant.

Establishing Credibility

- KNOW that you KNOW your stuff
- KNOW that you ARE an expert
- Speak loud enough
- Look at your audience
- Don't assume your audience knows
- Be in control at all times
- Watch for audience's non-verbal feedback and adjust your rate, content, or eye contact for them.

Presentation Day

- Check the room
- Bring your presentation on a reliable disk
- Check the presentation projection
- Decide how loud you must speak
- Decide where you and your audience will be located

Teamwork Considerations

- Work out all transitions
  - Between sections
  - Between team members
  - Practice as a team
- Speak with one voice
- Give the speaking member full attention. If you look distracted it will make your audience suspicious

Some other Guides for Technical Presentation

Study the masters

Before you even start building your presentation (or "deck"), study the masters: Steve Jobs and Al Gore. They're both famous for turning a potentially dry format a slide show on a stage into a powerful and inspiring performance. Just cue up a video on the web of Jobs or Gore presenting. Within a few minutes, you'll understand why both are known as "rock star" presenters. They move around the stage, making eye contact and speaking directly to attendees. They speed up their vocal delivery to build momentum, and slow it down to drive emphasis. Their slides are full of impressive graphics and facts that illustrate their points. They're passionate. They're confident. They're funny. Watch them work and analyze why they're so good. It's a great tool for creating your own presentation.

Ditch the podium

Here's your new mantra: walk, talk, and engage. Pay attention to the way Jobs and Gore walk across the stage as they address the crowd. They're dynamic. They use hand gestures. They stop, make a point, deliver a slide, and move on. They're constantly engaging the audience.

Design, design, design

Don't rush to the computer to start writing your presentation design it offline first. Work out the main message of your talk and build your presentation around it.

Using one piece of paper per slide, write some quick notes on what the audience should know after that slide has been presented. Once you have decided what the audience should know, you can go back and design *how* you are going to tell them. Engaging talks make use of stories, images and restraint. Weaving your set of slides into a logical narrative will fix your ideas in the minds of the audience long after the talk has finished. Images are more easily consumed than lists of text. In one experiment, two groups of students saw equivalent presentations and the group that saw more slides, with less text per slide, retained twice as much information as the other group.

Knowing what to leave out is key. What is not on a slide is as important as what is.

Another great way to plan and design your presentation is by creating a MindMap first. There are a lot of great open source/freeware applications you can download to do just that (Free mind, Cmap Tools, iMndmap, yED, Xmind). Making a MindMap will help you decide what you actually want to say in your presentation and how to best deliver it.

Practice, practice, practice

Once you've built your set of slides, it's time to practice. Sit in front of a mirror with your laptop at your side. Start from the beginning and work through your slides, presenting to yourself in the mirror. Make clear statements, but don't simply read what's on the slide this is a presentation, not a read-along. Create transitions that lead the audience into your next subject. Maintain a conversational and natural, yet confident tone. Be yourself don't speak or act like you think a presenter should. Imagine making eye contact with people in the audience when you're making a key statement.

Once you're comfortable in front of the mirror, ask your family if they will act as your audience. Use your laptop and a pocket projector to project your slides on a wall. Act as you would on stage: walk across the living room "stage," make eye contact, and speak directly to your audience. Then ask for feedback. Your wife may notice you're speaking too quickly; your son may point out that your slides are difficult to read. Though it may be slightly embarrassing, this kind of practice saves you from making bigger mistakes in front of a much larger, less sympathetic crowd.

Time everything

Find out what the allotted time is for your presentation. Then use a watch, or other timer to practice and repeat it every time you practice. Set an agenda so you know what time you need to be at certain slides; this allows one to micro-adjust during your delivery. If you go consistently over or significantly under your time limit, adjust your content or your delivery accordingly. Keep practicing and adjusting until you can deliver on target.

Start and end with a contact slide

The first and last slide in your deck should always have your name, e-mail address, phone number and appropriate web addresses. People should be able to contact you easily.

BYO equipment

Technical difficulties can drag down even the best presentations. Situations like, a projector isn't working, the client's laptop is on the fritz, or the file you sent via e-mail was "lost" can ruin an entire presentation.

Be self-sufficient. Bring your own laptop, a USB stick or SD card with your presentation on it, and carry a pocket projector. They are small and light enough to not weigh you down, and you can plug a USB stick directly into one and present without a laptop if you need to.

With all those elements in place, you can plug in, fire up your presentation, and project it without any additional assistance. (Remember to practice with the same gear you take to the presentation. This isn't the time for last-minute surprises.)

References

Argyle, M. (1988). Bodily Communication (2nd ed.) Madison: International Universities Press.

Argyle, M., Salter, V., Nicholson, H., Williams, M. and P. Burgess (1970). The communication of inferior and superior attitudes by verba and non-verbal signals. British journal of social and clinical psychology 9 : 222-231.

Chalapati Rao I.V., (1999). Communication and leadership, skills and strategies, Book links Corporation, Hyderabad.

Dasgupta, (1989). Diffusion Agricultural Innovations in village India, Wiley Eastern Limited, New Delhi.

Ganeshan, R. (2006). Reading Material for PG Courses, TNAU, Coimbatore.

Givens, D.B. (2000). Body speak: what are you saying? Successful Meetings (October) 51.

Gudykunst, W.B. and S. Ting-Toomey (1988). Culture and Interpersonal Communication. California : Sage Publications Inc.

Hargie, O. and D. Dickson (2004). Skilled Interpersonal Communication : Research, Theory and Practice. Hove: Routledge.

Kundu, C.L. & Tutoo D.N., (1985). Educational Psychology, Sterling publishers Private Ltd., Bangalore.

Paul, R. (2007). *Language disorders from infancy through adolescence: Assessment and intervention.* (3rd ed.) St. Louis, MO: Mosby/Elsevier. (See page 489. Cognitive uses of language of young school-age children. Based on chart developed by J. Tough.)

Post Graduate Diploma in Agricultural Extension Management (PGDAEM).

Prizant, B., Wetherby, A. M., Rubin, A. C. & Rydell, P. J. (2006). *The SCERTS model: A comprehensive functional approach for children with autism spectrum disorders. Vol. 1 Assessment*. Baltimore, MD: Paul Brookes Publishing. (See pgs. 251-310. SCERTS assessment forms.)

Quill, K. A. (2000). *Do-watch-listen-say: Social and communication intervention for children with autism*. Baltimore, MD: Paul Brookes Publishing. (See pgs. 54-69. Assessment of social and communication skills for children with autism.)

Ray, G L (2005). Extension Communication and Management, Express Graphics, Delhi 28.

Reddy, A. A. (1987). Extension Education, Sree Lakshmi Press, Bapatla.

Rogers, E. M., (1995). Diffusion of innovations Fourth Edition, Free Press, New York.

Rogers, E. M., and Shoemaker, F F (1971). Communication of Innovations, Free Press, New York.

Samuel A. Malone, (1997). Mind skills for managers, Gower.

Shiv Khera, (2002). You can win, winners don't do different things they do things differently, MacMillan, New Delhi.

Singh, K.N, C.S.S. Rao and S.N. Sahay, 1970. Research in Extension Education for accelerating development process published by Indian Society of Extension Education, New Delhi.

Swanson, B. E. 1984. Agricultural Extension a reference manual Second Edition, FAO, Rome, Oxford & IBH Publishing Co. Pvt. Ltd. New Delhi.

Vicker, B. (2008). Communicative functions or purposes of communication. The Reporter, 14 (1), 13-17.

Wiseman Gordon & Barker and Larry, (1974). Speech – Interpersonal communication, Chandler Publishing Company.

Agarwal, R. (1997). Innovations in entrepreneurship: major issues in new era of gloalisation. *Abhigyan*. Vol (15), pp. 65-71.

Aitken, H.G.J. (1963). The future of entrepreneurial research, explorations in entrepreneurial history. Second series, Vol (1), pp. 3-9.

Akhouri, M.M.P. (1975). Planning and monitoring entrepreneurship. Developing entrepreneurship. A hand book (Ed. Rao, T. V and Pareek, U). Learning system, New Delhi. pp. 336-347.

Akhouri, M.M. (1996). Conceptual background and rationale for entrepreneurship development training programmes. Paper presented in orientation course on entrepreneurship development in agriculture. December 17-31. IARI, New Delhi. pp. 10.

Ali, H. (1995). Gateways to market entry. *The Journal of entrepreneurship*. Vol (4), No. (I), pp. 49-69.

Appley, L.A. (1956). Management in action. American management association,

New York. pp. 19.

Atkinson, J.W. and Litwin, G.H. (1960). Achievement motive and test anxiety conceived motive to approach success and pp. 105 motive to avoid failure. *Journal of abnormal and social psychology*, Vol (60), No. (1), pp. 52-63.

Banerjee, M. and Talukdar, R.K. (1997). Variables influencing entrepreneurship of women entrepreneurs. Indian Journal of Extension Education. Vol. (33), No. (1&2). pp. 25-36.

Bankston, C.L. and Zhou, M. Q. (1996). Go fish: The Louisiana Vietnamese and ethnic entrepreneurship in extra active industry. *National Journal of Sociology*. Vol (10), No. (1), pp. 37-55.

Basavanna, M. (1971). A study of self-confidence as an attitude of self-concept. Ph.D. Thesis (Unpublished). Srivenkateswara University, Tirupathi, Andhra Pradesh.

Basu, S.K. and Moulik, T.K. (1979). Preference oriented selection criteria for rural entrepreneurs, identification and selection of small scale entrepreneurs. (Ed. Rao, T.V. and Moulik, T.K.). Indian Institute of Management, Ahmedabad, Gujarat.

Battacharjee, S.K. and Akhouri, M.M.P. (1975). Profile of small industry entrepreneurs. SEDME. Vol. (2), No. (1), pp. 73-85.

Battacharya, S.N. (1983). Entrepreneurship development in India and the South East Asian countries in agricultural and industrial sectors. Metropolitan book company, New Delhi.

Bema, J. (1960). Industrial entrepreneurship in Madras State. Asia Publishing House, London.

Bethal, L.L. Atwar, F.S. Smith, G.B.E. and Stockman, H.A. (1930). Industrial Organization and Management. Mc Graw Hill Book Company, New York, pp. 126.

Bhagat, G.R. and Singh, Y.P. (1995). Effectiveness of entrepreneurial development training, on the knowledge level of farm entrepreneurs. *Indian Journal of Extension Education*. Vol. (31), No. (1-4), pp. 102-104.

Bhatt, R.S. (1974). Growth of entrepreneurship in small and medium sectors. *The Indian Journal of Public Administration*. Vol (20), No. (3), pp. 453-465.

Bhattacharya, D. Sarkar, S. and Gangopadhyay, P. (1997). Decision making role of women in rural development. A quantitative study. Social science perspectives in agriculture. Edited by Das Gupta, D. and Mukhopadhyay, S.D. Santineketan, W.B. pp : 143-156.

Bisht, N.S. and Sharma, P.K. (1991). Entrepreneurship expectations and experience. Himalaya publishing house, New Delhi. pp. 3-4.

Bradford, L.A. and Glenn, L. J. (1953). Farm management analysis. John Wiley and Sons, Inc. New York.

Bull, I. and Willard, G. E. (1993). Towards a theory of entrepreneurship. *Journal of Business Venturing*. Vol. (8), pp: 183-195.

Busenitz, L.L. (1996). Research on entrepreneurial alertness: Sampling measurement and theoretical issues. *Journal of Small business management*. Vol. (34), No. (4), pp. 35-44.

Bygrave, W.D. and Hofer, C.F. (1991). Theorizing about entrepreneurship. *Entrepreneurship Theory and Practice*. Vol. (6), No. (2), pp. 13-22.

Cantillon, R. (1955). International encyclopedia of social sciences. Quoted by Cochran, T.C. (1968) under entrepreneurship. Macmillan, New York. Vol. (5), pp. 88-90.

Carland, J. W. (1984). Differentiating entrepreneurs from small business owners. *Academy of Management Review*. Vol. (9), No. (2), pp. 354-359.

Carrier, C. (1996). Entrepreneurship in small business: An exploratory study. *Entrepreneurship Theory and Practice*. Vol. (2), No. (1), pp. 5-20.

Christopher, K.J. (1969). Socio psychological factors influencing the adoption of innovationsof starting a small industry unit. SIET Institute, Hyderabad, pp. 68-71.

Christopher, K.J. (1971). Identification and detection of latent entrepreneurship. *Productivity*. Vol. (12), No. (1), pp. 78- 83.

Cohen, N. (1980). "The Five Ages of the Entrepreneur." *Venture*: 40-42.

Cole, A.H. (1949). Entrepreneurship and entrepreneurial history. In change and the entrepreneur. Prepared by the research centre in entrepreneurial history. Harward University Press, Cambridge.

Cole, A.H. (1959). Business enterprise in it social setting. Harward University Press, Cambridge.

Cole, A.H. (1968). Meso-economics: A contribution from entrepreneurial history. *Explorations in Entrepreneurial History*. Vol. (6), No. (1), pp. 78-86.

Cooper, A.C. Timothy, B and Woo, C. (1995). Entrepreneurial information search. *Journal of business venturing*. Vol. (10), No. (2), pp. 107-120.

Covin, J.G. and Slevin, D.P. (1991). A conceptual model of entrepreneurship as firm behaviour. *Entrepreneurship Theory and Practice*. Vol. (16), No. (1), pp. 7-25.

Dasgupta, A. (1969). Indian business management. Deptt. of business management and industrial relations, University of Delhi. pp. 60.

De, D. (1985). Status symbol and innovative entrepreneurship are predictors of farmers progressivism. *Journal of Extension System*. Vol. (1), No. (1), pp. 81-84.

De, D. (1986). Factors affecting entrepreneur characteristics of farmers. *Indian Journal of Social Work*. Vol. (XLVI), No. (4), pp. 541-546.

De, D. and Rao M. S. (2001). Entrepreneurial behaviour of farmers: An Axiomatic theory. Ganga Kaveri Publishing House, D. 35/77 Jangamwadi Math Varanasi.

Deivasena Pathy, P. (1986). "Entrepreneurial success: Influence of certain personal variables". *The Indian Journal of Social Work*. Vol. (XLVI), No. (4), pp. 547-553.

Derossai, F. (1971). The maxican entrepreneur. Development centre of the organisation for economic co-operation and development. Paris.

Desai, G.R. (1981). A critical analysis of the contribution of Extension guidance to economic performance of cotton farmers of Karnataka State. Thesis (Unpublished), U.A.S., Banglore.

Downing, J. (1991). Gender and growth of enterprises. *Small Enterprise Development.* Vol. (2), No. (1), pp. 4-12.

Drucker, P. (1954). The practice of management. Harper and Row, New York. pp. 17.

Drucker, F.P. (1985). Innovation and Entrepreneurship – Practice and Principles. Heinemann, London. pp. 31.

East West Centre, (1977). Entrepreneurial discovery and development progress of action research. East –West technology and development institute, Honolulu, Hawai. pp. 82-83.

Encyclopedia Britanica : *Micropedia* Vol. III : Ready reference and index.

Encyclopedia Britanica, Inc. Benton, W. Publishers, 1943-1973. Helen Hemingway, Chicago.

Encyclopedia of psychology, Search press, Ed. Eysenck, H.J. and Arnold, W. and Meili, R. London. Vol (2).

Encyclopedia of social sciences, (1933). Mc Millan Company, New York. Vol. (20), pp. 76-77.

English, H.B. and English, A. (1958). A comprehensive dictionary of psychological and psychoanalytical terms. Longmans. Green and Company, New York.

Fishbein, M. (1965). Attitude and the prediction of behaviour. In Fishbein, M (Ed.), Readings in attitude theory and measurement. John Wiley and Sons, New York. pp. 477-492.

Fraser, T.M. (1961). Achievement motivation as a factor in rural development, a report on research in Western Orissa. Haverford College, Haverford, Paris.

Gaikwad, J.G. Inamke, N.M. and Kadam, K. R. (1997). Involvement of farm women in decision making. Maharashtra. *J. Extn. Edn.* Vol. (16), pp. 359-361.

Gaikwad, V.R. (1978). Entrepreneurship: The concept and social context in developing entrepreneurship. A hand book of Rao, T. V. and Pareek, U. in collaboration with Deshpande, M. V. Mehta, P. and Nadkarni. M.S. Learning systems, New Delhi.

Gartner, W.B., Bird, B.J. and Starr, J.A. (1992). Acting as if Differentiating entrepreneurial from organizational behaviour. *Entrepreneurship theory and Practice.* Vol. 16 (3). pp. 13-31.

Ghosh, R. Gupta, M. and Dhar, S.N. (1998). Women and entrepreneurship in India. Entrepreneurship innovation edited by Kanungo, R.N. Sage Publications, New Delhi. pp. 156-175.

Gogoi, M. and Talukdar, R.K. (1997). Determining achievement motivation of agricultural scientists. *Indian Journal of Extension Education.* Vol. (33), No. (1&2),

pp. 64-67.

Gopakumar K (1995). Entrepreneurship in economic thought: A thematic review. *Journal of Entrepreneurship*, 4 (1) pp. 1-17.

Gordon, L. (1961). Private enterprise and international development. *The American Review*. Vol. (6), No. (1), pp. 5-13.

Gupta, C.B. and Srinivasan, N. P. (1992). Entrepreneurial development: Second edition. Sultanchand and sons educational publishers, New Delhi.

Harbison, F.N. and Hyers, C.A. (1964). Education man power and economic growth. In Nodder, B.W. (Eds.). Economic development in the topics, Meuthen, London.

Haredero, J .M. (1979). Agricultural entrepreneurship. Identification and selection of small scale entrepreneurs. (Ed. Rao. T.V. and Moulik, T.K.). Indian Institute of Management, Ahmedabad. pp. 151-159.

Harper, M. and Vyakaranam, S. (1988). Rural enterprise: Case studies for developing countries. Intermediate technology publications, London.

Hawley, F.B. (1882). Entrepreneurship in interaction. In Encyclopedia of the social sciences (1968). Mac Millan, New York. Vol. (5), pp. 87-90.

Hegan (1968). The economics of development. Irwin RD Inc. Illinois, New York

Hepple, L.M. (1959). Group organization and Leadership in rural life. Lucas bros, Columbia, Missouri.

Himachalam, A (1990). Developing entrepreneurial industrialist – An alternate approach. *Productivity*. Vol. 34 (2). pp. 281-287.

Homans, G.C. (1970). The relevance of psychology to the explanation of social phenomena. Borger, R. Cioffi, F. (Ed.) Explanation in the behavioural sciences. Cambridge University Press, Cambridge. pp. 313-326.

Hornaday, J.A. and Aboud, J. (1971). Characteristics of successful entrepreneurs. *Personal psychology*. Vol. (14), No. (6), pp. 66-78.

Hoselitz, B.F. (1957). Non-Economic factors in economic development in Tandoil, B.C. (Eds.). Environmental and entrepreneur. Cheigh publications, Allahabad, India.

Hundal, P.S. and Singh, S. (1975). Structure of personality characteristics and motive pattern of farmers. *Indian Journal of Psychology*: Vol. (50), No. (1), pp. 33-43.

Jaiswal, N.K. (1965). A study of factors associated with low level of adoption of improved agricultural practices. Ph.D. thesis, Division of Agricultural Extension, IARI, New Delhi.

Jenkins; J. C. (1983). A resource mobilization theory and the study of social movements. *Annual review of sociology*. Vol. (9), pp. 527-533.

Jhamtani, A. (1996). Dynamics of entrepreneurship. Orientation course on entrepreneurship development in agriculture. December 17-31, JARI, New Delhi, pp. 5-6.

Jones, S. and Cohen, M.B. (1990). The emerging business. Ronald Press, New York.

Joshi, S.S. and Kapur, T.R. (1973). Fundamentals of farm business management. Kalyani publishers, New Delhi.

Jyothi, V. (1990). Entrepreneurship development among women – study of entrepreneurs in Guntur district. M. Phil dissertation submitted to Nagarjuna University. Guntur, Andhra Pradesh.

Jyothi, V. and Prasad, G. (1993). A profile of potential rural women entrepreneurs. Small enterprises development management and extension *Journal* (*SEDME*). Vol. (20), No. (1), pp. 23-38.

Kanungo, R.N. and Bhatnagar, J. (1978). Achievement orientation and occupational values: A comparative study of Young French and English Canadians. *Canadian Journal of Behavioural Science*. Vol. (10), No. (3), pp. 202-213.

Kanungo, R.N. and Conger, J.A. (1995). Model orientation in leadership research and their implications for developing countries. *New approaches to employee management*. Vol (3), pp. 155-170.

Kanungo, R.N. and Mendonca, M. (1994). Fundamentals of organizational behaviour. Kendall Hunt publishing company, Iowa.

Keshari, P.K. and Kota, R. (1995). Determinants of the propensity to export: The case of small firms in the diesel engine and chemical industries in Gujarat. *The Journal of entrepreneurship*. Vol. (4), No. (1), pp. 35-48.

Khan, N.A. (1998). Agricultural marketing: Present position and future strategy, *Kurukshetra*. Vol. (47), No. (3), pp. 11-13.

Khan, W. M. (1998). Technological innovation in SMEs in Pakistan. Entrepreneurship and innovation edited by Kanungo, R.N. Sage publications, New Delhi. pp. 287-309.

Khanka, S.S. (1998). Making the entrepreneurial society. *Yojana*. Vol. (42), No. (2), pp. 9-11.

Khemmani, M. (1983). Turning thoughts towards training, Manita, phillippines, Asia and Pacific programme for development training and communication planning (DTCP), UNDP. Training tips No. 1 Editor: Andrew P. Bartlett.

Kilby, P. (1971). Entrepreneurship and economic development. Free press, New York.

Kirzner, I.M. (1979). Perception, opportunity and profit: Studies in the theory of entrepreneurship. The University Chicago press, Chicago.

Kittur, M.M. (1976). A study of adoption behaviour of marginal farmers in relation to their characteristics and value orientation in Bizapur district of Karnataka state. M.Sc. (Ag.) Thesis. Division of Agricultural Extension, U.A.S. Dharwar.

Knudson, W., Wysocki A., Champagne J. and H. Christopher Peterson, (2004). Entrepreneurship and innovation in the Agri-food System, *Amer. J. Agr. Econ.* 86:1330-1336.

Kogan, N. and Wallach, M.A. (1964). Risk taking. A study in cognition and personality.Holt, Rinehart and Winston, New York.

Kotler, P and Levy, S.J. (1969). Broadening the concept of marketing. *Journal of marketing*. Vol (16), No. (8), pp. 63-64.

Koontz, H. (1994). Essentials of management. (I edition). Mc. Graw. Hill publishing company, New York. pp. 39.

Kourilsky, M. (1980). Predictors of entrepreneurship in stimulated economy. *Journal of creative behaviour*: Vol. (14), No. 3. pp. 76-86.

Krishnan, S.B. Kumar, K. G. and Kanungo, R.N. (1998). Entrepreneurship development concept and context. *Entrepreneurship and innovation*. Sage publications, New Delhi. pp. 21.

Kumar, E.S. (1998). Theorizing in entrepreneurship: A review, critique and future directions. *Entrepreneurship innovation*. Edited by Kanungo, R.N. Sage publications, New Delhi, pp. 90-115.

Kumar, G.K. (1995). The entrepreneur in economic thought: A thematic overview. *The Journal of entrepreneurship*. Vol. (4), No. (1), pp. 1-15.

Kumar, K.G. (1998). Management skills for rural enterprises. *Entrepreneurship innovation*. Edited by Kanungo, R.N. Sage publications, New Delhi, pp. 338-357.

Kumar, V.S. (1998). Small business marketing: Need for specilized skills. Entrepreneurship innovation. Edited by Kanungo, R.N. Sage publications, New Delhi, pp. 325-337.

Lamb, R.K. (1965). The entrepreneur and the community in men in business. In Baumback, C.M. and Mancuso, JoR. (Eds.). 'Entrepreneurship and venture management. Prentice Hall inc, Engle wood cliffs, New Jersey.

Learned, K.E. (1992). What happened before the organization? A model of organization formation. *Entrepreneurship theory and Practice*. Vol. 17 (1). pp. 39-48.

Leeds, C.N. and Stainton, R.S. (1978). Management and business studies (Second edition). Macdonald and Evans.

Leibenstein, H. (1976). Beyond economic man: A new foundation for micro economics. Harward University, Cambridge, Massachusetts.

Lekha, K.C. Kalyan, W and Lavoie, D. (1998). Micro enterprise management by women in India. *Entrepreneurship innovation*. Edited by Kanungo, R.N. Sage publications, New Delhi, pp. 176-199.

Lim, C. (1977). Small industry in Malaysia: A socio economic profile of entrepreneurs. *Social action*. Vol. (27), No. (2), pp. 22-36.

Lumpkin, G.T. and Dess, G.G. (1996). Clarifying the entrepreneurial orientation construct and linking it to performance. *Academy of management review*. Vol. (21), No. (1), pp. 135-172.

Ma, Hao and Tan, Justin, (2006). Key components and implications of entrepreneurship: A 4 P framework, *Journal of Business Venturing*. 21:704-725.

Mall, C.P. (1993). Impact of financial incentives on entrepreneurial development in backward regions: A case study of Deoria district of V.P. *The Journal of entrepreneurship*. Vol. (2), 1 No. (2), pp. 199-207.

Mancuso, J. and Mascolo, M. (1987). Recognising achievement motivation. *Motivation and emotion*. Vol. (11), No. (3). pp. 323-330.

Manimala, M.J. (1998). Net working for innovation: Anecdotal evidences from a large sample study of innovative enterprises. *The Journal of entrepreneurship*. Vol. (7), No. (2), pp. 153-169.

Manimekalai, N. (1998). Gem cutting and rural entrepreneurship development. *Kurukshetra*. Vol. XLVI, No. (5), pp. 22-26.

Manimekalai, N. and Abdullah, M. (1998). Charcoal production and rural entrepreneurship development. *The Indian economic journal*. Vol. (45), No. (2), pp. 159-169.

Manjula, S. (1995). A study of entrepreneurial behaviour of rural women in Ranga Reddy district of Andhra Pradesh. M.Sc. (Home Science). Thesis. ANGRAV, Rajendra Nagar, Hyderabad.

Massie, J .L. (1982). Essentials of management (Third edition). Prentice Hall of India private limited, New Delhi.

Mcclelland, D.C. (1953). Achievement motive. Appleton, New York.

Mcclelland, D.C. (1965). Need achievement and entrepreneurship: A longitudinal study. *Journal of personality and social psychology*. Vol. (1), pp. 389-392.

Mcclelland, D.C. (1969). Motivating economic achievement. Free Press, New York.

Mishra, S.P. (1979). A study of farm entrepreneurship in a backward district of Bihar, Ph.D. Thesis (Unpublished), IARI, New Delhi.

Mishra, S.P. (1996). Training in entrepreneurship – A conceptual background. Paper presented in orientation course on entrepreneurship development in agriculture. December 17-31. IARI, New Delhi.

Mokry, B.W. (1988). Entrepreneurship and public policy. Quorum books, New York.

Nair, K.B. Sivaraman, M. and Kanungo, R.N. (1998). Rural entrepreneurship in a developing economy: A contingency approach. *Entrepreneurship innovation*. Edited by Kanungo, R.N. Sage publications, New Delhi. pp. 78-89.

Nandapurkar, G.G. (1980). A study of the entrepreneurial behaviour of small farmers. Unpublished Ph.D. thesis. U.A.S., Bangalore.

Nandi, A. (1973). Motives, modernity and entrepreneurial competence. *Journal of social psychology*. Vol. (91), No. (3), pp. 127-136.

Neill, R.E. and Rogers, E.M. (1963). Measuring achievement motivation among farmers. Department series A.E. 346. Ohio Agricul. Expt. Stn. Columbus.

Niles, M.C. (1956). The essence of management. Orient Longmans, Bombay. pp. 20.

Palmer, B.l. Ahiamadu, R.K. (1997). Leadership style and performance of public enterprises.*A Nigerian study*. Productivity. Vol. (38), No. (1), pp. 54-64.

Pandey, VK. and Sharma, K.C. (1996). Crop diversification and self sufficiency in food grains. *Indian Journal of agricultural economics*. Vol. (5), No. (4), pp. 644-651.

Pareek, U. and Nadkarni. M. (1978). Development of entrepreneurship. A conceptual model. In Rao, T.V and Pareek, U. (Eds.), Developing entrepreneurship: A hand book, learning systems, New Delhi.

Pareek, U. and Rao, T.V. (1995). Counseling and helping entrepreneurs. *The Journal of entrepreneurship*. Vol. (4), No. (1), pp. 19-34.

Patel, D.R. (1995). Entrepreneurship and small business development for women. *Kurukshetra*. Vol. (XLIII), No. (11), pp. 65-68.

Patel, G. V. (1970). A successful ventures in developing entrepreneurs. Lessons for Gujarat. A symposium paper on development of entrepreneurial talent. New Delhi.

Patel, M.M. Sanoria, Y.C. (1997). Correlates of entrepreneurial behaviour of sugarcane growers. Maharashtra J Extn. Edn. Vol. (16), pp. 344-346.

Paul, K.C. (1998). A study on entrepreneurial behaviour of vegetable growers in Krishna district of Andhra Pradesh. M.Sc. (Ag.) thesis, ANGRAU, Rajendra Nagar, Hyderabad.

Pearce, I.A. Kramer, T.R. and Robbins, K. (1997). 'Effect of managers' entrepreneurial behaviour on subordinates. *Journal of business venturing*. Vol. (12), No. (2), pp. 147-160.

Pollo, L.V. (1996). On economic theory of entrepreneurship : The strategic role of the entrepreneurial function. *International review of sociology*. Vol. (6), No. (2), pp. 191-213. PRADAN (2000) System setting document. Self Help Groups Pradan Programme, New Delhi.

Prakasam, D.S. (1998). Entrepreneur: An important change agent in every society. *Kurukshetra*. Vol, (46), No. (10), pp. 51- 52.

Prasad, R.M. (1983). Comparative analysis of achievement motivation of rice growers in three states in India. Ph.D. thesis (unpublished), U.A.S., Banglore.

Prasad, S.K. (1996). Promoting an entrepreneurship development programme. *Laghu Udyog*. Vol. (21), No. (2), pp. 3-5.

Rajak, M.P. (1998). Craft technology and entrepreneur. *Kurukshetra*. Vol. (XLVI), No. (5), pp. 17-19.

Ramachandran K and Ray Sougata (1998) A framework and entrepreneurial typology for developing a comprehensive theory of entrepreneurship. In Entrepreneurship and innovation: Models for development. Ed Kanungo R. N. Sage Publications New Delhi. Pp. 40-63.

Ramana, A. V. and Papaiah, R. (1998). Nuances of entrepreneurship development. *Indian management*. Vol. (37), No. (2), pp. 48-53.

Ramakrishnan, K. (1979). Interview techniques as a method of selection in identification and selection of small entrepreneurs (Eds. Rao, T.V. and Moulik, T.K.) Indian Institute of Management, Ahmedabad.

Ramana, A. V. (1999). Entrepreneurship and economic development. *Kurukshetra*. Vol. (48), No. (2), pp. (8 to 10 & 45).

Rani, C. (1986). Potential women entrepreneurs. A study. *SEDME*. Vol. (13), No. (3), pp. 13-32.

Rao, G.K.V. (1985). A prediction analysis of farming performance of farmers through their entrepreneurial behaviour factors. Ph.D. thesis, A.P.A.U. Rajendra Nagar, Hyderabad.

Rao, M.S. (2000). Axiomatic approach to theory building in entrepreneurial behaviour of farmers. Ph.D. thesis, Institute of Agricultural Sciences, B.H.U., Varanasi.

Rao, M.S. and Mishra, S.K. (1997). Farmers perceptions of environmental risk in pesticide use. *Maharashtra Journal of Extension Education*. Vol. (16), pp. 214-219. Rao, T. V (1975). Development of an entrepreneur: A behaviouristic model. *SEDME*. Vol. (1), No. (4), pp. 325-345.

Rao, T.V. (1975). Characteristics of entrepreneurs as perceived by entrepreneurial development agents. Paper presented at the work shop on identification and selection of entrepreneurs. Indian Institute of Management, Ahmedabad.

Rao, T.V. and Mehta, P. (1978). Psychological factors in entrepreneurship. In Rao, T.V. and Pareek, U. (Eds.). Developing entrepreneurship. A hand book. Learning systems, New Delhi.

Rao, T.V.S. (1989). A study on entrepreneurial characteristics of vegetable growers in Guntur district of Andhra Pradesh. Unpublished M.Sc. (Ag.) thesis, A.P.A.U., Rajendra Nagar, Hyderabad.

Ray, S. and Ramachandran, K. (1996). Towards a frame work for a comprehensive theory of entrepreneurship. *The Journal of entrepreneurship*. Vol. (5), No. (1), pp. 1-22.

Razin, E. and Langlois, A. (1996). Metropolitan characteristics and entrepreneurship among immigrants and Ethnic groups in Canada. *International Migration Review*. Vol. (30), No. (3), pp. 703-727.

Reddy, Y.V.R. (1992). Economics of Agro forestry system with particular reference to agro horticultural systems on dry land region. Int. sym. on natural resource management for a sustainable agriculture, New Delhi. Feb. 6-10.

Richart, C.B. Maurer, J.G. (1997). The moderating effect of environmental variables on the entrepreneurial and marketing orientation on entrepreneur -led firms. *Entrepreneurship Theory and Practice*. Vol. (22), No. (1), pp. 47-58.

Ricky, W.G. (1992). Essentials of management prentice Hall of India Private Limited, New Delhi. pp. 22-24.

Ripsas, S. (1998). Towards on interdisciplinary theory of entrepreneurship. *Small business economics*. Vol. (10), No. (2), pp. 103-115.

Rogers, E.M. and Svenning, L. (1969). Modernization among peasants: The impact of communication. Hort, Rinehart and Winston, New York.

Rogers, E.M. and Shoe maker, F.F. (1971). Communications of innovations, A cross cultural approach. The Free Press, A division of Macmillan publication company, Inc. New York.

Rogers, F.M. and Neill, R.E. (1966). Achievement motivation among colombian peasants. Michigan state University, East I Lansing, Michigan.

Rousmasset, J .A. (1977). Risk and uncertainty in agricultural development. Seminar report No. 15. Agricultural Development Council, New York.

Samanta, R.K. (1977). A study of some agro economic, socio psychological and communication variables associated with repayment behaviour of agricultural credit users of nationalised banks. Ph.D. thesis, Bidhan Chandra Krishi Vidyalaya, Nadia, West Bengal.

Samanta, R.K. (1989). Management in agriculture and rural development. UDH publishing house, Naisarak, New Delhi.

Samuelson, P.A. (1963). Discussion-problems of methodology. *American economic review*. Vol. (53), pp. 236.

Sara, C. (1998). Portfolio entrepreneurship in the farm sector: Indigenous growth in rural areas? *Entrepreneurship and regional development*. Vol. (10), No. (1), pp. 17-32.

Say, J.B. (1815). Quoted under entrepreneurship in international encyclopedia of the social sciences (1968). Macmillan, New York. Vol. (5), pp. 87-90.

Schumpter, J.A. (1961). The theory of economic development: An enquiry into profits, capital, interest and the business cycle. Cambridge Harward University Press. Cambridge.

Schumpter, J.A. (1970). The entrepreneur as innovator. Reading in management (Second edition). Mc. Graw Hill, New York. Schwartz, E.A. (1979). Entrepreneurship: New female frontier. *Journal of Contemporary business winter issue*. Vol. (15), pp. 150-163.

Scott, C.E. (1986). Why more women are becoming entrepreneurs. *Journal of small business management*. Vol. (24), No. (4), pp. 37-44.

Scott, M. (1997). Reengineering the rapidly growing entrepreneurial company. *Industrial management*. Vol. (39), No. (3), pp. 21-23.

Sengupta N (1998) Empowerment: A socio-psychological approach to Self Help Group formation. Prajnan. XXVI (4): 525-531

Shailendra, V. (1990). When the harvest is in developing rural entrepreneurship. Oxford and IBH publishing Co. Pvt. Ltd. New Delhi.

Sharma, A Mehta: P. and Sha S.K.(1996) Evaluation of task in diversified crop farming of Himanchal Pradesh. A case study of vegetable farmers; *Indian Journal of Agricultural Economics*; Vol. (51), No. (4); pp. 688.

Sharma, K.L. (1973). Growth of small scale entrepreneurship in India: A theoretical frame of analysis. *Interdisciplinary*. Vol. (10), No. (3), pp. 97-108.

Sharma, S.V.S. (1980). Characteristics of entrepreneurs. In developing entrepreneurship: Issue and problems. SIET (Small industry extension training). Institute, Hyderabad.

Singh, D.N. (1986). Developing entrepreneurship for economic development. Vikas publishing house private limited, New Delhi. pp. 83-92.

Singh, K. (1992). Women entrepreneurs. Ashish publishing house, New Delhi.

Singh, K.N. and Pal, K. (1974). A study on achievement motivation of farmers. *Interdiscipline*. Vol. (II), No. (I), pp. 31-40.

Singh, N.P. (1979). Achievement motivation among agricultural and business entrepreneurs of Delhi. *Journal of Social Psychology*. Vol. (81), pp. 146-149.

Singh, N.P. (1979). Tribal entrepreneurship, efforts and achievements. In Rao, T.V. and Moulik, T.K. (Eds.), identification and selection of small scale entrepreneurs. Indian Institute of Management, Ahmedabad.

Singh, N.P. and Gupta, R.S. (1985). Promotion women entrepreneurs. NIESBUD, New Delhi.

Singh, N.P. and Singh, K. (1971). Motivational components of agricultural and business entrepreneurs in India. *Indian Journal of Industrial relations*. Vol. (7), No. (1), pp. 31-52.

Singh, P.N. (1986). Factors influencing entrepreneurship. Developing entrepreneurship for economic development. Vikas publishing house private limited, New Delhi. pp. 46-92.

Singh, S. (1978). Achievement motivation, decision making orientation and work values of fast and slow progressing farmer in India. *The Journal of social psychology*. Vol. (106), pp. 153-160.

Singh, S. (1979). Motives, Personality characteristics and work values of fast and slow progressing farmers. *Behavioural science and rural development*. Vol. (2), No. (2), pp. 71-77.

Singh, T.P. and Kumar, K. (1975). Achievement motivation, socio- economic status and green revolution. *Indian Journal of Extension Education*. Vol. (1.1), No. (10), pp. 1-7.

Sinha, B.P. and Mehta, P. (1972). Size and land holding and farmers achievement motivation and change proneness. *Behaviour science and community development*. Vol. (6), No. (2), pp. 244-251.

Sinha, T.N. (1996). Human factors in entrepreneurship effectiveness. *The Journal of entrepreneurship*. Vol. (5), No. (1), pp. 23-39.

Sisodia, G.S. (1976). A study of the achievement motivation of farmers for mechanization. *Indian Journal of Extension Education*. Vol. (13), No. (3 & 4), pp. 72-75.

Stevenson, H.H. and Jarillo, I.C. (1990). A paradigm of entrepreneurship: Entrepreneurial management. *Strategic management Journal*, 10th summer special issue. Vol (11), pp. 17-27.

Strayton, R. (1986). The organization of course and conferences in ILO (1986 b) pp.

1-27. Sullivan, A.O. Munir, K. and Dougherty, D. (1998). Building organizational capacities for product innovation. Entrepreneurship innovation edited by Kanungo, R.N. Sage publications, New Delhi.

Sundari, S.S.K. (1995). Entrepreneurship development for rural women. *SDEM*. Vol. (1)., pp. 15-30.

Suresh, B. Kumar, G.K. and Kanungo, R.N. (1998). Entrepreneurship development: Concept and context. Entrepreneurship innovation. Edited by Kanungo, R.N. Sage publications, New Delhi.

Sutton, R.I. and Staw, B.M. (1965). What theory is not. *Administrative science*, Vol. (40), No. (3), pp. 371-384.

Swamy, B.A. (1988). Agricultural entrepreneurship in India. Ph.D. thesis, Andhra University, Waltair, Andhra Pradesh.

Terry, G.R. (1968). Principles of management. Irwin, R.D. Inc., Homewood, Illinois.

Thagamuthu, and Manimekhalai, N. (1989). Generation of employment for women through DWACRA. *Journal of rural development*. Vol. (8), No. (40), pp. 431-438.

Thomas, T. T. and Padmakar, K. (1993). Developing entrepreneurial industrialists. *An alternative approach productivity*. Vol. (34), No. (2), pp. 281-287.

Timmons, I.A. (1996). Characteristics and role demands of entrepreneurship. Paper presented in orientation course on entrepreneurship development in agriculture. December 17-31, IARI, New Delhi.

Tripathi, D. (1985). An integrated view of entrepreneurship. *Economic and political weekly review of management*. Vol. (20), No. (48), pp. 62-69. *Administrative science*. Vol. (31), No. (3), pp. 439-465.

Tyson, S. and York, A. (1989). Personal management. Made simple books, an imprint of Heinemann professional publishing limited. Oxford, England.

Vplankar, A. T. (1977). A study of self employed industrial entrepreneurs. *The Indian Journal of Social Work*. Vol. (38), No. (2), 137-147.

Vijaya, V. Kamalanabhan, T.J. (1998). A scale to assess of entrepreneurial motivation. *The Journal of entrepreneurship*. Vol. (7), No. (2), pp. 183-198.

Vinze, M.D. (1987). Women entrepreneurs in India. Mittal publications, New Delhi.

Vyas, V.S. (1996). Diversification in agriculture: Concept, rationale and approaches. *Indian Journal of Agricultural Economics*. Vol. (51), No. (4), 635-643.

Weber, M. (1930). Entrepreneurship and economic development. The Free Press, New York.

Webster, S. (1968). New world dictionary of the American language. Cleveland.

Westermark, N. (1951). The human factor and success in farming. *Acta agriculture scandanavica*. Vol. (1), pp. 146.

Westermark, N. (1961). The human factors and economic progress. *Indian Journal of Agricultural Economics*. Vol. (16), pp. 12-17.

Wick, J.H. (1998). Understanding entrepreneurial intentions: The attitudinal determinants. Entrepreneurship and innovation. Edited by Kanungo, R.N. Sage Publications, New Delhi, pp. 119-138.

Wilken, P.H. (1979). Entrepreneurship. A comparative and historical study. Ablex publishing, Norvood. pp. 60.

Yasunobu, K. (1997). Entrepreneur fanners in peninsular Malaysia : profile of Peladang Jaya. JIRCAS News letter. (December). No. (13). pp. 4-5.

Zahir, M.A. (1994). Entrepreneurship in Punjab. A case study. *Yojana*. Vol. (38), No. (8), pp. 24-28.

www.ingramcontent.com/pod-product-compliance
Lightning Source LLC
Chambersburg PA
CBHW051356240926
51852CB00001B/42
9789351307464